Cement: Composition, Properties and Applications

Cement: Composition, Properties and Applications

Edited by
Dominic Brady

WILLFORD PRESS
www.willfordpress.com

Published by Willford Press,
118-35 Queens Blvd., Suite 400,
Forest Hills, NY 11375, USA

ISBN: 978-1-64728-339-1

Cataloging-in-Publication Data

Cement : composition, properties and applications / edited by Dominic Brady.
    p. cm.
Includes bibliographical references and index.
ISBN 978-1-64728-339-1
1. Cement. 2. Cement composites. 3. Cement--Analysis. I. Brady, Dominic.
TN945 .C46 2022
666.94--dc23

For information on all Willford Press publications
visit our website at www.willfordpress.com

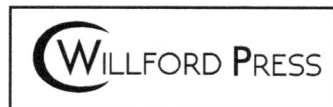

WILLFORD PRESS

Contents

Permissions

List of Contributors

Index

Preface

The world is advancing at a fast pace like never before. Therefore, the need is to keep up with the latest developments. This book was an idea that came to fruition when the specialists in the area realized the need to coordinate together and document essential themes in the subject. That's when I was requested to be the editor. Editing this book has been an honour as it brings together diverse authors researching on different streams of the field. The book collates essential materials contributed by veterans in the area which can be utilized by students and researchers alike.

Cement is a substance that is used for construction to set and harden other materials and to bind them together. It is usually mixed with aggregate to produce mortar for masonry and with gravel and sand to produce concrete. The two main forms of cement are geopolymer cement and portland cement. Geopolymer cement is made up of the mixture of water-soluble alkali metal silicates along with aluminosilicate mineral powders like metakaolin and fly ash. Portland cement is produced by heating limestone with other materials in a process known as calcination. On the basis of their hardening and setting mechanisms, the cement materials are classified into two categories – hydraulic and non-hydraulic cements. The setting and hardening of hydraulic cements include hydration reactions and require water. While, non-hydraulic cements are able to directly set under air as they react with carbon dioxide in the air. The topics included in this book on cement are of utmost significance and bound to provide incredible insights to readers. It is a valuable compilation of topics, ranging from the basic to the most complex advancements in the applications of cement. It will provide comprehensive knowledge to the readers.

Each chapter is a sole-standing publication that reflects each author's interpretation. Thus, the book displays a multi-facetted picture of our current understanding of application, resources and aspects of the field. I would like to thank the contributors of this book and my family for their endless support.

Editor

Analysis of Chemical Composition of Portland Cement in Ghana: A Key to Understand the Behavior of Cement

Mark Bediako[1] and Eric Opoku Amankwah[2]

[1]*CSIR-Building and Road Research Institute, Materials Engineering Division, Kumasi, Ghana*
[2]*Development Office, University of Education, Winneba, Kumasi Campus, Kumasi, Ghana*

Correspondence should be addressed to Mark Bediako; b23mark@yahoo.com

Academic Editor: Wei Liu

The performance of Portland cement in concrete or mortar formation is very well influenced by chemical compositions among other factors. Many engineers usually have little information on the chemical compositions of cement in making decisions for the choice of commercially available Portland cement in Ghana. This work analyzed five different brands of Portland cement in Ghana, namely, Ghacem ordinary Portland cement (OPC) and Portland limestone cement (PLC), CSIR-BRRI Pozzomix, Dangote OPC, and Diamond PLC. The chemical compositions were analyzed with X-Ray Fluorescence (XRF) spectrometer. Student's t-test was used to test the significance of the variation in chemical composition between standard literature values and each of the commercial cement brands. Analysis of variance (ANOVA) was also used to establish the extent of variations between chemical compositions and brand name of the all commercial Portland cement brands. Student's t-test results showed that there were no significant differences between standard chemical composition values and that of commercial Portland cement. The ANOVA results also indicated that each brand of commercial Portland cement varies in terms of chemical composition; however, the specific brands of cement had no significant differences. The study recommended that using any brand of cement in Ghana was good for any construction works be it concrete or mortar formation.

1. Introduction

Portland cement is without any argument among the most important and necessary materials in the world. Without it, the construction industry that utilizes huge tonnages of concrete annually would struggle to survive. Besides this, concrete is rated as the second most highly consumed product after water [1]. It is known that some developed countries depend on the construction industry as one of the main pillars for the growth of their economies. In developing economies, the construction industry provides many jobs for people in both the formal and the informal sectors. Any shortfall that stagnates the construction industry usually leads to serious economic slump.

The Ghanaian construction industry depends hugely on Portland cement for almost every type of construction including bridges, offices, and residential facilities [2]. It is estimated that approximately four million tonnes was utilized in 2014. The prediction is that cement consumption may hit a record high of about five million tonnes by 2020. In Ghana, until 2002, the cement industry was monopolized by Ghana cement manufacturers (Ghacem). After the breakdown of the monopoly, the cement industry has witnessed the influx of many other sources of cement products. Some of these Portland cement products are imported from China, India, and other western European countries. Currently the cement market is diverse and huge and therefore customers have the power to make choices. Available Portland cement products that builders depend on are normally the brands from Ghacem, West Africa cement manufacturers (Wacem), and the Dangote cement. The Ghanaian cement market in recent years has seen the influx of minor entrants like Fortress, and so forth.

The breakdown of cement monopoly which allowed the entrance of other brands of cement is currently creating a major problem for builders in making the best and preferred choice of cement for their constructional work. Engineers and other related building professionals are always confronted with the tough task of selecting the best cement brand in the Ghanaian cement market. The reason for this is because usually masons who normally use the cement products give out various complains to clients who are the main financiers. Sometimes masons complaints are justified; however, the details of their complaints really lack engineering basis.

Most Ghanaian engineers in construction which is largely made of mostly civil engineers make their preferred choice of Portland cement based on strength classification. However, other information such as chemical composition, mineralogical, and even physical properties could be used to corroborate with strength in making good decisions on the best cement in the market. This could be an important key for the selection of best performing cement. In this work, commonly used available Portland cement in the Ghanaian cement market was analyzed in terms of its oxide composition. The main aim of the study was to determine the extent of variation that exists between the commercially available cement and standard literature requirements. In achieving the main aim of the work, the study was guided by this hypothesis "Is there any major differences between the chemical compositions of Ghanaian commercial Portland cement and standard requirements from literature?"

2. Literature Review

Portland cement is the most commonly utilized cement in almost every part of the world. The understanding of the embodiment of Portland cement could lead to a more sustainable concrete and mortar design. It chemically reacts with water to attain setting and hardening properties when used in the construction of buildings, roads, bridges, and other structures. Portland cement was patented by Joseph Aspdin in 1824 and was named after the cliffs on the isle of Portland in England [3].

The production of Portland cement is made by the calcination of a mixture of a calcareous and an argillaceous material at a temperature around 1450°C [7]. Calcareous substances are of calcium oxide origin usually found in limestone, chalk, or oyster shells whereas argillaceous substances are of silicate and aluminate origin predominantly found in clays, shale, and slags [4]. The calcination process between well-proportioned argillaceous and calcareous substances leads to the production of clinker. Portland cement is obtained when the produced clinker is mixed together with a predefined ratio of gypsum and milled together in a ball mill.

The chemical composition of Portland cement involves both major and minor oxides [5]. The major oxides include CaO, SiO_2, Al_2O_3, and Fe_2O_3 whereas the minor oxides also include MgO, SO_3, and some alkali oxides (K_2O and Na_2O) and sometimes the inclusion of other compounds, P_2O_5, Cl, TiO_2, MnO_3, and so forth [5]. Each of the oxides performs unique work during cement hydration; however,

TABLE 1: Summary of chemical data for a selection of Portland cement.

Component	Minimum	Average	Maximum
SiO_2	18.40	21.02	24.50
Fe_2O_3	0.16	2.85	5.78
Al_2O_3	3.10	5.04	7.56
CaO	58.10	64.18	68.00
MgO	0.02	1.67	7.10
SO_3	0.00	2.58	5.35
Na_2O	0.00	0.24	0.78
K_2O	0.04	0.70	1.66
Equivalent alkalis	0.03	0.68	1.24
Free lime	0.03	1.24	3.68

each content of the oxide must be in the right quantity during proportioning of raw materials [6]. Lea [7] provided the required oxide composition of Portland cement (see Table 1). A deviation from standard specifications of the oxide composition may lead to unsoundness and sometimes failure of concrete structures. Many experienced authors have shown that cement oxides which fall very close to the average values are more suitable to maintain concrete integrity [8, 9].

During cement hydration CaO in conjunction with SiO_2, Al_2O_3, and Fe_2O_3 leads to hardening of Portland cement due to the formation of calcium aluminosilicates and aluminoferrite hydrate. With Portland cement, an increased presence of MgO (greater than 2%) may be detrimental to the soundness of cement, especially at late ages. High percentage of SO_3 tends to cause unsoundness of cement. For the Americans in their standard, ASTM C618 limits SO_3 to 4% and 5% whilst the Indian standard limits SO_3 to 2.75%. Alkalis at higher levels and in the presence of moisture gives rise to reactions with certain types of aggregates to produce gel which expands and gives rise to cracking in mortars and concretes.

Sometimes Loss on Ignition (LOI) is classified as a component of chemical composition. LOI indicates the amount of unburnt carbon in the material. However, in some instances it may not necessarily be a measure or indication of carbon content. It may be burning away of residual calcite, bound water molecules, and clay materials [10]. High LOI content may be detrimental to concrete and mortar. It is also known that a high value of LOI results in increased water requirement and dosage of super plasticizer usage in mortar and concrete [11]. Maximum LOI values for both American and Indian standards for common pozzolanic material are 10% and 12%, respectively.

3. Materials and Experimental Procedure

The Portland cement analyzed was five main available commercial cement brands in Ghana which included Ghacem OPC (Class 42.5N) and PLC (Class 32.5R), BRRI Pozzomix, Dangote brand (Class 42.5R), and Diamond brand (Class 42.5N). These brands of cement were obtained in 50 kg bags from a cement distribution outlet in Kumasi, the second biggest city of Ghana. A representative sample of about 20 g was taken from the bulk 50 kg bags as received from the factory for the chemical analysis.

The chemical compositions were performed with the X-Ray Fluorescence (XRF) by the name Spectro X-Lab 2000, at the Ghana geological survey in Accra. The XRF machine uses a polarized energy dispersion. About 4 g of the cement sample was mixed with about 0.09 g of wax. The mixture was milled in a milling machine (RETSCH Mixer Miller (MM 301)) for about three minutes to produce a homogeneous mixture, obtaining a particle size of about 60 μm. The mixture was placed in a dice and placed under the press pellet equipment (SPECAC hydraulic press). The equipment produced a pellet and was then placed in the Spectro X-Lab instrument. The major and other minor oxides were described in graphical histogram presentation. The chemical compositions of each sample were performed three times. The average values of each brand cement were analyzed against the average composition of cement provided by Lea [7] using Student's t-test at alpha (α) value of 0.05. The hypothesis made for the t-test was

$$Ho: \mu_1 = \mu_2,$$
$$Ha: \mu_1 \neq \mu_2, \tag{1}$$

where μ_1 and μ_2 are the mean values of average composition by Lea [7] and commercial Portland cement, respectively.

After Student's t-test, analysis of variance (ANOVA) was used to determine the extent of variation that existed in commonly used Portland cement in Ghana. The hypothesis that was established was that

$$Ho: \mu_1 = \mu_2 = \mu_3 = \mu_4 = \mu_5,$$
$$Ha: \mu_1 \neq \mu_2 \neq \mu_3 \neq \mu_4 \neq \mu_5, \tag{2}$$

where μ_1, μ_2, μ_3, μ_4, and μ_5 are the mean values of Ghacem OPC, Ghacem PLC, CSIR-BRRI Pozzomix, Dangote OPC, and Diamond PLC. The alpha (α) value used for the ANOVA test was 0.05.

4. Results and Discussions

Figures 1(a)–1(e) present the chemical compositions of the commercial cement available in Ghana. The figure showed the major and the minor oxides present in the Portland cement. A comparison with each cement brand indicated variations in the chemical compositions existing between them. From Figure 1, the predominant oxide compositions were CaO followed by SiO_2, Al_2O_3, and then Fe_2O_3 in that order. The minor oxides included MgO, Na_2O, K_2O, MnO, TiO_2, P_2O_5, and SO_3. The compositions of the various oxides in the commercial Portland cement shown in Figure 1 fall within the requirements of cement oxide compositions provided by Lea [7] and Neville [12] who also obtained similar compositions but at different percentages in their studies.

4.1. Student t-Test. Table 2 presents the predictive (P) values and remarks of Student's t-test performed between the average chemical composition values provided by Lea [7]

TABLE 2: Predictive values and remarks of Student's t-test between average and various cement brands.

Brand	P value	Remark
Ghacem OPC	0.96	Failed
Ghacem PLC	0.93	Failed
BRRI Pozzomix	0.91	Failed
Dangote OPC	0.96	Failed
Diamond OPC	0.98	Failed

and each of the commercial Portland cement brands. All the P values indicated that the test failed to reject the hypothesis that there is any significant effect between the average literature composition values and the commercial Portland cement. This indicated that commercial Portland cement in Ghana is well within and without any major deviation from accepted generalized standard specifications.

4.2. Analysis of Variance. Table 3 presents the results of the ANOVA test performed among commercial Portland cement brands. The results gave a predictive value of approximately $3.927E - 19$ for the rows which represented the chemical compositions whereas that for the columns was approximately 0.85 representing the various brands of the commercial Portland cement. The predictive values indicated that there exists a significant difference with respect to the chemical compositions among cement brands. The variation in chemical composition may be attributed to the differences in the proportioning of raw materials and the nature of production used to produce Portland cement. However, with respect to Portland cement brands, any of them could be used for construction since there were no significant differences.

5. Conclusions

The chemical compositions of commonly used Portland cement in Ghana were analyzed with both spectroscopic analysis and statistical tools. Generally, Student's t-test results confirmed that, with regards to chemical composition, all commonly used cement in the country has no deviation from standard requirements prescribed from the literature. This therefore shows that any of the commercial cement brands is very suitable for construction or concrete works. However, the Anova output indicated that each brand of commercial Portland cement has individual variations with respect to chemical composition. This is due to the differences that exist with individual factory proportioning of raw materials for Portland cement production.

Conflict of Interests

The authors declare that there is no conflict of interests regarding the publication of this paper.

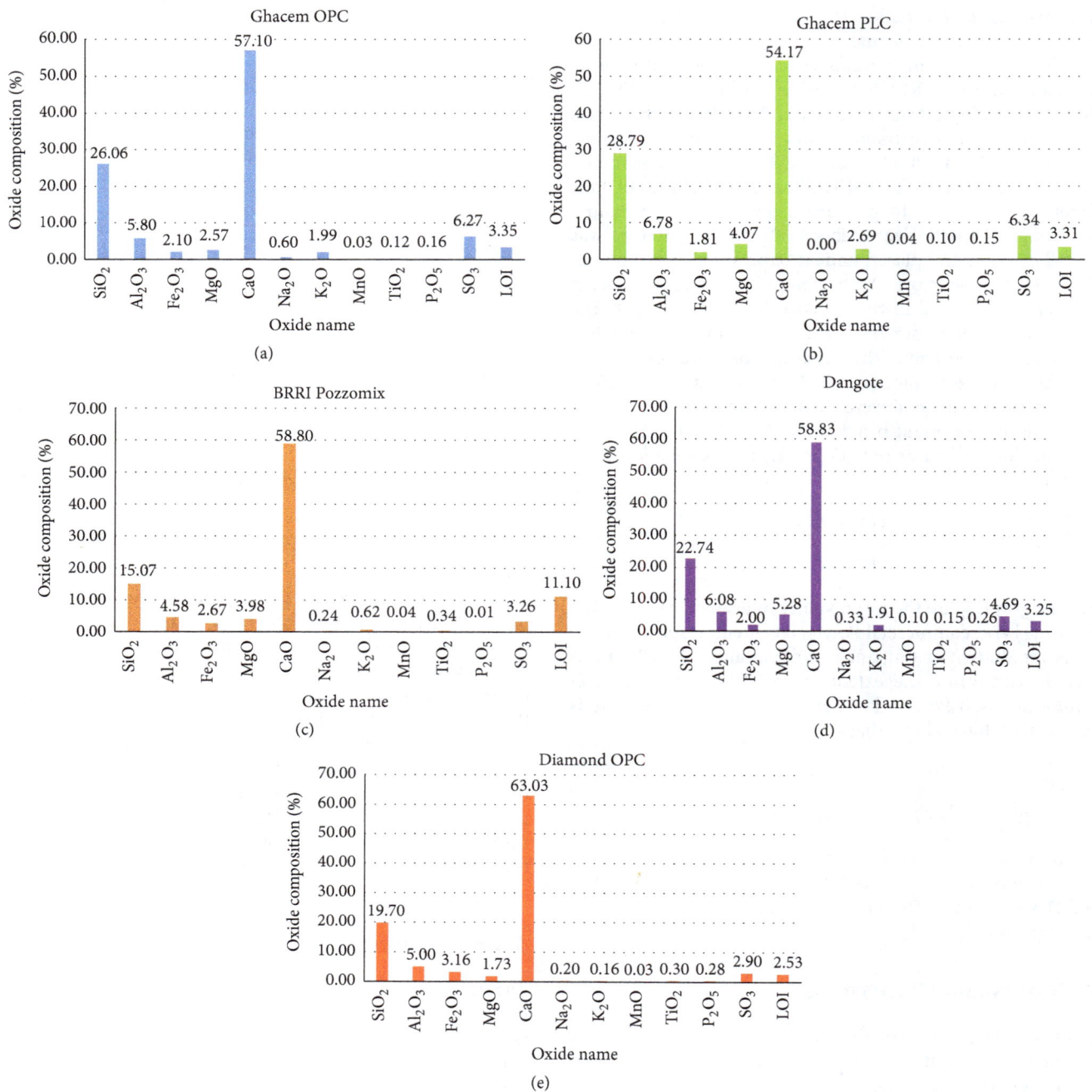

(a)

(b)

(c)

(d)

(e)

FIGURE 1: Oxides composition of commercial Portland cement in Ghana.

TABLE 3: ANOVA results of commercial Portland cement.

Source of variation	SS	df.	MS	F	P value	$F_{crit.}$
Rows	11320.3788	8	1415.05	167.0854	$3.927E-19$	2.35508149
Columns	6.684288889	3	2.2281	0.263088	0.8512781	3.00878657
Error	203.2561111	24	8.469			
Total	11530.3192	35				

SS: sum of squares; df.: degree of freedom; MS: mean square; F: F stat; $F_{crit.}$: F critical.

References

[1] C. M. Hanson, "Concrete: the advanced industrial material of the 21st century," *Metallurgical & Materials Transactions A*, vol. 26, pp. 1321–1341, 1995.

[2] M. Bediako, S. K. Y. Gawu, and A. A. Adjaottor, "Suitability of some Ghanaian mineral admixtures for masonry mortar formulation," *Construction and Building Materials*, vol. 29, pp. 667–671, 2012.

[3] S. H. Kosmatka, B. Kerkhoff, and W. C. Panarese, *Design and Control of Concrete Mixtures*, Portland Cement Association, Skokie, Ill, USA, 14th edition, 2002.

[4] M. S. Mamlouk and J. P. Zaniewski, *Materials for Civil and Construction Engineers*, Prentice Hall, Upper Saddle River, NJ, USA, 2006.

[5] T. Punmatharith, M. Rachakornkij, A. Imyim, and M. Wecharatana, "Co-processing of grinding sludge as alternative raw material in portland cement clinker production," *Journal of Applied Sciences*, vol. 10, no. 15, pp. 1525–1535, 2010.

[6] D. N. Huntzinger and T. D. Eatmon, "A life-cycle assessment of Portland cement manufacturing: comparing the traditional process with alternative technologies," *Journal of Cleaner Production*, vol. 17, no. 7, pp. 668–675, 2009.

[7] F. M. Lea, *The Chemistry of Cement and Concrete*, Arnold Publishers, London, UK, 3rd edition, 1970.

[8] J. F. Young, S. Mindess, R. J. Gray, and A. Bentur, *The Science and Technology of Civil Engineering Materials*, Prentice-Hall, Upper Saddle River, NJ, USA, 1998.

[9] H. F. W. Taylor, *Cement Chemistry*, Thomas Telford, London, UK, 2nd edition, 1997.

[10] S. H. Kosmatka and M. L. Wilson, *Design and Control of Concrete Mixtures*, Portland Cement Association, Stokie, Ill, USA, 2011.

[11] V. Sata, C. Jaturapitakkul, and K. Kiattikomol, "Influence of pozzolan from various by-product materials on mechanical properties of high-strength concrete," *Construction and Building Materials*, vol. 21, no. 7, pp. 1589–1598, 2007.

[12] A. Neville, *Neville on Concrete*, ACI, Farmington Hills, Mich, USA, 2003.

2

Rheological Method for Alpha Test Evaluation of Developing Superplasticizers' Performance: Channel Flow Test

Jae Hong Kim, Jin Hyun Lee, Tae Yong Shin, and Jin Young Yoon

School of Urban and Environmental Engineering, Ulsan National Institute of Science and Technology, Ulsan 44919, Republic of Korea

Correspondence should be addressed to Jae Hong Kim; jaekim@unist.ac.kr

Academic Editor: Doo-Yeol Yoo

Advance in high-range water-reducing admixture revolutionizes the workability and constructability of conventional vibrated concrete as well as self-consolidating concrete. Its need from construction fields has increased, and consequently a variety of new-type polycarboxylates, base polymers for the admixture, are being formulated in these days. Synthesizing new polymers needs a quick, but reliable, test to evaluate its performance on concrete. The test is also asked for selecting the best applicable brand of them before a test concrete will be mixed. This paper proposes a "channel flow test" and its usage for the purpose. The proposed procedure for the test includes the mix proportion of a test mortar, the test method, and rheological interpretation of the test results.

1. Introduction

Polycarboxylate- (PCE-) based high-range water-reducing admixture (HRWRA) was firstly introduced by Nippon Shokubai, Inc. (cement dispersant, JP 842,022; S59-018338; 1981), and then it became one of the most important polymers for chemical admixtures [1–3]. The merits of PCE, a raw material for HRWRA, can be found on its superior performance based on steric repulsion and its variability to control the time of its functioning. Changing its polymeric structure, such as trunk chain length and grafts' configuration, allows us to control the degree and speed of its absorption on cement grains, which results in controlling the time-dependent fluidity of cement-based materials.

A variety of polymeric structures have been developed to respond to needs from construction fields. A simple test to evaluate its performance in the middle of development is accordingly required in a polymer-synthesizing lab. Testing its application on concrete mixtures is certainly necessary at the end. The former can be called "alpha test," evaluating it in the producing lab, and the latter is "beta test" to evaluate its final blend with the field materials such as Portland cement and aggregates. A mini slump flow test [3–5] using a Hager-man cone has been widely used for the alpha test, but it has several limitations on considering the aggregate effect and the

sensitivity on fluidity [6]. The cone used for the mini slump flow test was originally designed for a thick mortar. Its flow is triggered by dropping the sample using a so-called flow table.

This paper proposes the use of a channel flow test as an alpha test of PCEs. The volume of a sample increases compared to the mini slump flow test, which expects to get a stable measure of the test result. Finding the mix proportion for a test mortar is also one of the important tasks. The fluidity of the test mortar should represent that of concrete to be applied. Finally, a model functions to evaluate the rheological properties based on the results of channel flow test of the test mortar.

2. Channel Flow Test for a Test Mortar

2.1. Apparatus. The channel flow test is designed to sensitively measure the enhanced fluidity of a mortar sample. A volume of 100 mm × 100 mm × 100 mm (its total volume is 1 L) of a mortar sample is placed in the cubic space surrounded by a gate and walls, as shown in Figure 1. The higher volume of a sample, compared to the mini slump flow test (180 mL), increases the reliability of the measurements. Lifting up the gate induces the mortar flow by its self-weight. When the flow stops, the final length of the channel flow and the time to get the final length or 500 mm approach are measured. The channel guided to one-direction flow gives higher sensitivity

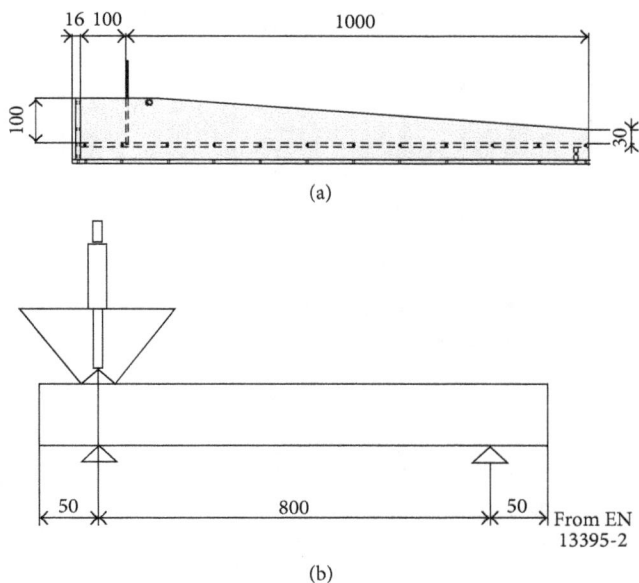

FIGURE 1: Equipment for (a) channel flow test and (b) grout flow test (dimensions are in mm).

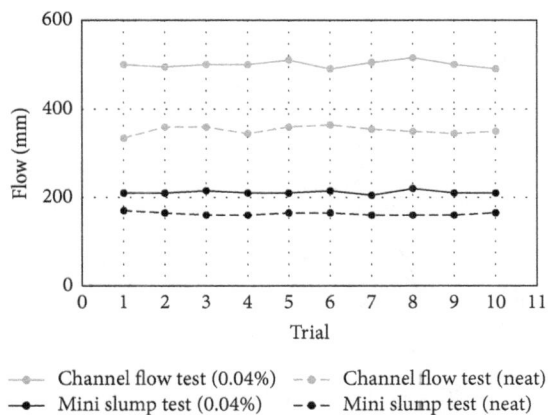

FIGURE 2: The result of resolution and repeatability.

than the radial flow by the mini slump flow test. The one-directional flow is based on the idea of the grout flow test (EN 13395-2. Products and systems for the protection and repair of concrete structures. Test methods. Determination of workability. Test for flow of grout or mortar). While the grout flow test permits the drop of a mortar sample from the charging hopper, a sample in the channel flow test is not subjected to such a dynamic motion. Note that the cross section of the grout flow apparatus is 130 mm in width by 75 mm in height and a one liter sample of grout is required for the test.

2.2. Resolution, Sensitivity, and Stability. The prototype of the channel flow test was priorly applied to only indicate the fluidity of mortar samples, where its application was acceptable [7, 8]. More pretests using various cement paste and mortar mixes support the fact that the range of 300 mm to 700 mm is acceptable for the channel flow. A sample showing less than 300 mm channel flow is so thick that it collapses when the gate opens. The collapse of a sample is not wanted to evaluate its flow behavior. On the other hand, a sample showing more than 700 mm channel flow is susceptible to segregation. Assuming a thin layer is composed at the end of the high channel flow, a sample showing the channel flow of 700 mm would be 14.3 mm thick (1 L-volume divided by the planer section 100 mm × 700 mm). The front-end thickness is lower than the calculation in practice because the top surface is inclined. Fine aggregates have the maximum grain of 5 mm, which is higher than the one-third of the thin layer thickness. The aggregate segregation is likely to occur in such a case. Note that the effective range of the mini slump flow is considered as 200 mm to 400 mm with the same reason.

A resolution test is accomplished in comparison to the mini slump flow test. Figure 2 shows the results of 10 replicated mixes, where two samples were used: one was a neat cement paste prepared by the water-to-cement ratio (w/cm)

of 0.40 and the other having the same w/cm incorporated a HRWRA (0.04% dosage per cement mass). The superplasticzed cement paste gave the channel flow of 500 ± 13 mm, which is approximately ±3% variation on their average. Its mini slump flow was 210 ± 8 mm (±4% variation approximately). The neat cement paste showed the channel flow of 350 ± 15 mm (±4% variation) and the mini slump flow of 165 ± 5 mm (±3% variation). On the other hand, comparing the HRWRA effect on both flows indicates the sensitivity of the tests. While the mini slump flow showed 27% increase in its measurement (from 165 mm to 210 mm), the channel flow did 43% increase (from 350 mm to 500 mm). Therefore, it can be said that the channel flow test has better sensitivity on the fluidity of a sample and its resolution is in the same range, within ±4%, of the mini slump flow.

The viscosity of a sample is generally related to the time of spreading. One measure to consider the spreading time is the time to get 500 mm spread, similar to the slump flow test. The other measure for the viscosity evaluation is the time to get the final spread, which showed better correlation on the mini slump flow test [4]. The measuring stability of each time was compared in Figure 3. A total of 4 replicated mixes having w/cm = 0.35 and sand-to-cement ratio of 1.5 by mass were tested. A commercially available HRWRA (0.06% dosage per cement mass) was applied. For a single test, both times for 500 mm and the final spread were measured together. As can be compared in Figure 3, the times for the final spread and 500 mm spread showed 10 s variation (±22%) and 3 s variation (±23%), respectively. The value for 500 mm spread was more stable, even though the resolutions for each of the measurements are similar in percentage. Therefore, the use of the time for 500 mm spread is recommended because the viscosity is very sensitive to the value of spreading time.

2.3. Rheological Analysis for the Results. Developing a relationship between the rheological properties and the results of a field test allows us to have quantitative understanding on the flow behavior. A theoretical analysis [9] concluded an inverse relationship between the yield stress and the slump flow of a concrete mix. The relationships for the mini slump flow test and marsh cone test were summarized and compared [5],

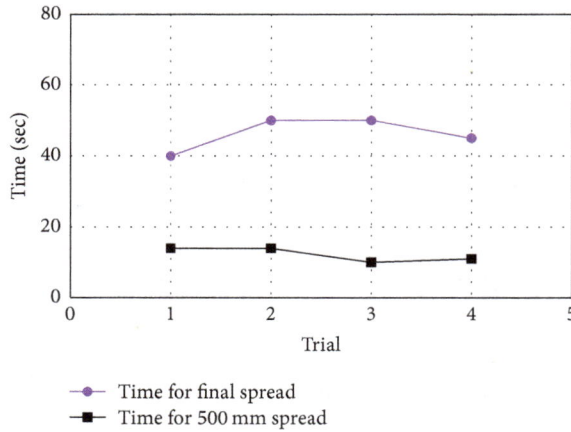

FIGURE 3: Repeatability on the measurement of the spreading time.

where the volume-of-fluid (VOF) simulation was adopted. For example, the yield stress (τ_y) and the plastic viscosity (η_p) can be obtained from the measurement of the mini slump flow test [4]:

$$\tau_y = 0.00660 D_f^{-5.81}, \tag{1}$$

$$\eta_p = \tau_y \left(0.00641 T_f - 0.00194 \right), \tag{2}$$

where D_f and T_f are the mini slump flow, in meter, and the time to get the final mini slump flow, in second, respectively.

The volume-of-fluid (VOF) technique was also applied to simulate the channel flow test. A half-symmetric model was composed with 8-node hexagonal elements. The average mesh size was 10 mm, which is small enough to have an accuracy in the flow simulation of cement-based materials [10]. Thus, the number of elements was 2,700 for modeling an 800 mm long channel. The time increment for explicit computation was 0.1 s, and the channel flow was simulated for 65 s. Input variables for the flow simulation were yield stress and plastic viscosity assuming a mortar sample is a Bingham model fluid. Its density was set to 2,200 kg/m³. Figure 4 shows an example result of flow simulation: (a) snapshots of the VOF simulation and (b) time spread curve for a Bingham fluid having 15 Pa yield stress (τ_y) and 30 Pa·s plastic viscosity (η_p). The front of the channel flow was determined by the volume fraction of each element.

In order to establish database of the flow simulation, each spread curve was fitted by a logarithmic function:

$$C(t) = \alpha + \beta \ln(t), \tag{3}$$

where α and β are the parameters related to the final length of spread and its damping, respectively. As shown in Figure 4(b), the logarithmic function showed a satisfactory fitting than an exponential function. The coefficient of determination was higher than 0.95 for most of spread curves. Assigning a cut-off velocity (V_f) of 1 mm/s allows us to catch the time for the Bingham fluid stops: its flow front stops

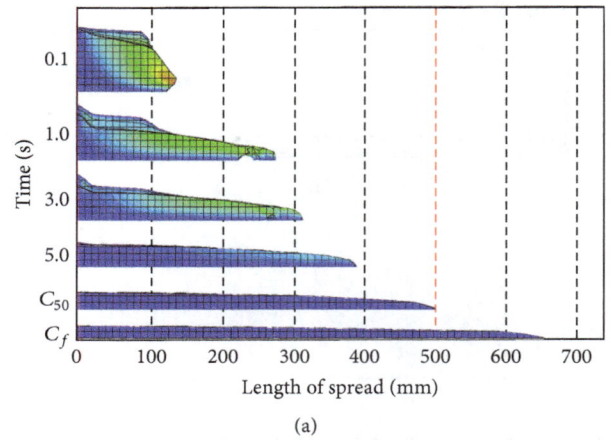

(a)

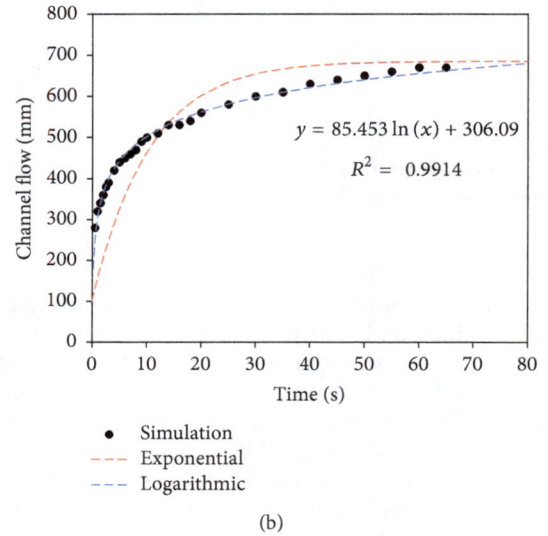

(b)

FIGURE 4: Simulation contour and spread curve of channel flow simulation.

moving at the time of β/V_f from the derivative of (1). The final spread of the channel flow test is then calculated by

$$C_f = \alpha + \beta \ln \left(\frac{\beta}{V_f} \right), \tag{4}$$

which is a measure of the channel flow (C_f). In addition, the time for its 500 mm flow is

$$C_{50} = \exp \left(\frac{500 - \alpha}{\beta} \right). \tag{5}$$

The flow simulations were conducted for the range of 1 to 60 Pa yield stress and 0 to 100 Pa·s plastic viscosity. A total of 48 combinations were parameterized by (3). Finally, the database composed of τ_y, η_p, C_f, and C_{50} developed using (4) and (5).

Figure 5(a) shows the correlation between the rheological properties and the results of the channel flow test. The channel flow (C_f) was inversely proportional to the yield stress of

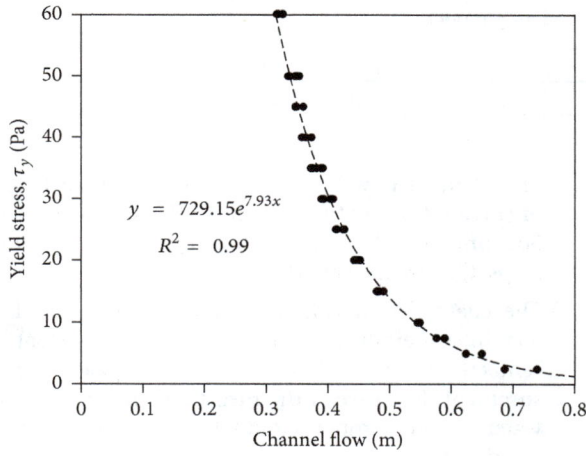

(a)

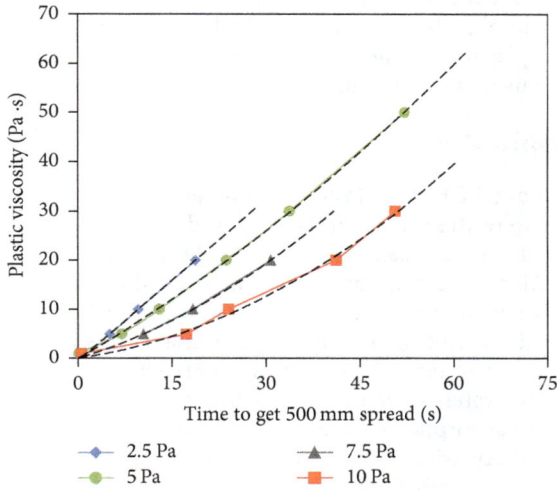

(b)

FIGURE 5: Correlation for the rheological evaluation.

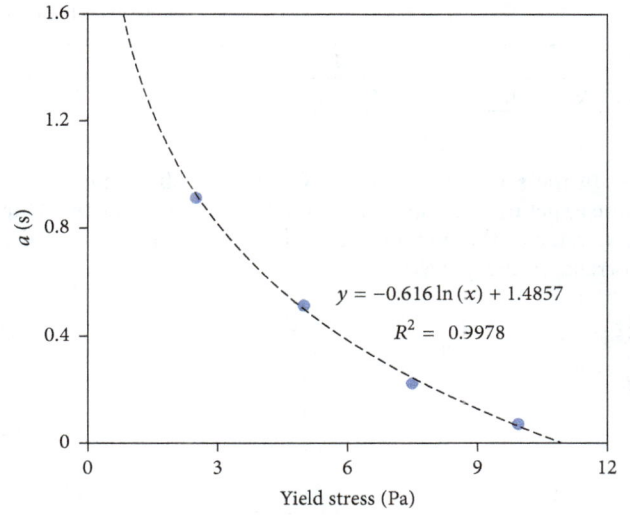

(a)

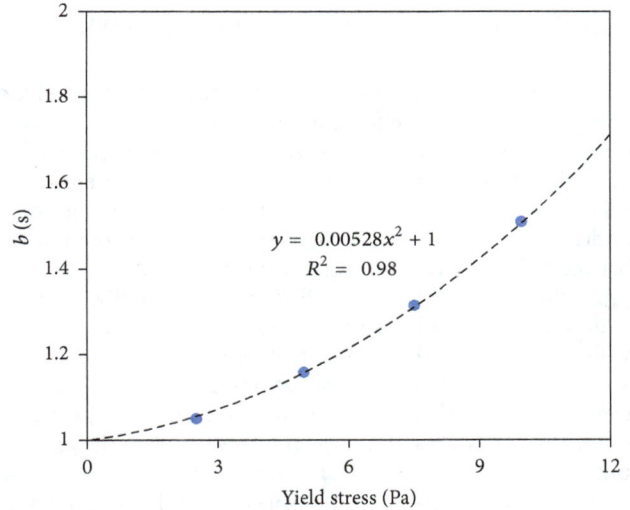

(b)

FIGURE 6: Regression analysis for the model parameters a and b.

a Bingham fluid, which is similar to the tendency for the slump flow. The inverse proportionality was modeled by

$$\tau_y = k_1 \exp\left(-\frac{C_f}{k_2}\right), \tag{6}$$

where k_1 and k_2 are fitting coefficients in the unit of "Pa" and "mm," respectively. Their values determined by a linear regression are listed in Table 1. In addition, as shown in Figure 5(b), the 500 mm flow time had a power-law relation with the plastic viscosity when the yield stress is fixed. Encapsulating the effect of the yield stress in the model parameters gave

$$\eta_p = a \cdot \left(\frac{C_{50}}{k_6}\right)^b, \tag{7}$$

where k_6 is a coefficient determining the unit of the plastic viscosity, here in "s." The model parameters, a and b, were then functions of the yield stress:

$$a = -k_3 \ln\left(\frac{\tau_y}{k_4}\right),$$

$$b = \left(\frac{\tau_y}{k_5}\right)^2 + 1, \tag{8}$$

where a and b were dimensionless parameters, and the units of k_3, k_4, and k_5 were determined to "Pa·s," "Pa," and "Pa," respectively, for nondimensionalization. Excepting the cases having a high yield stress ($\tau_y \geq 10$ Pa), the coefficients of determination were larger than 0.99 for a linear regression with respect to a and b. Figure 6 shows the results of the linear regression for the η_p-C_{50} relationship. The high yield stress cases also follow the trends in (8) with marginal errors.

TABLE 1: Coefficients for the correlating model.

k_1	k_2	k_3	k_4	k_5	k_6	k_7	k_8	k_9	k_{10}	k_{11}
729 Pa	126 mm	0.616 Pa·s	11.2 Pa	13.8 Pa	1.0 s	3.0 s	1568 Pa	46.1 s	11.6 Pa	1 Pa·s

In the same way as the η_p-C_{50} relationship, that for the time to get the final spread (η_p-C_{tf}) could be correlated. One may refer to the previous study [11]. Here, the result of the correlation is reported:

$$\eta_p = k_{11} \exp\left(\frac{C_{tf} - d}{c}\right),$$

$$c = -k_7 \ln\left(\frac{\tau_y}{k_8}\right), \tag{9}$$

$$d = k_9 \exp\left(\frac{-\tau_y}{k_{10}}\right),$$

where the value of each coefficient is listed in Table 1.

2.4. Mix Design of a Test Mortar. PCEs are usually applied for having the fluidity of self-consolidating concrete (SCC), and then a test mortar needs to represent the rheophysical state of SCC. The rheophysical state of a concentrated suspension is usually described by solid volume fraction [12]. Suspension models such as Krieger-Dougherty equation also indicate that the volume fraction of aggregates is critical [13, 14]. The aggregate volume fraction for SCC is intentionally reduced to decrease its yield stress, and its water-to-cement ratio is also reduced to have better stability. For example, an SCC's mix proportion by mass ratio is given by 0.33 : 1 : 1.63 (cementitious binder, water, and fine and coarse aggregates, resp.). The volume fraction of aggregates is 0.65 in the mix. In comparison, normal concrete is proportioned by 0.56 : 1 : 2.74 and the volume fraction of aggregates is 0.71. Finally, the mix proportion for a test mortar is regularized by 0.35 : 1 : 1.5 for considering the rheophysical state of SCC.

Using a bag of standard sand (ISO 679. Methods of testing cements. Determination of strength) makes it easy to prepare a sample. The mass of the bag is 1.35 kg, and then those of water and cement become 0.315 kg and 0.9 kg, respectively. Approximately 1.12 L of a test mortar is then produced. Note that ISO 679 provides the mix proportion of 0.5 : 1 : 3 for measuring the cement strength. The mix proportion for evaluating the PCE effect needs a higher water-to-cement ratio (0.35) and relatively smaller content of sand (1.5).

The optimal procedure to evaluate the performance of PCE is finally established:

(1) A test mortar sample is prepared with the cement-to-sand ratio of 1.5 and the water-to-cement ratio of 0.35. Approximately 1.12 L will be obtained with 0.315 kg-water, 0.9 kg-cement, and 1.35 kg-standard sand. The dosage of PCE needs to be adjusted to get the channel flow of 300 to 700 mm.

(2) The raw materials are mixed for 5 min using a planetary mixer.

(3) The channel flow test is followed. The final length of spread, C_f, the time taken for the sample to have 500 mm spread, C_{50}, and the time when the flow stops, C_{tf}, are measured.

(4) The channel flow test is repeated at an interval of 30 min, generally for 2 h, where it is important to remix the sample for 1 min before each repeating measurement. In between the measurements, the loss of water in the sample should be prevented using a plastic cover.

(5) The measured values of C_f and C_{50} at 0, 0.5, 1.0, 1.5, and 2.0 h are reported. Finally, the yield stress and plastic viscosity of a test mortar can be computed using the correlating equations.

3. Application

A total of 6 PCEs were tested. Monomers for their polymerization were the same, acrylic acid and polyethylene oxide. All of them were categorized in the MPEG-type [1]. Their molecular structures, however, varied by different process of polymerization, which results in showing various performance. Following their designed performance, the PCEs were divided into water-reducing type or consistency-maintaining type. The water-reducing PCEs, labeled by LA, LB, and LC, show fast adsorption on cement grains, and then a mix shows high fluidity with a small dosage. In terms of total solid content, a water-reducing PCE was incorporated in 0.12% of cement mass. The consistency-maintaining PCEs, labeled by LD, LE, and LF, were in 0.16% of cement mass. They were adsorbed rather slowly and the samples incorporating them show less fluidity.

Table 2 shows the results of the channel flow test accompanied with the mini slump flow test. The channel flow test was applied into the test mortar following the optimal procedure in Section 2.4, but the mini slump flow test was done with a cement paste having the water-to-cement ratio of 0.34. The dosage for each PCE was the same for both tests.

The channel flows for the water-reducing PCEs were similar to each other at 0 h, and they decreased over time. Only 71%, 54%, and 73% of channel flow of LA12, LB12 and LC12, respectively, were maintained for 2 h. However, the corresponding mini slump flows showed discrepancy: LB12 showed much less fluidity even at 0 h. Loss of mini slump flow was much smaller than that of channel flow. Their application in concrete mixes supports the trend of the channel flow, where the loss of slump flow was observed with all the water-reducing PCEs.

In the case of consistency-maintaining PCEs, LE16 and LF16 had rather lower initial channel flows, and then each of them had nearly doubled in 2 h. Note that the values in the parentheses exceed the limit of the measuring range, which was 300–700 mm for the channel flow and 200–400 mm

TABLE 2: Performance test results.

Label	Mini slump flow of pastes (mm)			Channel flow of mortars (mm)		
	0 h	1 h	2 h	0 h	1 h	2 h
LA12	400	390	360	635	555	450
LB12	290	260	260	645	430	350
LC12	360	350	340	640	610	470
LD16	280	300	310	520	360	320
LE16	260	370	400	(235)	430	510
LF16	310	(440)	(440)	300	530	560

TABLE 3: Rheological evaluation on the results of channel flow test.

Label	C_f (mm)	C_{50} (s)	τ_y (Pa)	η_p (Pa·s)
LA12	635	71	4.7	10.1
LB12	645	62	4.4	5.58
LC12	640	63	4.5	6.14
LD16	520	36	11.8	3.72
LE16	235	5	113	1.88
LF16	300	7	67.4	2.07

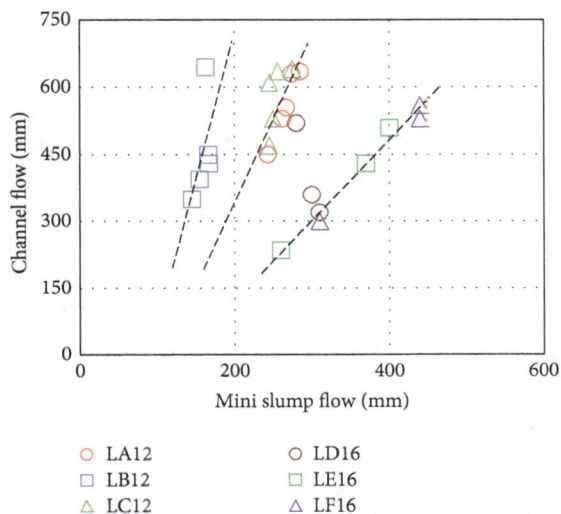

○ LA12	○ LD16	
□ LB12	□ LE16	
△ LC12	△ LF16	

FIGURE 7: Comparison of channel flow and mini slump flow.

for mini slump flow as previously stated. The results of the mini slump flow corresponded with the increasing trend of fluidity and comparatively worse performance with LE16. LD16 showed a similar increase in the mini slump flow. However, its channel flow initially gave the highest values and it lost the fluidity within 1 h.

Figure 7 comparatively shows the channel flow of the test mortars and the mini slump flow of the test pastes. Each linear line provides a correlation between two measurements. The correlations for the water-reducing PCEs again indicate poor sensitivity of the mini slump flow of the test pastes: it did not reveal the loss of fluidity. Even though the results of consistency-maintaining PCE samples are on a single line, the right line in the figure, the other PCE samples are far from the coherent trend. Therefore, it can be concluded that the mini slump flow of the test pastes does not represent the flow behavior of the corresponding mortar samples. The use of a test mortar, accompanied with the channel flow test, is believed to better explain the fluidity of concrete mix.

More investigation on the performance of PCE could be made by rheological interpretation. For the initial measurement of channel flow test, (4) to (6) calculated the rheological properties of the test mortars. Table 3 reports the calculated properties. Even though the water-reducing PCEs show the same fluidity ($C_f \approx 640$ mm and $\tau_y \approx 4.5$ Pa), the plastic viscosity is different: 10.1, 5.58, and 6.14 Pa·s for LA12, LB12, and LC12, respectively. A PCE having lower plastic viscosity is preferable for higher workability of concrete.

4. Discussion

The channel flow of a test mortar is also dependent on the characteristics of constituting materials such as cement and aggregates. Therefore, in order to compare the performance of PCEs, it is required to use the same materials for producing a test mortar. The variation according to the constituting materials is summarized here for future reference.

4.1. Batch of Portland Cement. Even if Portland cement, type I, is manufactured by an identical company, the mineral contents of cement can be different depending on the manufacturing time and process of production [15]. Table 4 shows the oxide composition of Portland cement from three different batches. The channel flows of LC12 using the three cements were in the range of 490 ± 150 mm (517.7 ± 5.91 in). The variation is approximately ±31%, which is beyond the resolution of the test (±4%) and then not an error factor.

4.2. Types of Binder. Blended cement is generally used to increase the fluidity [16] and the resistance against segregation. In this paper, a ternary binder (40% cement A, 50% ground-granulated blast furnace slag, and 10% fly ash) was compared. The test mortar using the ternary binder was segregated with LC12, while the channel flow of 100% cement A-LC12 was 550 mm (21.7 in), as reported in Table 4. When the PCE dosage was decreased, the channel flow of 100% cement A-LC10 was 340 mm (13.4 in) and that using the ternary binder was 900 mm (35.4 in). In addition, the channel

TABLE 4: Oxide composition of three batches.

Label	CaO	SiO$_2$	Al$_2$O$_3$	Fe$_2$O$_3$	SO$_3$	MgO	K$_2$O	Na$_2$O	TiO$_2$	C_f for LC12
Cement A	64.0%	19.3%	4.4%	3.8%	3.4%	2.4%	1.3%	0.4%	0.3%	550 mm
Cement B	64.6%	18.0%	4.8%	3.3%	3.7%	3.6%	1.1%	0.2%	0.3%	640 mm
Cement C	64.9%	17.9%	4.7%	3.3%	3.7%	3.6%	1.1%	0.2%	0.3%	345 mm

TABLE 5: The rheological properties of each mortar sample.

Type	TSC/C	τ_y (Pa)	η_p (Pa·s)
BE	0.18%	26.13	2.96
	0.21%	19.03	3.18
	0.24%	12.80	1.97
	0.27%	6.27	1.28
LC	0.10%	49.26	3.59
	0.12%	9.32	3.53
	0.14%	7.35	4.84

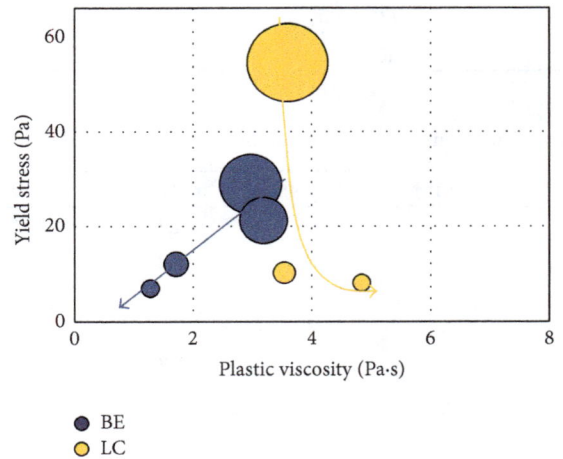

• BE
○ LC

FIGURE 8: Rheograph showing the effect of PCE dosage.

flow of LC10 increases up to 780 mm (30.7 in) when the water-to-cement ratio increased from 0.35 to 0.40. Thus, the use of ternary binder is very effective in increasing the flow of cement-based materials.

4.3. Types of Sand. The standard sand is required for preparing a test mortar. The use of the other sand dramatically changes the channel flow even though its quality is "good" and its grading is comparable with the standard sand. A comparison was made with good-quality river sand. The average particle size and fineness modulus of river sand are 0.45 mm (0.0178 in) and 2.60, respectively, while those of standard sand were 0.77 mm (0.0303 in) and 2.51. Their difference was highlighted on the absorption rate: 0.79% versus 2.20% for the standard sand and the river sand, respectively. Adding 7.11 g (0.0157 lb; 0.79%-absorption) water to compensate the water adsorption of standard sand increased 10 mm channel flow of a test mortar, where the reference was 550 mm (21.7 in) for LC12, 100% cement A. However, the mortar proportioned with river sand did not flow in spite of adding 19.8 g (0.0437 lb; 2.20%-absorption) water. Inconsistency on the aggregate changes to a great extent the rheophysical state of a test mortar.

4.4. Dosage of PCE. The effect of PCE dosage can be compared with the results of channel flow test. For the mortar samples with a different dosage, the channel flow was firstly conducted. Table 5 shows the test results, where BE and LC are different-type PCEs. The dosage is reported in total solid content (TSC) per cement by mass. The rheograph, originally developed for SCC [17], was developed in Figure 8, where the PCE dosage is represented with the size of each symbol. A smaller dosage is depicted with a bigger circle symbol, which illustrates higher sensitivity to the dosage dependence. Adding more BE sample decreases both yields stress and

plastic viscosity of a mix. However, LC sample dominantly decreases the yield stress first, and then its excessive dosage (0.12% to 0.14%) increases the plastic viscosity, while its yield stress holds a constant value.

5. Conclusions

This paper has introduced the alpha test to evaluate the performance of PCE or HRWRA before a test concrete is mixed. The alpha test needs to be quickly finished and less loaded in terms of labor. The mini slump flow test for a test paste, currently widely used, has the limit on its sensitivity and representing the aggregate effect. The channel flow test with a test mortar is therefore proposed. The final spread and the time to get the stoppage of the spread are measures of the test, and a model converting them into rheological properties is also developed. The mix proportion of a test mortar is taken to represent the aggregate volume fraction of SCC. Future work will focus on ways to stack performance data of various PCEs or HRWRAs and quantitatively evaluate their performance grade.

Competing Interests

The authors declare that they have no competing interests.

Acknowledgments

This research was supported by a grant from Smart Civil Infrastructure Research Program funded by Ministry of

Land, Infrastructure and Transport (MOLIT) of Korea government (13SCIPA02) and Korea Agency for Infrastructure Technology Advancement (KAIA).

References

[1] J. Plank, E. Sakai, C. W. Miao, C. Yu, and J. X. Hong, "Chemical admixtures—chemistry, applications and their impact on concrete microstructure and durability," *Cement and Concrete Research*, vol. 78, pp. 81–99, 2015.

[2] E. Sakai, A. Ishida, and A. Ohta, "New trends in the development of chemical admixtures in Japan," *Journal of Advanced Concrete Technology*, vol. 4, no. 2, pp. 211–223, 2006.

[3] S. Hanehara and K. Yamada, "Rheology and early age properties of cement systems," *Cement and Concrete Research*, vol. 38, no. 2, pp. 175–195, 2008.

[4] N. Tregger, L. Ferrara, and S. P. Shah, "Identifying viscosity of cement paste from mini-slump-flow test," *ACI Materials Journal*, vol. 105, no. 6, pp. 558–566, 2008.

[5] A. Bouvet, E. Ghorbel, and R. Bennacer, "The mini-conical slump flow test: analysis and numerical study," *Cement and Concrete Research*, vol. 40, no. 10, pp. 1517–1523, 2010.

[6] K. Yamada, T. Sugamata, and H. Nakanishi, "Fluidity performance evaluation of cement and superplasticizer," *Journal of Advanced Concrete Technology*, vol. 4, no. 2, pp. 241–249, 2006.

[7] J. H. Kim, H. J. Yim, and S. H. Kwon, "Quantitative measurement of the external and internal bleeding of conventional concrete and SCC," *Cement and Concrete Composites*, vol. 54, pp. 34–39, 2014.

[8] J. H. Lee, J. H. Kim, and M. K. Kim, "Fine aggregates size effect on rheological behavior of mortar," *Journal of the Korea Academia-Industrial cooperation Society*, vol. 16, no. 8, pp. 5636–5645, 2015.

[9] N. Roussel and P. Coussot, "'Fifty-cent rheometer' for yield stress measurements: from slump to spreading flow," *Journal of Rheology*, vol. 49, no. 3, pp. 705–718, 2005.

[10] J. H. Kim, H. R. Jang, and H. J. Yim, "Sensitivity and accuracy for rheological simulation of cement-based materials," *Computers and Concrete*, vol. 15, no. 6, pp. 903–919, 2015.

[11] T. Y. Shin, J. H. Lee, J. H. Kim, and M. K. Kim, "Correlation between channel-flow test results and rheological properties of freshly mixed mortar," *Journal of the Korea Academia*, vol. 17, no. 7, pp. 237–244, 2016 (Korean).

[12] P. Coussot and C. Ancey, "Rheophysical classification of concentrated suspensions and granular pastes," *Physical Review E*, vol. 59, no. 4, pp. 4445–4457, 1999.

[13] I. M. Krieger and T. J. Dougherty, "A mechanism for non-newtonian flow in suspensions of rigid spheres," *Journal of Rheology*, vol. 3, no. 1, pp. 137–152, 1959.

[14] L. Struble and G.-K. Sun, "Viscosity of Portland cement paste as a function of concentration," *Advanced Cement Based Materials*, vol. 2, no. 2, pp. 62–69, 1995.

[15] R. P. Ferron, A. Gregori, Z. Sun, and S. P. Shah, "Rheological method to evaluate structural buildup in self-consolidating concrete cement pastes," *ACI Materials Journal*, vol. 104, no. 3, pp. 242–250, 2007.

[16] J. H. Kim, H. J. Yim, and R. D. Ferron, "In situ measurement of the rheological properties and agglomeration on cementitious pastes," *Journal of Rheology*, vol. 60, no. 4, pp. 695–704, 2016.

[17] O. H. Wallevik and J. E. Wallevik, "Rheology as a tool in concrete science: the use of rheographs and workability boxes," *Cement and Concrete Research*, vol. 41, no. 12, pp. 1279–1288, 2011.

3

Temperature Characteristics of Porous Portland Cement Concrete during the Hot Summer Session

Liqun Hu, Yangyang Li, Xiaolong Zou, Shaowen Du, Zhuangzhuang Liu, and Hao Huang

Key Laboratory of Special Area Highway Engineering of Ministry of Education, Chang'an University, Shaanxi, Xi'an 710064, China

Correspondence should be addressed to Liqun Hu; hlq@chd.edu.cn

Academic Editor: Hossein Moayedi

Pavement heats the near-surface air and affects the thermal comfort of the human body in hot summer. Because of a large amount of connected porosity of porous Portland cement concrete (PPCC), the thermal parameters of PPCC are much different from those of traditional Portland cement concrete (PCC). The temperature change characteristics of PPCC and the effects on surrounding environment are also different. A continuous 48-hour log of temperature of a PCC and five kinds of PPCC with different porosity were recorded in the open air in the hot summer. The air temperatures at different heights above concrete specimens were tested using self-made enclosed boxes to analyze the characteristics of near-surface air temperature. The output heat flux of different concrete specimens was calculated. The results show that the PPCC has higher temperature in the daytime and lower temperature in the nighttime and larger temperature gradient than the PCC. The air temperature above PPCC is lower than that of PCC after solar radiation going to zero at night. The total output heat flux of PPCC is slightly smaller in the daytime and significantly smaller at night than that of PCC. The results of tests and calculations indicate that PPCC contributes to the mitigation of heating effect of pavement on the near-surface air.

1. Introduction

Porous Portland cement concrete (PPCC), which is also known as pervious concrete, is a porous material, mixed, and molded by optimal percentages of water, cement, coarse aggregate, and additives. The range of air void content of PPCC is about 15% to 25% [1]. In recent years, with the improvement of the functional diversification requirements, engineers have paid more attention to the mechanical performance, water permeability, antisliding, and noise reduction of PPCC. Studies have shown that the PPCC not only could meet basic requirements of strength, but also have functions of breathability, permeability, antisliding, and noise reduction [2–6].

Pavement temperature periodically changes with the influence of solar radiation, air temperature, and wind. It is also significantly impacted by the properties of pavement materials. Therefore, temperature change of different pavements varies vastly even in the same environmental conditions. Li et al. found that the albedo of pavement surface had a momentous impact on the surface temperature.

The increase of albedo can sharply reduce the temperature of pavement surface in summer days [7]. Gui et al. analyzed the average maximum and minimum surface temperatures of pavement material with different thermophysical properties and found that the factor causing the highest decrease in the pavement maximum temperature is albedo, followed by thermal diffusivity, thermal conductivity, emissivity, and volumetric heat capacity, while the reduction in the pavement minimum temperature is most affected by emissivity, albedo, volumetric heat capacity, conductivity, and diffusivity in that order. Thermal conductivity, thermal diffusivity, and heat capacity have positive effects only on the pavement maximum temperature, but not on the minimum temperature with increasing values [8]. The surface and bottom temperatures of asphalt mixture slabs with different air void contents under steady state were analyzed. Hassn et al. found that, with the increase of air void content, the surface and bottom temperature of asphalt mixture slabs increased, and the maximum temperature of asphalt mixture slabs also increased [9]. Because PPCC have more air void content than traditional PCC, there are obvious differences between them. Studies by Haselbach et al. and Zhang et al. have shown that PPCC has rougher surface and lower albedo than traditional PCC pavement, so that PPCC can absorb more solar radiation [10, 11]. Kevern et al. collected temperature data of pervious pavement and traditional pavement at different depths, respectively, and found that the temperature at middle-level (8 cm below the surface) in both concrete types was always higher than the air temperature. The temperature at middle-level in pervious pavement was higher than that of traditional pavement about 5°C during daytime. However, in the night, the temperature at middle-level in both pavements would decrease and reach similar temperature [12]. However, research on the temperature change characteristics of PPCC pavement is still limited and further study is needed.

The change of pavement temperature has an impact on surrounding environment and the near-surface air temperature and affects the thermal comfort of human body. For the sake of environmental protection, researchers pay more attention to this area and conduct more research in this area. Benrazavi et al. observed thermal performances of two kinds of polished granite, cement concrete, and asphalt concrete pavements in three different environments, namely, open space, near water, and under shade. They found that there was constant heat exchanging between surface and air. In addition, surface temperature of pavement had important effects on the change of near-surface air temperature [13]. Through analyzing the change characteristics of air temperature at 1.5 m above the surface of impervious asphalt and concrete pavements, pervious brick pavers, and grass, Guan et al. found that surface temperatures showed a positive and moderately strong correlation to air temperatures when surface temperatures increased. However, the air temperature above the different materials did not mimic the trend shown in surface temperatures [14]. Lin et al. also found that surface temperature of asphalt concrete pavement showed a positive correlation to air temperature in summer and thought that the heat output of pavement significantly impacted the air temperature and mean radiant temperature [15]. Rosenfeld et al. found that the local air temperature decreased by 0.6°C when the pavement-surface albedo increased from 0.05 to 0.30 [16]. In the aspect of the impact on ambient temperature of PPCC, the research of Qin and Hiller showed that pervious pavement could increase surface albedo and evaporation, which can promote the cool effect of pavement [17]. The research of Li et al. and Kevern et al. had shown that pervious concrete pavements store less energy than traditional pavements, which had a lower thermal effect on near-surface air and could alleviate urban heat island effect [7, 12]. However, research in this area is still relatively insufficient.

PPCC have a smaller density and a good permeability because of 15% to 20% porosity. When used as a pavement material, PPCC is significantly different from PCC in thermal properties such as radiation, convection, and conduction due to rough surface and internal air. In order to understand the temperature change characteristics of PPCC and the impact of the temperature change on the surrounding environment, the data of surface and internal temperature of five kinds of PPCC and a PCC as reference at hot weather in summer were logged. Their impacts on the near-surface air temperature were studied by comparison of PPCC and PCC. The heat outputs of different types of concrete were calculated based on the model of pavement-surface energy balance. The results of the study would be helpful to further understand mutual influence law between PPCC and external environment.

2. Materials and Methods

2.1. Materials. Six types of concrete specimens were designed in this study, which were one kind of PCC and five kinds of PPCC. Different aggregates with single particle size were used to produce the PPCC. The dimensions of specimens were 30 cm (length) × 30 cm (width) × 15 cm (thickness). Parameters of those concrete are shown in Table 1. Albedo is the ratio of the reflected radiant flux to the incident solar radiant flux, which characterizes the ability of the pavement to absorb and reflect solar radiation [8]. Albedo and coefficient of thermal storage were calculated according to the references [9, 11, 18].

2.2. Temperature Monitoring of Concrete. The temperature of pavement changes under the effect of environment factors. Because PPCC contains a certain amount of porosity, the temperature change characteristics of PPCC have great difference from PCC. In order to obtain the temperature change characteristics of PPCC, the data of temperatures on the surface, internal (at the depths of 5 cm and 10 cm), and the bottom of specimens were logged. Six pits of 30 cm in length, 30 cm in width, and 15 cm in depth were dug in the natural clay soil ground. After compaction of soil at the bottom of the pits, graded crushed stone of 15 cm was filled and compacted as base course in the pits, and the six specimens were placed on the graded crushed stone layer to simulate pavement structures. The schematic figure of slabs specimens is presented in Figure 1(a).

Thermal resistance temperature sensors of PT100 with precision of 0.1°C were embedded in each specimen at the depths of 0 cm, 5 cm, 10 cm, and 15 cm from the surface. DT85G produced by Datataker in Canada was applied as Data Collector, whose temperature data were automatically recorded in every 15 min. In order to avoid the influence of the surrounding air on the test results, specimens were tightly wrapped with foam boards. A Weather Station was also placed closed to the specimens to log the data of weather conditions. The photograph of the experimental setup is shown in Figure 1(b).

2.3. Near-Surface Air Temperature Monitoring. The change of pavement temperature has impact on the near-surface air temperature and affects the human body's comfort. In order to study the influence of the heated pavement on the near-surface air temperature, the specimens were heated by exposing under the solar radiation from 9:00 to 15:15, and then the near-surface air temperature was investigated using five temperature sensors that were settled at the heights of 25 cm, 50 cm, 75 cm, 100 cm, and 130 cm from the surface of specimen. Because the size of the specimens is small, it is difficult to test the influence of the specimens on the

TABLE 1: Parameters of concrete.

Types of concrete	Particle size/mm	Porosity/%	Apparent density/kg·m^{-3}	Albedo	Emissivity	Thermal conductivity/W·m^{-1}·°C^{-1}	Specific heat/J·kg^{-1}·°C^{-1}	Coefficient of thermal storage/W·m^{-2}·°C^{-1}
PCC$_{0.075-16.0}$	0.075–16.0	0.8	2100	0.4320	0.96	1.40 [9, 19]	1050 [9, 18]	14.99
PPCC$_{2.36-4.75}$	2.36–4.75	15.1	2048	0.3557	0.95	1.28	920	13.25
PPCC$_{4.75-9.50}$	4.75–9.50	18.2	1974	0.3401	0.94	1.26	900	12.77
PPCC$_{9.50-13.2}$	9.50–13.2	20.1	1928	0.3305	0.94	1.23	870	12.26
PPCC$_{13.2-16.0}$	13.2–16.0	21.8	1887	0.3219	0.93	1.22	860	12.01
PPCC$_{16.0-19.0}$	16.0–19.0	24.9	1812	0.3063	0.92	1.20	840	11.53

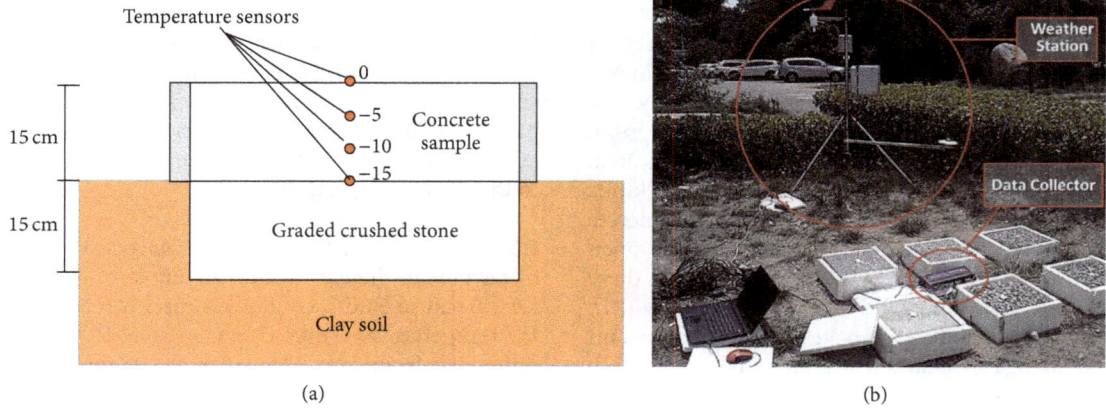

(a)   (b)

FIGURE 1: Temperature monitoring system: (a) schematic figure of slabs specimens; (b) photograph of the experimental setup.

near-surface air temperature. In addition, the near-surface air temperature is easily affected by the surrounding air temperature. In order to enlarge the impact of the heated specimens on the near-surface air temperature and avoid the influence of the surrounding air, foam board boxes of 150 cm (height) × 30 cm (length) × 30 cm (width) were placed above the specimens. The schematic figure of near-surface air temperature monitoring device is shown in Figure 2(a). The temperature data were automatically logged by Datataker in every 15 min. The photograph of near-surface air temperature monitoring device is shown in Figure 2(b).

2.4. Calculation of Heat Output of Pavement. The influence of pavements on surrounding environment is closely related to the heat input and output of pavements. To further illustrate the effect of different concrete specimens on surrounding environment, the characteristics of heat output of different concrete were calculated and analyzed according to the changes of the surface temperature of the specimens.

Figure 3 shows the model of pavement-surface energy balance. Pavement-surface energy balance equation is shown as follows:

$$H_\Delta = (q_s - H_r) + H_c + H_l + H_{\text{cond}}, \tag{1}$$

where H_Δ is the change of pavement energy; q_s is the solar radiation; H_r is the short-wave solar radiation reflected by road surface; H_c is the convective heat between surface and

air; H_l is the long-wave net radiation; and H_{cond} is the pavement-surface-downward heat conduction.

Heat flux from pavement surface to outside air was calculated, which included the convective heat between pavement surface and the outside air, long-wave net radiation, and the short-wave solar radiation reflected by road surface. The total output heat flux, which is H_o, is the sum of the three types of heat flux. H_c, H_l, H_r [15], and H_o are calculated as follows:

$$H_c = h_\infty \left(T_s - T_\infty \right), \tag{2}$$

$$H_l = \Psi_{\text{sky}} \varepsilon \sigma \left(T_s^4 - T_{\text{sky}}^4 \right), \tag{3}$$

$$H_r = \tilde{a} \times q_s, \tag{4}$$

$$H_o = H_c + H_l + H_r, \tag{5}$$

where h_∞ is the convective heat coefficient of air, $h_\infty = 0.664 k_\infty \text{Pr}_\infty^{0.3} v_\infty^{-0.5} L^{-0.5} U_\infty^{0.5}$ [8], and k_∞ is the thermal conductivity of air; Pr_∞ is the Prandtl number; v_∞ is the kinematic viscosity; L is the length; U_∞ is the wind speed; T_s is the temperature of surface; Ψ_{sky} is the sky view factor; the value is set to be one because the pavement structure is completely open to the sky. ε is the infrared emissivity of the surface, σ is the Stefan–Boltzmann constant, 5.67×10^{-8} Wm^{-2} K^{-4}; T_{sky} is the sky temperature, which can be estimated using $T_{\text{sky}} = T_\infty (0.004 T_{\text{dew}} + 0.8)^{0.25}$ [8]; T_∞ is the atmospheric dry-bulb temperature; T_{dew} is the dew point, unit in K; $\tilde{a}$ is the albedo.

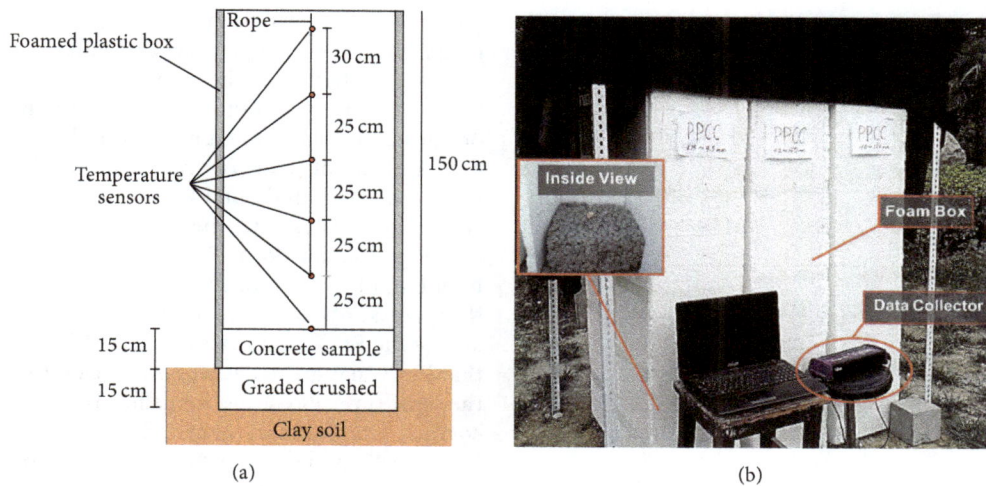

FIGURE 2: Near-surface air temperature monitoring device: (a) schematic figure of near-surface air temperature monitoring device; (b) photograph of near-surface air temperature monitoring device.

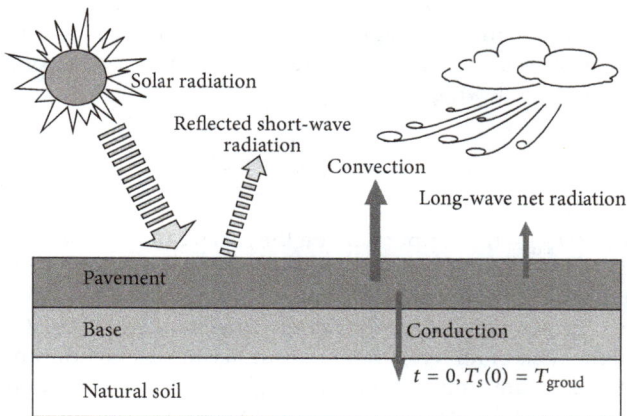

FIGURE 3: The model of pavement-surface energy balance.

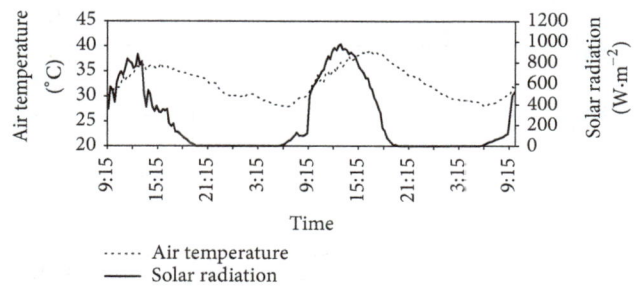

FIGURE 4: Air temperature and solar radiation.

3. Results and Discussion

3.1. Temperature Change Characteristics of Concrete. Data of air temperature and solar radiation at the test site were logged continuously from 9:15 on June 18, 2016, to 9:15 on June 20, 2016, by Weather Station. The curves of air temperature and solar radiation are shown in Figure 4. The weather at the test site on June 18, 2016 was mainly cloudy. The maximum temperature of 36.5°C was reached at 15:30 and the largest solar radiation of 881 W/m^2 was reached at 12:45. Solar radiation continued until 19:30. The weather at the test site on June 19, 2016 was mainly sunny. The maximum temperature of 39.2°C was reached at 16:30 and the largest solar radiation of 979 W/m^2 was reached at 13:00. Solar radiation continued until 19:45. The peaks of temperature were reached later and then the peaks of solar radiation were reached. The peaks of the temperature and the solar radiation on June 19 were higher than those on June 18. The solar radiation on June 19 had a longer duration.

Temperature data of six types of concrete at different depths were logged from 9:15 on June 18, 2016, to 9:15 on June 20, 2016, by the Data Collector. The curves of temperature are shown in Figure 5. As presented in Figure 5, the curves at different depths fluctuated similarly as the curves of air temperature. The temperature of each concrete on surface (depths of 0 cm) changed basically synchronously with the air temperature. Both of them reached the maximum at about 15:00 and reached the minimum at around 5:00 in the next day. However, the changes of temperature at the depths of −5 cm, −10 cm, and −15 cm lagged behind the changes of air temperature and the lag obviously enlarged with the depth. In addition, the lag of temperature of PCC at each measuring point was more significant than that of PPCC. The curves of temperature of PPCC fluctuated more drastically than those of PCC.

PPCC has a smaller specific heat capacity due to its porous structure, and the specific heat capacity of PPCC decreases with the increase of porosity [9, 18, 20]. The specific heat capacity of PPCC approximately is lower by 12.4% to 20.0% than that of PCC. Specific heat capacity has an influence on heat absorption and release for materials, thereby affecting the temperature changes of materials. Moreover, the coefficient of thermal storage has a similar effect on materials. The

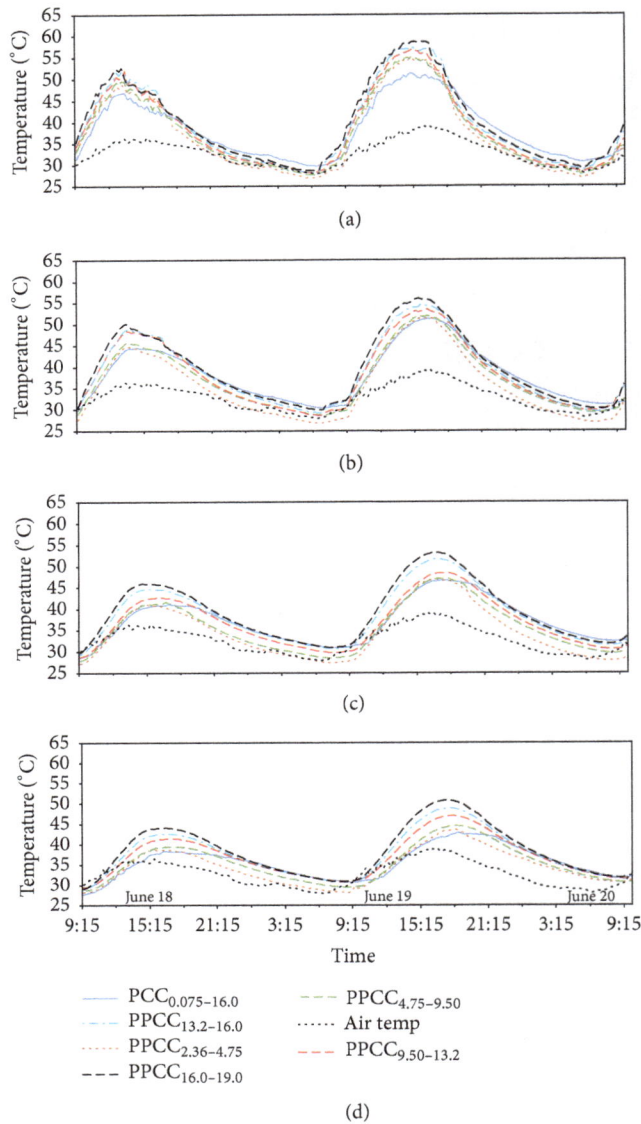

FIGURE 5: Temperature values of concrete at different depths: (a) 0 cm; (b) −5 cm; (c) −10 cm; and (d) −15 cm.

temperature of material with smaller specific heat capacity and coefficient of thermal storage is more susceptible to external environment, which is the reason for the fluctuation of temperature changes of PPCC that is significantly greater than that of PCC under the same test conditions.

The comparisons of different concrete specimens at different depths for daytime maximum temperature and nighttime minimum temperature are shown in Figures 6(a) and 6(b), respectively. It can be seen that the temperatures of PPCC were basically higher than the temperatures of PCC at corresponding positions during daytime. In the night, the minimum temperatures of PPCC were lower than the minimum temperatures of PCC at corresponding positions. As shown in Figure 6, the temperature changes of different PPCC related to the particle sizes of aggregates. The PPCC

with the larger particle size of aggregate had the relatively higher maximum temperature and minimum temperature.

Solar radiation is the main factor to promote cement concrete surface warming. Generally, the higher the albedo of surface is, the more the solar radiation is reflected to reduce rising of temperature caused by absorption of solar radiation. Five kinds of PPCC contain a large number of coarse aggregates and the surface texture is rough after forming. The albedo of PPCC is smaller by 17.7% to 29.1% than that of PCC, with the albedo of PPCC increasing with the increase of aggregate particle size. In addition, the surface area of the PPCC increases because of the uneven surface, so that PPCC has greater acceptance area for solar radiation. At the same time, PPCC has a smaller specific heat capacity and coefficient of thermal storage, which make the heating rate of PPCC faster and the maximum temperature of PPCC higher than that of PCC in the daytime. The temperatures of PPCC lower than that of PCC during nighttime are closely related to smaller specific heat capacity and coefficient of thermal storage of PPCC, and a lot of interconnected pores of PPCC contribute to the rapid release of heat.

The vertical temperature difference in the concrete due to the effect of the external environment is called temperature gradient. The temperature gradient can be calculated according to the following:

$$\text{Temperature gradient} = \frac{(T_{sur} - T_{bot})}{H}, \quad (6)$$

where T_{sur} = the temperature on the surface of specimen, °C; T_{bot} = the temperature at the bottom of specimen, and °C; H = the thickness of specimen, cm.

The temperature gradients of concrete were calculated by the maximum temperature of surface in the daytime and the minimum temperature in the nighttime. The temperature gradients of different concrete are shown in Figure 7.

It can be seen from Figure 7 that the temperature gradients of the PPCC were 0.7 to 1.2°C/cm when the surfaces reached the maximum temperature during the day. The temperature gradient increased with the particle size of aggregate. The temperature gradient of PCC was 0.7°C/cm and significantly less than that of PPCC. When the surface temperature of the specimens respectively reached the minimum temperature at night, the temperature gradient of PCC was smaller than those of 5 kinds of PPCC. But the temperature gradient at night is much smaller than that during the day.

Thermal conductivity is an important parameter to measure internal heat transfer capability of objects. PPCC contains more amount of coarse aggregate than PCC does. The point contacts between aggregate particles weaken the heat conduction leading to thermal conductivity of PPCC being less than that of the compacted PCC. The thermal conductivity of PPCC is lower by 8.5% to 14.3% than that of PCC in this research and the thermal conductivity of PPCC decreases with the increase of aggregate particle size. The smaller the thermal conductivity is, the greater the temperature difference between surface and internal different positions usually is. Therefore, small thermal conductivity in

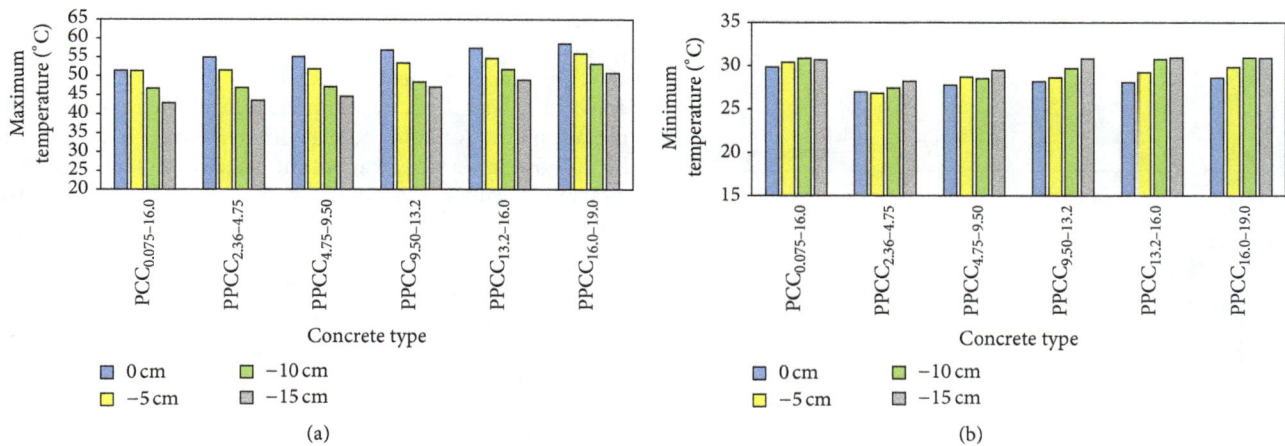

FIGURE 6: Maximum temperature and minimum temperature of different concrete: (a) maximum temperature in the daytime; (b) minimum temperature in the nighttime.

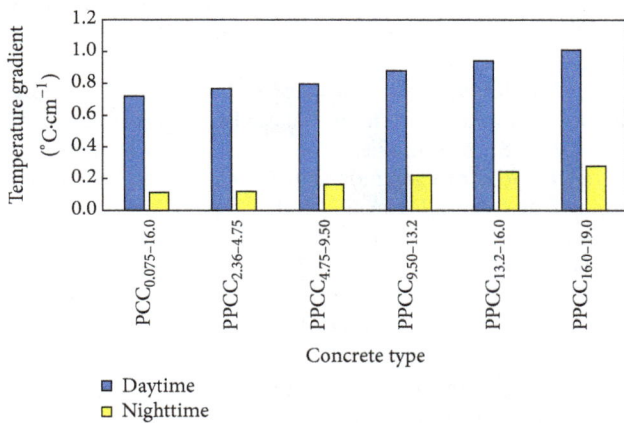

FIGURE 7: The temperature gradients of different concrete.

PPCC increases the difference between surface temperature and bottom temperature. This is the main reason for the temperature gradient of PPCC being greater than that of PCC.

3.2. Analysis of Characteristics on Change of Near-Surface Air Temperature of Concrete.

The near-surface air temperatures were tested on July 17, 2016 and data of air temperature at 25 cm, 50 cm, 75 cm, 100 cm, and 130 cm above surface of specimens started to be logged at 15:45. The air temperature was 33.9°C at 15:45. The initial surface temperature of $PCC_{0.075-16.0}$, $PPCC_{2.36-4.75}$, $PPCC_{4.75-9.50}$, $PPCC_{9.50-13.2}$, $PPCC_{13.2-16.0}$, and $PPCC_{16.0-19.0}$ was 40.5°C, 41.7°C, 42.0°C, 42.7°C, 43.0°C, and 42.5°C, respectively. The curves of near-surface air temperature and surface temperature are shown in Figure 8.

From Figure 8, it can be seen that the hot specimens started to heat the air in box after specimens were covered by foamed plastic box. The air temperature at 25 cm, 50 cm, 75 cm, 100 cm, and 130 cm above specimen surface increased with time firstly and reached the peaks about at 60 min after the start of test. Then, the air temperatures at the different heights dropped slowly after the peaks and levelled off over time.

The surface temperature and thermal storage of specimen in the box are the key factors for influence on the near-surface air temperature. Because solar radiation was blocked by the box, the temperature of specimen surface decreased gradually. Within initial 240 min, the surface temperature of the PPCC was higher than the surface temperature of PCC, and air temperatures at the different heights above PPCC in the boxes were higher than that of PCC in the corresponding positions. The air temperature in each box reached the peak at 16:45 (60 min after test beginning). The near-surface air temperatures above different concrete are presented in Figure 9. It can be seen from Figure 9(a) that the air temperatures at different position above PPCC specimens were higher than those above the PCC specimen in the corresponding position. Compared with different PPCC, the near-surface air temperatures at corresponding positions increased with the particle size of aggregates. In addition, the near-surface air temperature decreased with height.

After about 240 min of test beginning, the surface temperature of PPCC specimens was lower than that of PCC gradually, which is due to specific heat capacity and coefficient of thermal storage of PPCC is small compared with the PCC and PPCC is easier to cool when it loses the heat source of solar radiation. From Figure 9(b), it can be seen that the air temperatures at different positions above five kinds of PPCC specimens were lower than those above PCC specimen at the corresponding position at 6:45.

The results of tests indicated that PPCC increases the near-surface air temperature during a short time after solar radiation being shaded in hot summer; however, the situation conversely changes over time. Because the capacity of heat releasing of PPCC is less than that of PCC, the air temperature above PPCC is lower than that above PCC after the peak of air temperature.

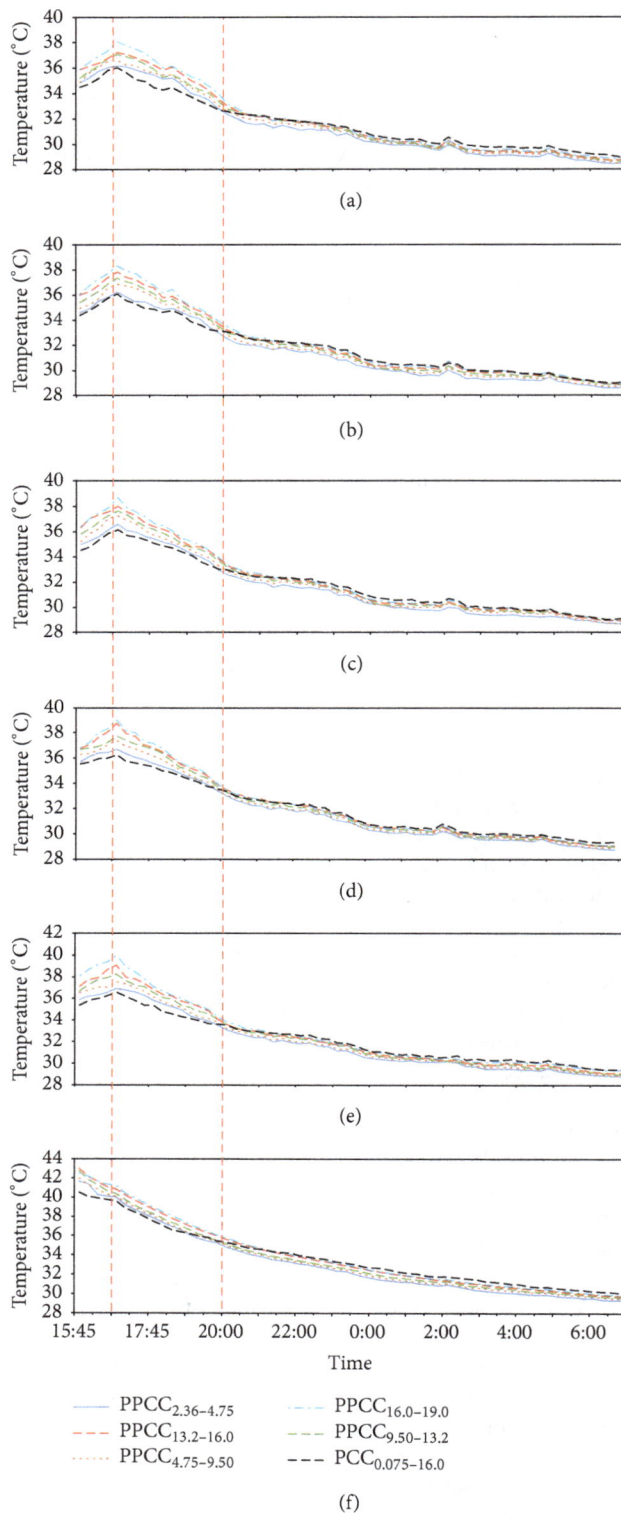

FIGURE 8: The curves of surface temperature and near-surface air temperature: (a) 130 cm; (b) 100 cm; (c) 75 cm; (d) 50 cm; (e) 25 cm; (f) 0 cm.

3.3. Heat Output Analysis. The temperature of pavement is influenced by environmental conditions. At the same time, the heat output from the pavement impacts the air temperature, especially near-surface air temperature. There are three types of heat output from pavement: the first is convective heat flux, the second is reflected short-wave solar radiation heat flux, and the last is long-wave net radiation heat flux. Those three type heat output values of different kinds of concrete within 48 h are calculated according to (2)–(4). The calculation results are shown in Figures 10(a), 10(b), and 10(c), respectively.

As shown in Figure 10, in the daytime, the convective heat and long-wave net radiation of PPCC are higher than those of PCC. The reflected short-wave solar radiation of PPCC is lower than that of PCC. The heat flux has obvious difference between different PPCC. The convective heat and long-wave net radiation increase and the reflected short-wave solar radiation decreases with the increase of particles size of aggregate. In addition, because of the impact of the wind change in the daytime, the curves of convective heat have obvious sawtooth wave.

The convective heat and long-wave net radiation are mainly related to the surface temperature of specimens. Therefore, the sequence of convective heat and long-wave net radiation of different PPCC are basically in accordance with surface temperature of PPCC, respectively. Reflected short-wave solar radiation is mainly associated with the albedo of material. Surface temperature of PCC is lower due to its higher albedo. However, more solar radiation will be reflected to the surrounding air by PCC at the same time, which is the reason that the short-wave solar radiation of PCC is significantly higher than that of PPCC. The calculation results also show that the heat output of reflected short-wave solar radiation is much higher than the heat output of convective heat and long-wave net radiation.

At night, because of solar radiation disappearing, the reflected short-wave solar radiation values of several concrete types are all zero. Therefore, the surface temperature of PCC specimens is higher than those of five kinds of PPCC gradually. And the change trends of the convective heat and long-wave net radiation of several concrete types are contrary to those in the daytime. That is, the convective heat and long-wave net radiation of five kinds of PPCC are lower than those of PCC. And the coarser aggregate particles size of PPCC is, the larger corresponding convective heat and long-wave net radiation are.

The impacts of concrete on surrounding air depend on its total output heat flux. The total output value of heat flux is the sum of the convective heat flux, the reflected short-wave solar radiation, and the long-wave net radiation [15]. The curve of total output heat flux versus time is shown in Figure 11.

It can be seen from Figure 11 that, in the daytime, the total output heat flux of PCC is slightly larger than those of PPCC. For example, the max of total output heat flux of $PCC_{0.075-16.0}$, $PPCC_{2.36-4.75}$, $PPCC_{4.75-9.50}$, $PPCC_{9.50-13.2}$, $PPCC_{13.2-16.0}$, and $PPCC_{16.0-19.0}$ on June 19 is 618.6, 591.8, 590.1, 603.1, 616.9, and 610.8 W/m^2, respectively. The average of total output heat flux of those six concrete types is 320.9, 295.6, 292.6, 303.2, 311.6, and 317.6 W/m^2, respectively. The calculations show that total output heat flux of the six concrete types and the effect on the

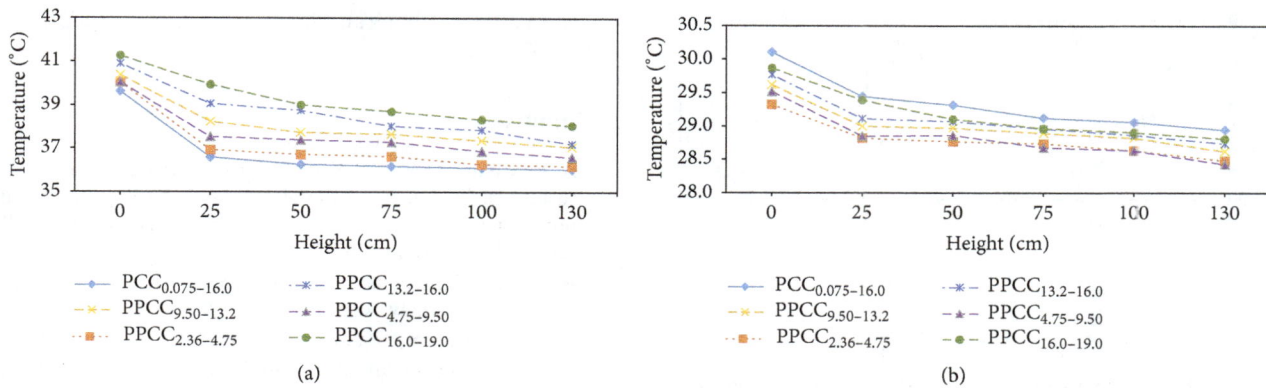

FIGURE 9: The near-surface air temperature above different concrete: (a) 16:45; (b) 6:45.

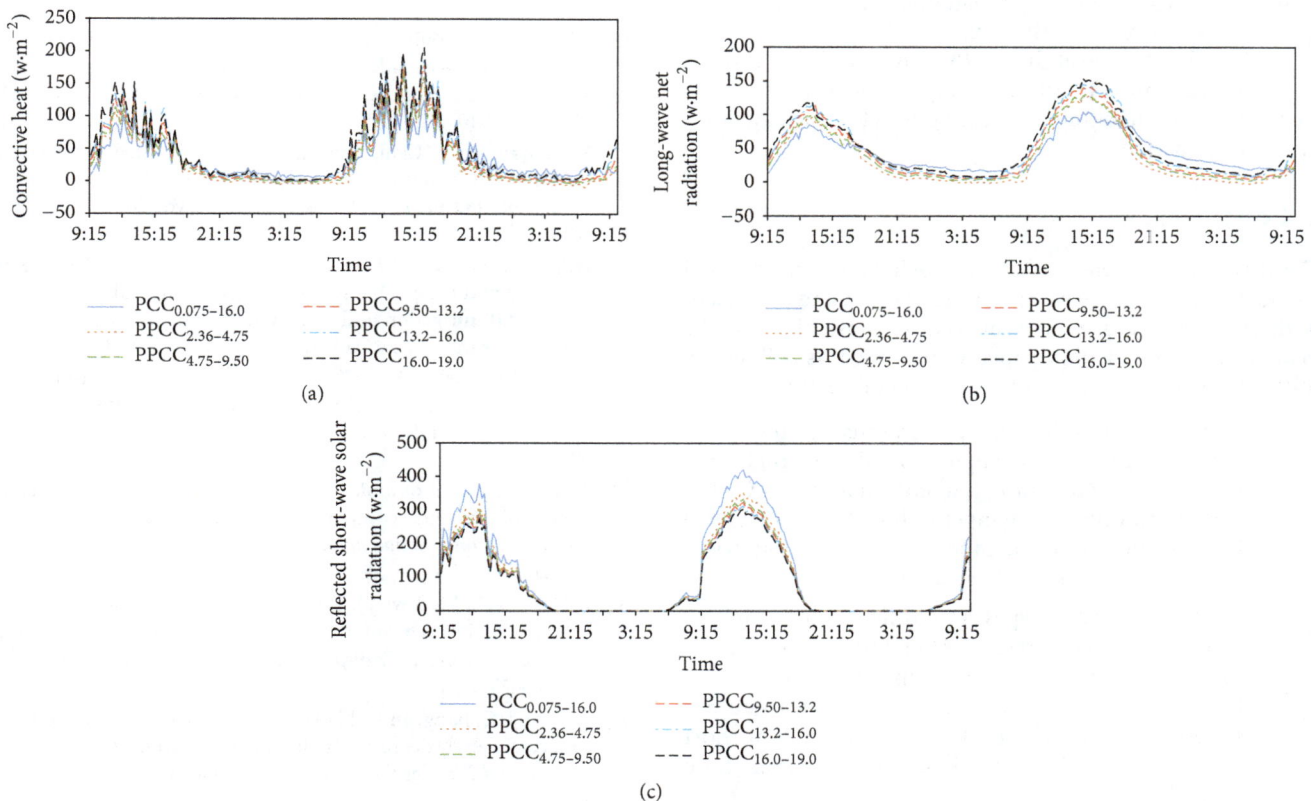

FIGURE 10: Surface heat flux of different concrete during 48 h in June 2016: (a) convective heat; (b) long-wave net radiation; (c) reflected short-wave solar radiation.

near-surface air temperature are little difference, although the surface temperatures of PPCC are higher than that of PCC in the daytime.

At night, the total output heat flux of concrete only relates to the convective heat flux and the long-wave net radiation in the nighttime because there is no solar radiation. The total output heat flux of all kinds of concrete decreases sharply. Taking $PCC_{0.075-16.0}$ and $PPCC_{16.0-19.0}$ as examples, the average of total output heat flux during the nighttime of the two kinds of concrete is 17.5% and 11.7% of that in the

daytime. Additionally, the total output heat flux of PCC is obviously larger than those of PPCC in the nighttime. Taking the calculations of the total output heat flux in the nighttime of June 19 as an example, the average of total output heat flux of $PPCC_{2.36-4.75}$, $PPCC_{4.75-9.50}$, $PPCC_{9.50-13.2}$, $PPCC_{13.2-16.0}$, and $PPCC_{16.0-19.0}$ decreases by 86.0%, 68.6%, 58.6%, 43.0%, and 33.3% compared with 56.0 W/m^2 of $PCC_{0.075-16.0}$, which shows that PPCC has less effect on the outside air temperature compared with the PCC at night. This can also explain from the view of heat output why the near-surface air temperature

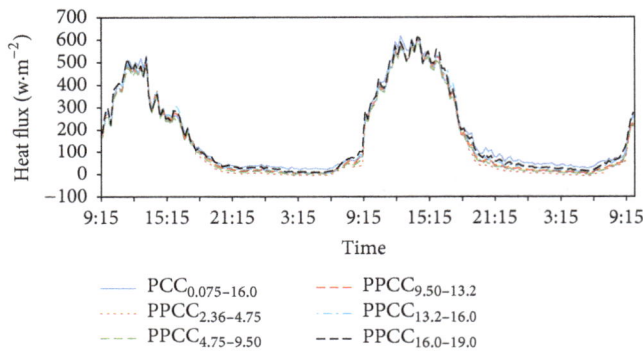

FIGURE 11: Total output heat flux of surface of different concrete during 48 h in June 2016.

above PPCC is generally lower than that above PCC after the specimens are covered by the box.

The total output heat flux in the daytime and nighttime is considered comprehensively; compared with PCC, PPCC contributes to mitigating the heating effect of pavement on the near-surface air.

4. Conclusions

The thermal parameters of PPCC, such as albedo, thermal conductivity, specific heat capacity, emissivity, and coefficient of thermal storage, are quite different to those of PCC. The main conclusions in this paper were drawn as follows by contrast experiments and calculation in hot weather.

(1) The PPCC has higher temperature in the sunny daytime and lower temperature in the nighttime and larger temperature gradient than the PCC. The surface and internal temperatures of the PPCC rise faster in the sunny daytime and decrease also faster in the nighttime than those of the PCC.

(2) Different concrete types have different influence on near-surface temperature. After the concrete specimens being exposed to the same solar radiation to heat up and then covered with boxes, the near-surface air temperatures of the PPCC are higher in a short period of the beginning and then are lower than those of the PCC.

(3) The total output heat flux of PPCC is slightly smaller than that of PCC during the day and is significantly smaller at night. PPCC could contribute to reducing the heating effect of pavement on the near-surface air.

Competing Interests

The authors declare that they have no competing interests.

Acknowledgments

This work was funded by Ministry of Science and Technology of China (2014BAG05B04). This project was also supported by the Special Fund for Basic Scientific Research of Central College, Chang'an University, China (310821153308, 310821153202).

References

[1] A. K. Chandrappa and K. P. Biligiri, "Pervious concrete as a sustainable pavement material—research findings and future prospects: a state-of-the-art review," *Construction and Building Materials*, vol. 111, pp. 262–274, 2016.

[2] A. Yukari, *Development of Pervious Concrete*, University of Technology, Sydney, Australia, 2009.

[3] B. J. Wardynski, R. J. Winston, and W. F. Hunt, "Internal water storage enhances exfiltration and thermal load reduction from permeable pavement in the North Carolina mountains," *Journal of Environmental Engineering*, vol. 139, no. 2, pp. 187–195, 2012.

[4] J. Mullaney and T. Lucke, "Practical review of pervious pavement designs," *Clean—Soil, Air, Water*, vol. 42, no. 2, pp. 111–124, 2014.

[5] J. J. Chang, W. Yeih, T. J. Chung, and R. Huang, "Properties of pervious concrete made with electric arc furnace slag and alkali-activated slag cement," *Construction and Building Materials*, vol. 109, pp. 34–40, 2016.

[6] M. Santamouris, "Using cool pavements as a mitigation strategy to fight urban heat island&A review of the actual developments," *Renewable and Sustainable Energy Reviews*, vol. 26, pp. 224–240, 2013.

[7] H. Li, J. T. Harvey, T. Holland, and M. Kayhanian, "The use of reflective and permeable pavements as a potential practice for heat island mitigation and stormwater management," *Environmental Research Letters*, vol. 8, no. 1, pp. 15–23, 2013.

[8] J. Gui, P. E. Phelan, K. E. Kaloush, and J. S. Golden, "Impact of pavement thermophysical properties on surface temperatures," *Journal of Materials in Civil Engineering*, vol. 19, no. 8, pp. 683–690, 2007.

[9] A. Hassn, M. Aboufoul, Y. Wu, A. Dawson, and A. Garcia, "Effect of air voids content on thermal properties of asphalt mixtures," *Construction and Building Materials*, vol. 115, pp. 327–335, 2016.

[10] L. Haselbach, M. Boyer, J. T. Kevern, and V. R. Schaefer, "Cyclic heat island impacts on traditional versus pervious concrete pavement systems," *Transportation Research Record*, no. 2240, pp. 107–115, 2011.

[11] R. Zhang, G. Jiang, and J. Liang, "The albedo of pervious cement concrete linearly decreases with porosity," *Advances in Materials Science and Engineering*, vol. 2015, Article ID 746592, 5 pages, 2015.

[12] J. T. Kevern, L. Haselbach, and V. R. Schaefer, "Hot weather comparative heat balances in pervious concrete and impervious concrete pavement systems," *Journal of Heat Island Institute International*, vol. 7, no. 2, 2012.

[13] R. S. Benrazavi, K. Binti Dola, N. Ujang, and N. Sadat Benrazavi, "Effect of pavement materials on surface temperatures in tropical environment," *Sustainable Cities and Society*, vol. 22, pp. 94–103, 2016.

[14] K. K. Guan, "Surface and ambient air temperatures associated with different ground material: a case study at the University of California, Berkeley," *Environmental Science*, vol. 196, pp. 1–14, 2011.

[15] T.-P. Lin, Y.-F. Ho, and Y.-S. Huang, "Seasonal effect of pavement on outdoor thermal environments in subtropical Taiwan," *Building and Environment*, vol. 42, no. 12, pp. 4124–4131, 2007.

[16] A. H. Rosenfeld, H. Akbari, J. J. Romm, and M. Pomerantz, "Cool communities: strategies for heat island mitigation and smog reduction," *Energy and Buildings*, vol. 28, no. 1, pp. 51–62, 1998.

[17] Y. Qin and J. E. Hiller, "Understanding pavement-surface energy balance and its implications on cool pavement development," *Energy and Buildings*, vol. 85, pp. 389–399, 2014.

[18] T. Nakayama and T. Fujita, "Cooling effect of water-holding pavements made of new materials on water and heat budgets in urban areas," *Landscape and Urban Planning*, vol. 96, no. 2, pp. 57–67, 2010.

[19] J. J. Muñoz-Criollo, P. J. Cleall, and S. W. Rees, "Factors influencing collection performance of near surface interseasonal ground energy collection and storage systems," *Geomechanics for Energy and the Environment*, vol. 6, pp. 45–57, 2016.

[20] Y. Qin, "A review on the development of cool pavements to mitigate urban heat island effect," *Renewable and Sustainable Energy Reviews*, vol. 52, pp. 445–459, 2015.

4

Influence of Organic Esters on Portland Cement Hydration and Hardening

Duan-le Li,[1] **Da-peng Zheng,**[1] **Dong-min Wang** ⓘ**,**[1] **Ji-hui Zhao,**[2] **Cheng Du,**[1] **and Cai-fu Ren**[1]

[1]*School of Chemical and Environmental Engineering, China University of Mining and Technology, Beijing, China*
[2]*Department of Civil Engineering, Tsinghua University, Beijing, China*

Correspondence should be addressed to Dong-min Wang; wangdongmin-2008@163.com

Academic Editor: Francisca Puertas

This paper investigated the effect of organic compounds with ester groups on the hydration and hardening of cement. The effects of five kinds of organic compounds with ester groups (ethyl acetate, dimethyl oxalate, glyceryl triacetate, trimethyl phosphate, and triethanolamine borate) on hydration heat, hydration degree, setting time, mechanical properties, microstructure, and pore structure of hardened cement slurry were studied. The test results showed that esters can make the end time of cement hydration induction longer and delay the occurrence of the second exothermic peak. Also, the effect of five kinds of esters on the hydration and hardening of cement was basically followed by TG> TB> DMO> EAC> TMP. In terms of molecular structure, for organic compounds containing only ester groups, the higher the number of ester groups, the greater the effect on the hydration of cement. The introduction of other functional groups (such as phosphate or borate) will influence the effect of the esters.

1. Introduction

In recent decades, the new progress in the preparation and application of cement concrete materials has benefited from the development and effective application of chemical admixtures. At present, the chemical admixture has become a cement-based material, which is an important regulatory component or modification method. Addition of a small amount of admixture, especially organic admixture, can effectively adjust the performance of cement-based materials, such as improving work performance or rheology, speeding up or delaying the hydration rate and coagulation time, and improving the physical and mechanical properties and durability of materials [1–5]. The performance of the admixture has an important relationship with the functional groups in its molecule. At present, the functional groups or functional groups contained in the organic admixtures used in the concrete materials mainly include hydroxyl groups, carboxyl groups, amino groups, and sulfonic acid groups [6–9].

In addition, it should be noted that the ester groups and phosphonate groups are also important functional groups.

On the one hand, the ester-based organics can be used as antifoaming agents. On the other hand, many studies have reported that the esterification modification method is an important means for improving or modifying the performance of hydroxyl- or carboxyl-containing admixtures [1, 10]. In view of the low grinding ability of triethanolamine in cement fine-grinding stage and the disadvantage on the late strength growth of cement, it is studied that the triethanolamine modified by carboxylic acid has a better effect on the grinding and dispersing, and the strength of cement at early and late stage has a significant growth [1, 11, 12]. The molecular structure and dosages of the admixtures play an important role in the hydration of cement, such as shorting the hydration time, reducing the hydration heat, and improving the mechanical strength [13–16]. However, it is not clear at present that the increase in the performance of the admixture is caused by changes in the molecular structure or by the action of the newly introduced ester functional groups after modification. Also, there is no research at this point.

In order to explore whether the ester groups have an important effect on the hydration and hardening of cement, five common organic compounds, ethyl acetate, dimethyl

TABLE 1: The chemical compositions and mineral compositions of cement wt%.

CaO	SiO$_2$	Al$_2$O$_3$	Fe$_2$O$_3$	MgO	SO$_3$	R$_2$O	f-CaO	LOI	C$_3$S	C$_2$S	C$_3$A	C$_4$AF
62.13	20.76	4.58	3.27	3.13	2.80	0.57	0.76	1.56	57.3	18.9	6.5	11.3

TABLE 2: The physical properties of cement test results.

Specific surface area/(m^2·kg^{-1})	Standard consistency, %	Setting time/min		Compressive strength/MPa	
		Initial	Final	3 d	28 d
347	27.2	155	215	28.3	51.2

TABLE 3: Basic information of five kinds of ester group.

Esters	Abbreviation	Chemical formula	Structure	Number of ester groups
Ethyl acetate	EAC	C$_4$H$_8$O$_2$		1
Dimethyl oxalate	DMO	C$_4$H$_6$O$_4$		2
Glyceryl triacetate	TG	C$_9$H$_{14}$O$_6$		3
Trimethyl phosphate	TMP	C$_3$H$_9$O$_4$P		3
Triethanolamine borate	TB	C$_6$H$_{12}$BNO$_3$		3

oxalate, glyceryl triacetate, trimethyl phosphate, and triethanolamine borate, were selected for comparative experiments in this study. The effects of five kinds of organic compounds on hydration heat, hydration degree, setting time, mechanical properties, microstructure, and pore structure of hardened slurry were studied in this study.

2. Materials and Methods

2.1. Materials. Ordinary Portland cement, complying with the requirements of Chinese National Standard GB 175-2007 (common Portland cement), was used for all experiments here. The chemical composition and mineral composition of cement are shown in Table 1, and the physical properties of cement are shown in Table 2. The five esters selected for the experiment were ethyl acetate (EAC), dimethyl oxalate (DMO), glyceryl triacetate (TG), trimethyl phosphate (TMP), and triethanolamine borate (TB), and their basic information is shown in Table 3.

2.2. Hydration Heat of Cement Paste. In this study, the hydration heat of cement paste was tested using an eight-channel micro-heat instrument (TAM-Air).

2.3. Hydration Degree of Cement. The hydration degree of cement was characterized by chemical-bound water and

calcium hydroxide content. The prepared cement samples were maintained under the standard curing conditions to the corresponding age (3 d and 28 d), then the central portion was removed and the hydration was terminated with absolute ethanol. The samples were burned at 1000°C in the muffle furnace for 2 h to constant weight, and the mass loss of the samples was chemical-bound water. In addition, TG/DTA6300 thermogravimetric analyzer was used to test samples, and due to the decomposition of calcium hydroxide, the sample will have a significant mass loss at 400°C–500°C, and the content of calcium hydroxide in the sample can be calculated.

2.4. Morphology and Pore Structure of Hydration Product. The morphology of cement hydration products was observed by JSM-6700F scanning electron microscopy, and the pore structure was measured by AutoPore IV 9500 mercury porosimeter.

2.5. Workability and Mechanical Properties of Cement. The standard consistency water consumption and setting time of cement are in accordance with GB/T 1346-2001 "cement standard consistency of water consumption, setting time, stability test method," and the cement mortar strength is in accordance with GB/T 17,671-1999 "cement mortar strength test method" (ISO method).

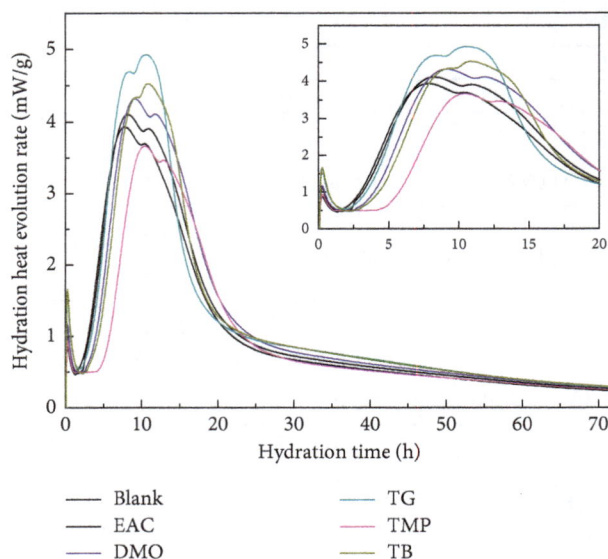

FIGURE 1: Effect of ester on the rate of heat release of cement hydration.

TABLE 4: Characteristic parameters of cement hydration.

Sample	The end time of the induction period (h)	The appear time of the second exothermic peak (h)	Second exothermic peak exothermic rate (J/g·h)	Total heat release (J/g)		
				24 h	48 h	72 h
Pure cement	1.3	7.9	14.2	295.0	374.0	416.6
+0.05% EAC	1.7	8.3	14.8	311.3	396.8	443.5
+0.05% DMO	2.2	9.2	15.6	325.6	418.1	468.4
+0.05% TG	1.8	10.6	17.7	330.0	430.1	482.2
+0.05% TMP	4.2	10.4	13.2	266.7	345.2	389.0
+0.05% TB	2.2	10.9	16.3	311.8	411.2	463.7

3. Results and Discussion

3.1. Effect of Ester on Hydration Heat of Cement. The amount of all kinds of esters was set to 0.05% by mass. Figure 1 shows exothermic rate curves during the hydration of pure cement paste and cement paste containing five kinds of ester. The variation of the characteristic parameters is shown in Table 4. It can be seen from Figure 1 and Table 4 that cement hydration exothermic rate and heat release are significantly changed after the addition of ester. First of all, the addition of five kinds of ester makes cement hydration induction time longer and postpone the appear time of the second exothermic peak. Also, with the increase of the number of ester groups, the appear time of the second exothermic peak of cement paste delays. The addition of TG and TB resulted in a delay of 2.7 and 3.05 hours for the appear of the second

exothermic peak. The results showed that the ester had a delayed effect on the hydration reaction rate of cement at the early stage (<10 h, hydration induction period and hydration acceleration period).

However, in addition to TMP, the other four kinds of esters improve the second exothermic peak of cement. When adding 0.05% TG to cement, the peak is raised from 14.15 to 17.74 J/g·h. The degree of effect was as follows: TG (3 ester groups)> TB (3 ester groups)> DMO (2 ester groups)> EAC (1 ester group). It indicates that the four esters will delay the early hydration reaction but will also promote the degree of response during the acceleration phase. TMP has a delayed effect on cement hydration and reduces the second exothermic peak, the reason may be that its molecular structure contains phosphorus.

The total amount of heat release in different hydration stages also shows a similarity to the second exothermic peak,

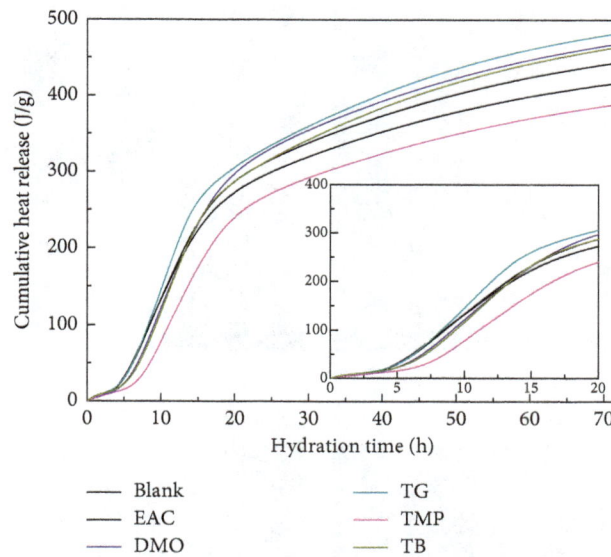

FIGURE 2: Effect of ester on the total heat release of cement hydration.

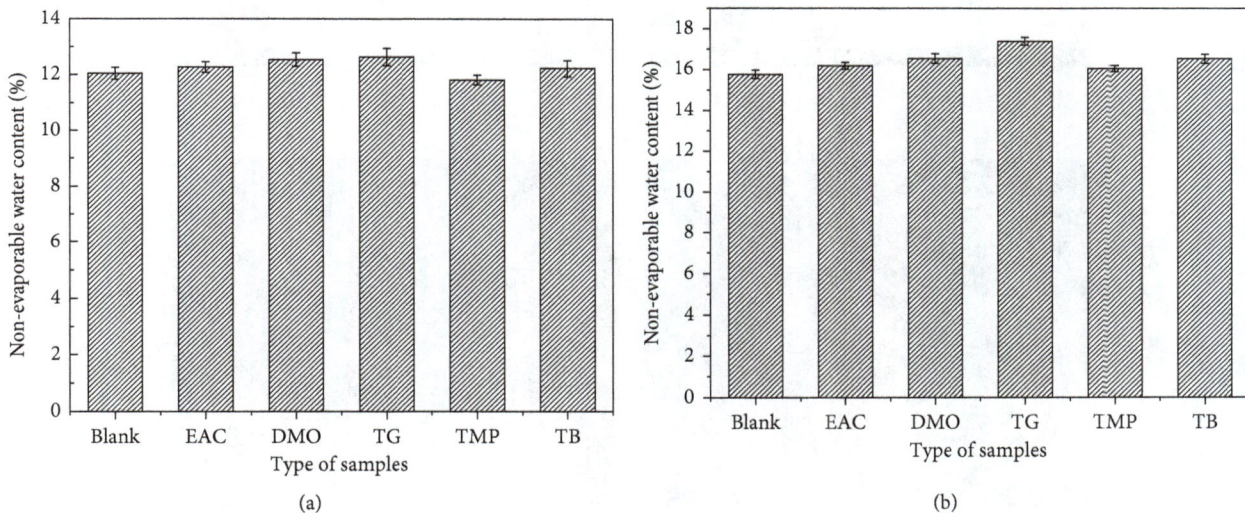

FIGURE 3: Effect of ester on nonevaporation water content of cement paste: (a) 3 d and (b) 28 d.

as can be seen from Figure 2, TG, TB, DMO, and EAC can improve the total amount of heat release in 24 h, 48 h, and 72 h, and the greater the number of ester groups, the stronger the effect.

3.2. Effect of Ester on Cement Hydration Degree.
Nonevaporable water content is usually used to characterize the degree of cement hydration. In the previous section, the hydration heat analysis speculated that, in addition to TMP, the other four kinds of organic compounds can promote the early hydration of cement. This section uses the results of non-evaporative hydration to verify that the above assumptions are correct. Figure 3 shows two histograms of the

nonevaporable water content of the cement slurry with different contents of ester at 3 d and 28 d. It can be seen from Figure 3(a) that TG, TB, DMO, and EAC, four kinds of organic compounds, can improve the nonevaporative water content of cement at 3 d. TG can increase the nonevaporable water content by 4.9%, while TMP will reduce the non-evaporative water content of cement by 1.9%, this is consistent with the laws of hydration heat. It can be analyzed from Figure 2(b) that the five kinds of ester can obviously improve the content of nonevaporable water content of cement at 28 d. Based on the above analysis, the ester can improve the hydration degree of the cement, and the effect of promotion from large to small is: TG> DMO> TB> EAC>

(a)

(b)

(c)

(d)

(e)

(f)

FIGURE 4: The hydration products morphology of different hardened cement at 3 d. (a) Blank. (b) +0.05% EAC. (c) +0.05% DMO. (d) +0.05% TG. (e) +0.05% TMP. (f) +0.05% TB.

TMP. TG has the best effect and can increase the non-evaporable water content of cement by 10.3% at 28 d.

Compared with the molecular structure of five kinds of organic compounds, EAC, DMO, and TG belong to the same series of organic matter with increasing number of ester groups, while TB and TMP contain three ester groups as in TG, but both contain boronic acid groups and phosphoric acid groups, respectively. It can be inferred that the higher the number of ester groups contained in the organic matter containing only the ester functional groups, the greater the effect on the degree of hydration of the cement. Also, the introduction of other functional groups

(such as phosphate or borate) will have some influence on the effect of ester organic matter.

3.3. Effects of Ester on the Morphology of Cement Hydration Products. The effects of ester on the morphology of hydrated products of cement at 3 d and 28 d are shown in Figures 4 and 5, respectively. It can be seen from Figure 4 that a large amount of hydrated products such as ettringite (AFt), CSH gel, and $Ca(OH)_2$ have been produced in the six kinds of cement slurries at the age of 3 d, but the structure is relatively loose and porous, indicating that the overall degree of

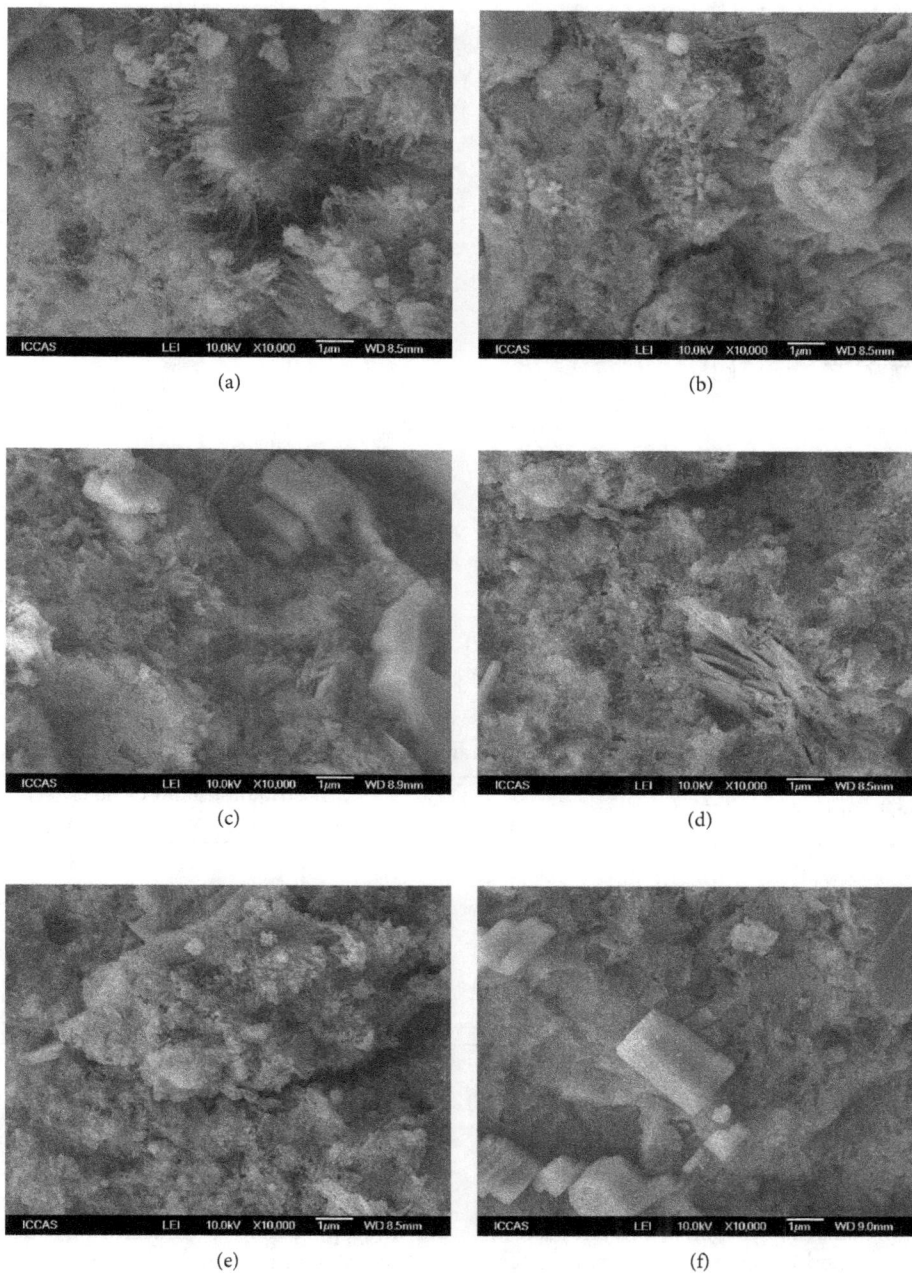

FIGURE 5: The hydration products morphology of different hardened cement at 28 d. (a) Blank. (b) +0.05% EAC. (c) +0.05% DMO. (d) +0.05% TG. (e) +0.05% TMP. (f) +0.05% TB.

hydration is not deep. Compared with the blank group, the hydration products of cement-hydrated products with EAC, DMO, TG, and TB were relatively dense, and the large amount of C–S–H gel was formed to encapsulate ettringite, and there are more $Ca(OH)_2$ generation. It was suggested that the addition of these four kinds of organic compounds promoted the formation of hydration products of cement at 3 d, in which TG and DMO had the most obvious effect. In contrast, the structure of cement with TMP is very loose, and the formation of hydration products is less, indicating that TMP has an inhibitory effect on the hydration of cement.

It can be seen from Figure 5 that CSH gel has been connected with one in each group of cement paste at 28 d and AFt is wrapped in it. The size and amount of $Ca(OH)_2$ increased, and the density was significantly higher than the 3 d age. The density of the cement paste in each group was as follows: TG> DMO> TB> EAC> TMP> blank group. The results showed that the five kinds of esters could promote the formation of hydration products of cement at 28 d, and the effect was more obvious than that of 3 d. Thus, it is further explained that these five kinds of organic compounds promote the cement hydration reaction and improve the degree of hydration at 28 d.

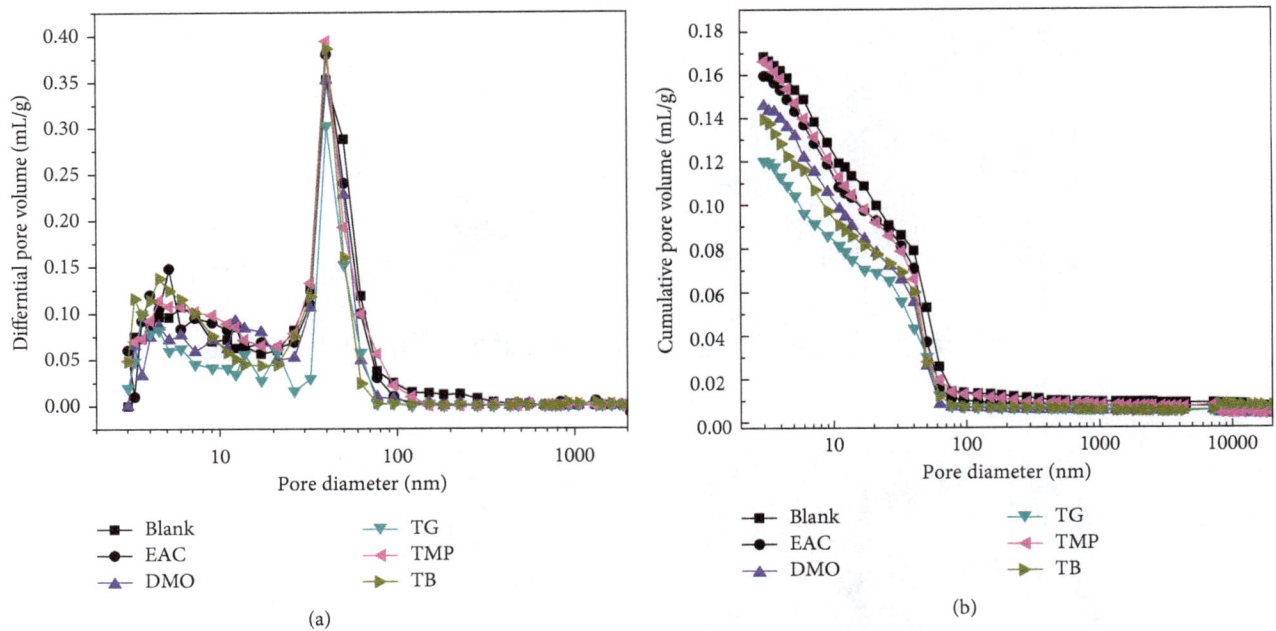

FIGURE 6: Pore volume distribution at different ages of hardened cement. (a) Differential distribution. (b) Cumulative distribution.

TABLE 5: Pore structural parameters of hardened cement.

Sample	Total pore volume (mL/g)	Distribution ratio of holes (%)			
		<20 nm	20–50 nm	50–200 nm	>200 nm
Blank	0.17	40.9	27.6	24.1	7.4
EAC	0.16	42.0	34.9	17.7	5.4
DMO	0.15	46.5	30.7	13.7	9.1
TG	0.12	42.7	32.4	19.4	5.5
TMP	0.17	44.8	38.5	10.2	6.5
TB	0.14	44.6	35.6	15.0	4.8

3.4. Effects of Ester on the Pore Structure of Cement. SEM can only qualitatively observe the density of cement hydration product structure, and MIP can visually analyze the pore structure of cement hydration products quantitatively. The effect of ester on the pore size distribution of cement paste at 28 d is shown in Figure 6, and the characteristic parameters of pore structure are shown in Table 3.

It can be seen from Figure 6 that the pore size of the hardened slurry is basically distributed in the range of less than 1000 nm. The five kinds of cement slurries with esters are not the same as the blanks and are about 40 nm. The most probable pore size of the five kinds of cement paste with organic matter was almost the same as that of the blank, and the total pore volume was significantly smaller than that of the blank group. TG significantly reduced the total pore volume of cement slurry, which was 28.7% lower than that of blank. TB and DMO decreased the total pore volume by 17.4% and 13.5%, respectively. The effect of TMP on the total pore volume was the smallest, and the order of action is as follows: TG> TB> DMO> EAC> TMP.

According to Renhe et al. [17], the pore in cement paste can be divided into four categories: much detrimental pores

(>200 nm), detrimental pores (50–200 nm), less harmful pore (20–50 nm), and innocuous pores (<20 nm). It can be seen from Table 5 that the five kinds of organic compounds have improved the content of innocuous pores and less harmful pore and significantly reduced the content of detrimental pores and much detrimental pores. This indicates that the five kinds of esters play a role in optimizing the pore size distribution of cement slurry, which is helpful to improve the macroscopic properties such as mechanical properties and durability of cement.

3.5. Effects of Ester on Mechanical Properties and Setting Time of Cement. The setting time of the cement slurry with different ester-based organic materials is shown in Figure 7. It can be seen from the results that the esters prolonged the initial setting time and final setting time of the cement, and the initial finishing time of the cement gradually prolonged with the increase of the ester. It is well explained by the results of the previous hydration heat that the five kinds of esters make the cement hydration induction end time longer and postpone the second exothermic peak, that is, the esters

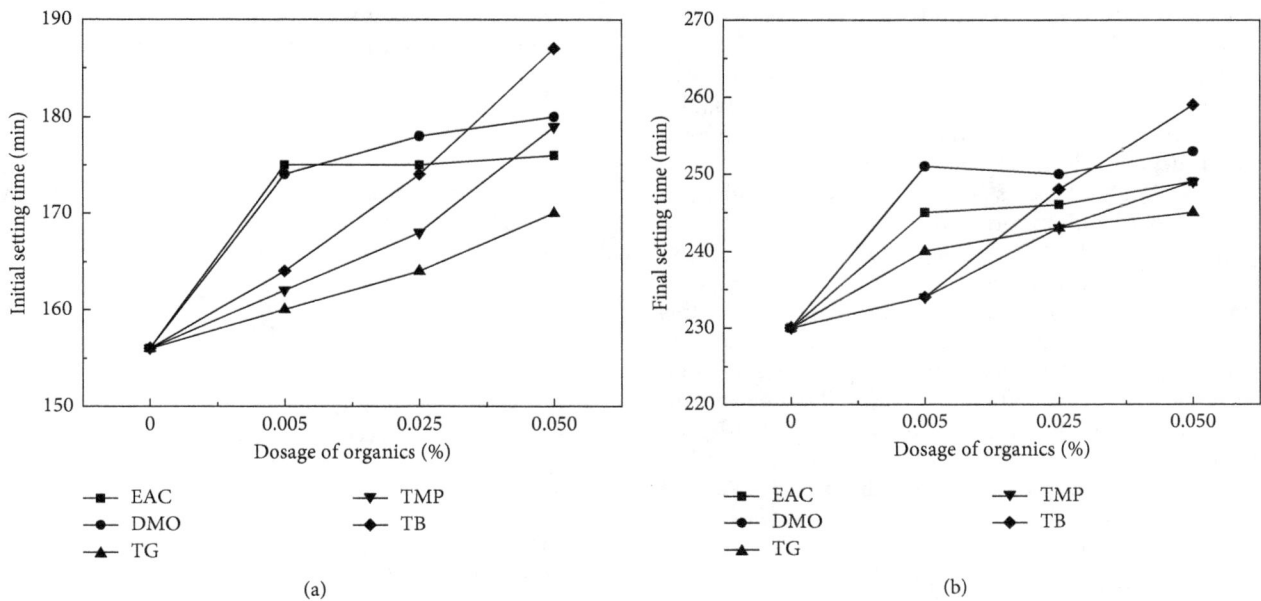

FIGURE 7: Effect of ester organics on the setting time of cement. (a) Initial setting time. (b) Final setting time.

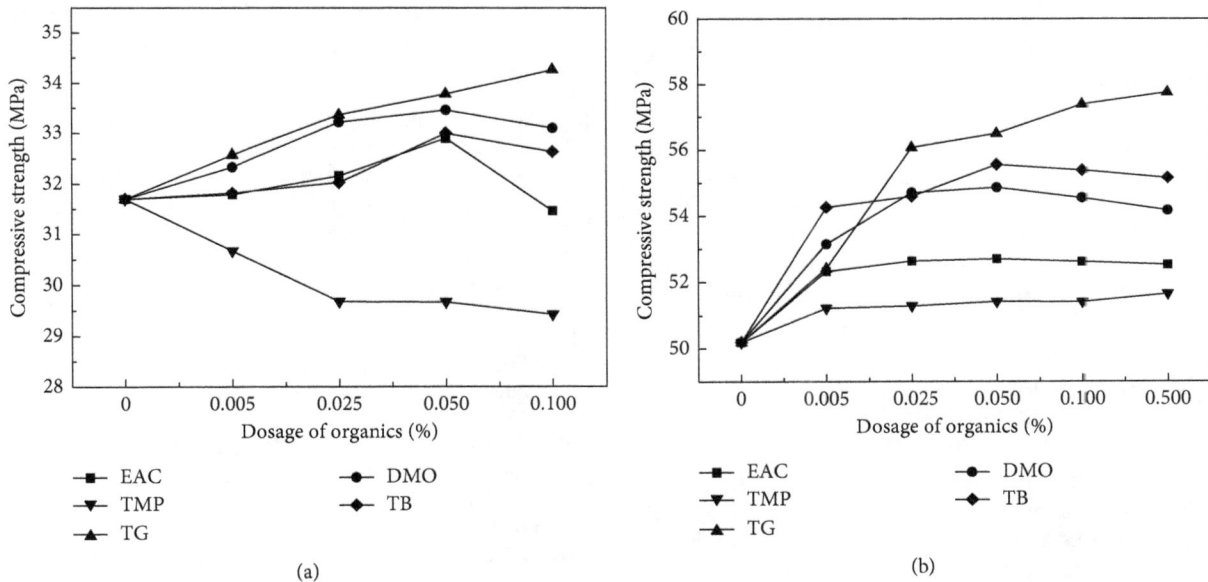

FIGURE 8: Effect of esters on the compressive strength of cement mortar. (a) 3d. (b) 28d.

have a delayed effect on the hydration reaction rate at the early stage of cement (hydration induction period and the acceleration period). When dosage is 0.05%, the delay effect of TB on the coagulation time is most obvious. As can be seen from Figure 7, the initial setting time was delayed from 156 minutes to 187 minutes and the final setting time is delayed from 230 minutes to 259 minutes. Also, TG has the minimum retardation effect.

The effect of different content of ester on the compressive strength of cement mortar is shown in Figure 8. The 3 d

strength of cement increased with the increase of EAC, DMO, TG, and TB contents. However, the strength of cement with EAC, DMO, and TB began to decrease after the content of 0.05%, indicating that the best dosage of these three esters is 0.05%. In contrast, the strength of cement mixed with TMP gradually decreased with the increase of organic matter content, which indicated that TMP was unfavorable to the development of cement 3 d strength. This can be explained by the abovementioned hydration heat, hydration degree, and hydration product structure, and it was further verified that TMP organic matter delayed the hydration and hardening of cement. The compressive strength at 3 d of each cement sample with 0.05% dosage of admixture are as follows: TG> DMO> TB> EAC> blank> TMP, which is consistent with the results of hydration heat, hydration degree, and hydration product.

The five kinds of esters have improved the hydration degree of cement and promoted the formation of hydration products and optimized the pore structure. Therefore, it can be concluded that the esters will increase the strength of cement at 28 d. The results in Figure 8(b) confirm this speculation, and as the amount of ester increased, the strength of cement increased significantly (the TB group decreased slightly at the content of 0.1%). When the content is 0.05%, the order of the strength of cement hardening slurry is TG> TB> DMO> EAC> TMP> blank, in which TG and TB increase the cement strength by 6.3 MPa and 5.4 MPa, respectively.

4. Conclusions

Based on the experimental results, the following conclusions can be drawn:

(1) Ester can make the end time of cement hydration induction longer and delay the occurrence of the second exothermic peak, indicating that it has a delayed effect on the hydration reaction rate of cement at early stage (hydration induction period and acceleration period).

(2) TG, TB, DMO, and EAC increased the second exothermic peak and hydration degree of cement and promoted the formation of hydration products and the optimization of the pore structure. Macroscopic performance is that mechanical properties of cement at 3 d and 28 d improved significantly. TMP has a retarding effect on cement and reduces the second exothermic peak of cement and inhibits the formation of early hydration products, which reduces the strength of cement at 3 d, but it improves the strength of cement at 28 d.

(3) The effect of five kinds esters on the hydration and hardening of cement was basically followed by TG> TB> DMO> EAC> TMP. In terms of the molecular structure, for organic compounds containing only ester groups, the higher the number of ester groups, the greater the effect on the hydration of cement. The introduction of other functional groups (such as phosphate or borate) will influence the effect of the esters.

Conflicts of Interest

The authors declare that there are no conflicts of interest regarding the publication of this paper.

Acknowledgments

The work described in this paper was fully supported by the scientific and technological projects of Shanxi Province, most through the Research Institute of Concrete and Ecomaterials in China University of Mining and Technology (Beijing) (Project no. MC2014-04).

References

[1] P. J. Sandberg and F. Doncaster, "On the mechanism of strength enhancement of cement paste and mortar with triisopropanolamine," *Cement and Concrete Research*, vol. 34, no. 6, pp. 973–976, 2004.

[2] N. Chikh, M. Cheikh-Zouaoui, S. Aggoun, and R. Duval, "Effects of calcium nitrate and isopropanolamine on the setting and strength evolution of Portland cement pastes," *Materials and Structures*, vol. 41, no. 1, pp. 31–36, 2008.

[3] D. Heinz, M. Goebel, H. Hilbig, L. Urbonas, and G. Bujauskaite, "Effect of TEA on fly ash solubility and early age strength of mortar," *Cement and Concrete Research*, vol. 40, no. 3, pp. 392–397, 2010.

[4] J. Cheung, A. Jeknavorian, L. Roberts, and D. Silva, "Impact of admixtures on the hydration kinetics of Portland cement," *Cement and Concrete Research*, vol. 41, no. 12, pp. 1289–1309, 2011.

[5] B. Łaźniewska-Piekarczyk, "The influence of chemical admixtures on cement hydration and mixture properties of very high performance self-compacting concrete," *Construction and Building Materials*, vol. 49, pp. 643–662, 2013.

[6] W. Jianfeng, *Study on Structure-Performance Relationship and Molecular Structure Design of Cement Grinding Aids*, China University of Mining and Technology (Beijing), Beijing, China, 2013.

[7] J.-P. Perez, A. Nonat, S. Garrault-Gauffinet, S. Pourchet, and M. Mosquet, "Influence of triisopropanolamine on the physical-chemical and mechanical properties of pure cement pastes and mortars," in *Proceedings of 11th International Congress on the Chemistry of Cement (ICCC)*, pp. 454–463, Durban, South Africa, May 2003.

[8] E. Gartner and D. Myers, "Influence of tertiary alkanolamines on Portland cement hydration," *Journal of the American Ceramic Society*, vol. 76, no. 6, pp. 1521–1530, 1993.

[9] K. Xiangming, L. Zhenbao, Y. Juan, L. Hui, and W. Dongmin, "Influence of triethanolamine on elemental concentrations in aqueous phase of hydrating cement pastes," *Journal of the Chinese Ceramic Society*, vol. 41, no. 7, pp. 981–986, 2013.

[10] W. Qisheng and Z. Ying, "Preparation and research of triethanolamine with dibasic organic acid-modified cement grinding aids," *Bulletin of the Chinese Ceramic Society*, vol. 33, no. 3, pp. 697–707, 2014.

[11] Z. Heren and H. Ölmez, "The influence of ethanolamines on the hydration and mechanical properties of Portland cement," *Cement and Concrete Research*, vol. 26, no. 5, pp. 701–705, 1996.

[12] J. P. Perez, A. Nonat, S. Garrault, S. Pourchet, and M. Mosquet, "Influence of triisopropanolamine on the physico-chemical and mechanical properties of pure cement pastes and mortars," in *Proceedings of the 13th French-Polish Seminar on Reactivity of Solids*, pp. S35–S42, Cluny, France, September 2003.

[13] K. Yamada, T. Takahashi, S. Hanehara, and M. Matsuhisa, "Effects of the chemical structure on the properties of the poly carboxylate-type superplasticizer," *Cement and Concrete Research*, vol. 30, no. 2, pp. 197–207, 2000.

[14] F. Winnefeld, S. Becker, J. Pakusch, and T. Gotz, "Polymer structure/concrete property relations of HRWRA," in *Proceedings of the 8th CANMET/ACI International Conference*

on Recent Advances in Concrete Technology, pp. 159–177, Montreal, Québec, Canada, May-June 2006.

[15] F. Winnefeld, A. Zingg, R. Holzer, R. Figi, J. Pakusch, and S. Becker, "Interaction of polycarboxylate-based superplasticizers and cements: influence of polymer structure and C3A-content of cement," in *Proceedings of 12th International Congress on the Chemistry of Cement (ICCC)*, Montreal, Canada, July 2007.

[16] A. O. Habib, I. Aiad, T. A. Youssef, and A. M. Abd El-Aziz, "Effect of some chemical admixtures on the physico-chemical and rheological properties of oil well cement pastes," *Construction and Building Materials*, vol. 120, pp. 80–88, 2016.

[17] Y. Renhe, L. Baoyuan, and W. Zhongwei, "Study on the pore structure of hardened cement paste by SAXS," *Cement Concrete Research*, vol. 20, no. 3, pp. 385–393, 1990.

5

Portland Cement Hydration Behavior at Low Temperatures: Views from Calculation and Experimental Study

Zhuangzhuang Liu,[1,2] Wenxiu Jiao,[2] Aimin Sha,[1,2] Jie Gao,[2]
Zhenqiang Han,[2] and Wei Xu[2]

[1]Key Laboratory for Special Area Highway Engineering, Ministry of Education, Xi'an, China
[2]School of Highway, Chang'an University, Xi'an, China

Correspondence should be addressed to Zhuangzhuang Liu; liuzhuangzhuang1986@gmail.com

Academic Editor: Aboelkasim Diab

Environmental condition affects the property of construction materials. This study gives an initial understanding of Portland cement hydration under low temperatures from the views of laboratory experiments (including electrical resistivity, degree of hydration (DoH), and maturity) as well as thermodynamic calculation. The hydrates of Portland cement at the given period were detected with X-ray diffraction (XRD), and their microstructure was observed by scanning electron microscope (SEM). Experiment result (i.e., DoH and electrical resistivity) indicated that the hydration of Portland cement was delayed by low temperature without hydration stopping at −5°C. Based on a basic kinetics model, the thermodynamic calculation predicted that the final hydrate differs in dependence on environmental temperatures. The mechanical behavior trend of Portland cement paste affected by low temperatures potentially is linked to the appearing of aluminate compounds and reduction of portlandite.

1. Introduction

Temperature affects the performances of Portland cement which is the most widely used materials in infrastructure constructions [1]. Meanwhile, the matrix of cement binder acts a greatly important role in Portland cement-based composites (i.e., pastes, mortars, concrete, stabilized stones, and treated soils). The performance of hardened Portland cement (e.g., mechanical behavior, and durability) is linked tightly with early age chemical hydration and hardening, while the relationships between hydration processes, hydrates produced, microstructures, and mechanical behaviors have been proved in previous studies [2–5]. In service life of infrastructures, cement-based materials have to be faced with severe environmental conditions, such as super-low temperatures [6–8]. In these conditions mechanical behaviors (e.g., compressive strength, flexural strength, elastic modulus, and Poisson ratio) of hardened Portland cement concrete will be enhanced by the super-low temperature, for example, −70~−10°C [7].

Differently, if cement-based materials (pastes, mortars, concrete, etc.) are experienced with low temperature, especially negative temperature (<0°C) during initial hydration stage, cement hydration will be greatly affected [9–11]. In this case, the hydrated products, phase conversion, for example, from ettringite (AFt) to monosulfate (AFm), and pores solution will be affected by low temperatures [5, 12]. In some limited conditions, the matrix would be even damaged. Thus, strategies in early ages should be applied to avoid the damages in matrix under cold weathers [13, 14]. To this end, a lot of studies were conducted to learn the Portland cement hydration under low temperatures [10, 11, 15–18], even though there is still lack of understanding in depth to the low temperature effects on cement hydration and hardening characteristics.

Better understanding of Portland cement hydration can absolutely promote the performance of cement composites under low temperatures, especially for the application at cold climates. Thus, this study aims to investigate the hydration process of Portland cement, including hydrates, microstructures, and mechanical behavior evolution. To further understand the low temperature effects on hydration process, thermodynamic approach is also employed to calculate the hydrates of Portland cement paste. Basically, this research provides a basic knowledge about the hydration process of Portland cement under low temperatures as a part of a systematic study.

2. Experimental Program

2.1. Raw Materials. A typical commercial ordinary Portland cement (OPC, produced by Jidong Cement Plant, Xi'an, China) was used in this study with oxide components detailed in Table 1 (PO42.5). What should be noted is that the oxide components measured by XRF here did not reflect the real component in Portland cement due to 5~10% filler replacement in clinkers. The mineral phases in OPC (via XRD) and the particle size distribution are shown in Figure 1. The technical properties of PO42.5 used in this study (provided by the producer) are as follows: specific surface area (Blaine) = $360 \, m^2/kg$, density = $3.02 \, g/cm^3$, initial setting time = 2.8 hrs, and final setting time = 4.7 hrs.

2.2. Methods and Instruments

2.2.1. Laboratory Experiments. The cement pastes were mixed according to ASTM C305-14 [20] and then were removed into molds ($40 \times 40 \times 160$ mm) or plastic containers; afterward, they were cured in chambers under $-5°C$, $0°C$, $5°C$, $8°C$, and $20°C$ (RH = 90%). It should be noted that before mixing raw materials (i.e., Portland cement and water), molds and bowls should be precooled in chambers consistent with their following curing temperature. For example, if the paste sample is going to be cured at $0°C$, the water, cement, bowls, and molds should be precooled under $0°C$ for two hours until their surface reaches $0°C$. In experimental study, the water-cement (w/c) ratio of prisms samples for strength measurement was set as 0.45, while the w/c of the paste stored in sealed plastic containers was given as 0.5 for complete reaction.

The pastes in container were treated following the solvent exchange method (with isopropanol) [21] and then were measured with XRD (Bruker, D8 Advanced, Cu-Kα) and SEM (Hitachi, S4800).

In order to describe the hydration process of Portland cement pastes, degree of hydration (DoH), maturity, and electrical resistivity were measured in this research. The DoH of Portland cement pastes was determined as (1) based on Powers model [22].

$$\text{DoH} = \frac{m_{105} - m_{950}}{0.25 \times m_{105}}, \quad (1)$$

where DoH is the degree of hydration (% by weight), m_{105} is the initial mass of sample pretreated in muffle furnace (6 hours) at $105°C$, and m_{950} is the final mass of sample heated at $950°C$.

Electrical resistivity of cement paste could be utilized to analyze the cement hydration process well [19]. Thus, the resistivity curve during initial Portland cement hydration was detected via CCR-II (produced by BC Tech, Shenzhen City, China). The equipment and sample were shown in Figure 2. To prevent moisture evaporation and temperature fluctuation, a plastic cover was set on the test-board, while the temperature was controlled by air-conditioner (general temperature) or chamber (lower temperatures).

The maturity was calculated according to the following equation [10, 11, 23]:

$$M = \sum_{0}^{t} (T_c - T_0) \cdot \Delta t, \quad (2)$$

where M is the maturity of Portland cement paste, T_c is the sample temperature ($°C$) measured by CCR-II (detailed above) or curing temperature ($-5°C$, $0°C$, $5°C$, $8°C$, and $20°C$), T_0 is the base temperature ($-10°C$, generally), and Δt is the time interval during curing stage (h).

2.2.2. Thermodynamic Calculation. Gibbs free-energy minimization criteria were used to calculate equilibrium phase assemblages and ionic speciation of chemical systems, such as the Portland cement paste. The modeling and software were detailed in our previous study [18], in which GEMS-PSI (software) and CEMDATA7.1 (database) were employed to calculate the hydrates of OPC. What should be noted is that in this study a basic kinetic function [24] of Portland cement hydration was modified by a solution equilibrium constant [18, 25]. In the thermodynamic simulation, input data included the following: C_2S = 11.1 g/100 g, C_3S = 62.9 g/100 g, C_3A = 6.0 g/100 g, C_4AF = 11.5 g/100 g, gypsum = 4.6 g/100 g, K_2O = 1.1 g/100 g, and Na_2O = 0.3 g/100 g. Also, 10000 days were adapted as the final hydration term in the simulation. The theoretical calculation from thermodynamics could provide deeper explanation to mechanical behaviors.

3. Results and Discussion

3.1. Electrical Resistance at Early Ages. Figure 3 shows the resistivity curve of Portland cement during hydration. After being mixed with water the ions (e.g., Ca^{2+}, K^+, Na^+, OH^-, and SO_4^{2-}) dissolved into the water forming an electrolytical solution [26], and then the hydrates would expend the ions in solution or occupy the space of solution; thus the resistivity of paste could be utilized to observe the hydration stages during hydration. It was reported that the cement hydration could be divided into five stages: (1) dissolution stage; (2) dynamic balance stage; (3) setting stage; (4) hardening stage; and (5) hardening deceleration stage [19]. Based on the previous study [19], the curve of resistivity and differential resistivity could point out the initial setting and final setting time well. In Figure 3, the normalized data of resistivity shows an obvious bottom which matches the start of initial setting. Also, the starting position of the hardening deceleration stage could be confirmed at the top of differential electrical resistivity curve. The curves in Figure 3 show that the temperature data and resistivity data were in a good agreement with each other. Therefore, the temperature data in CCR-II measurement can also be considered to describe the cement hydration. This

TABLE 1: Oxide composition of Portland cement used in this study[*].

Oxide	Na_2O	MgO	Al_2O_3	SiO_2	P_2O_5	SO_3	K_2O	CaO	TiO_2	MnO	Fe_2O_3	CuO	ZnO	Rb_2O	SrO	BaO	PbO	Cr_2O_3
PO42.5	0.30	1.30	5.20	18.00	0.05	3.00	1.10	65.80	0.40	0.07	4.80	0.02	0.10	0.00	0.09	0.08	0.02	0.00

[*]Data collected by X-ray fluorescence (XRF).

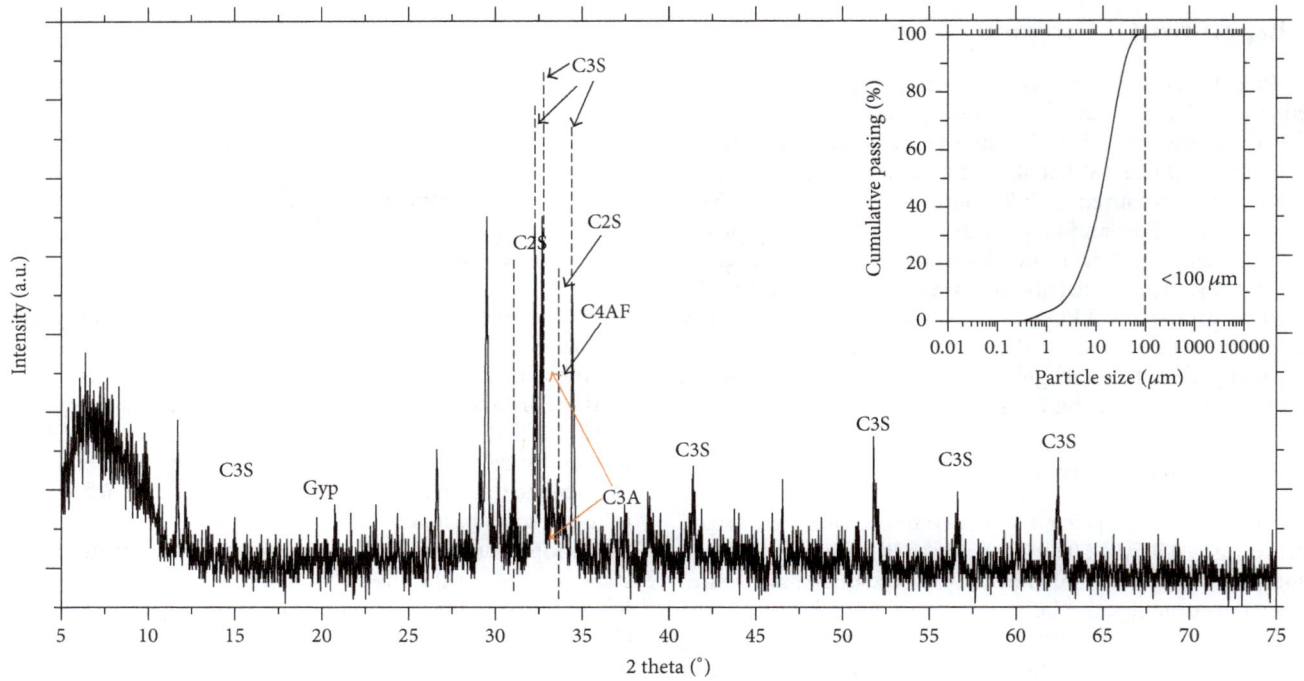

FIGURE 1: Particle size and minerals phases of raw materials.

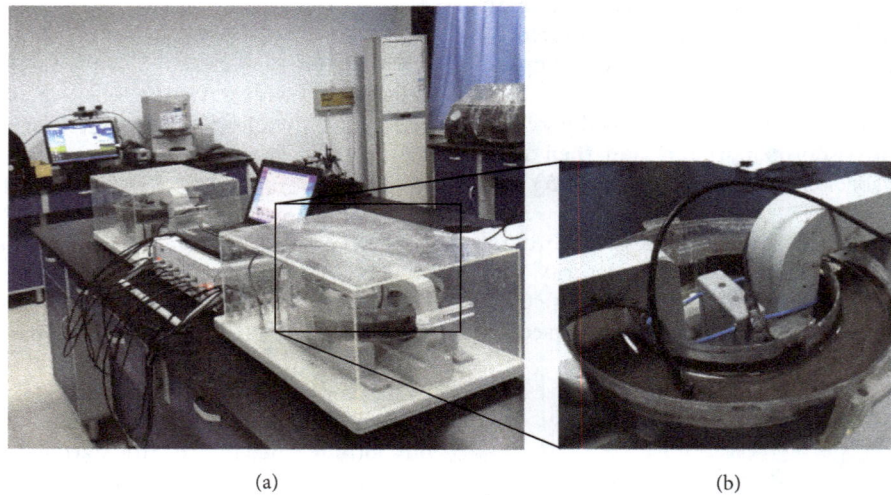

(a)   (b)

FIGURE 2: Electrical resistivity measurement: device and sample setting.

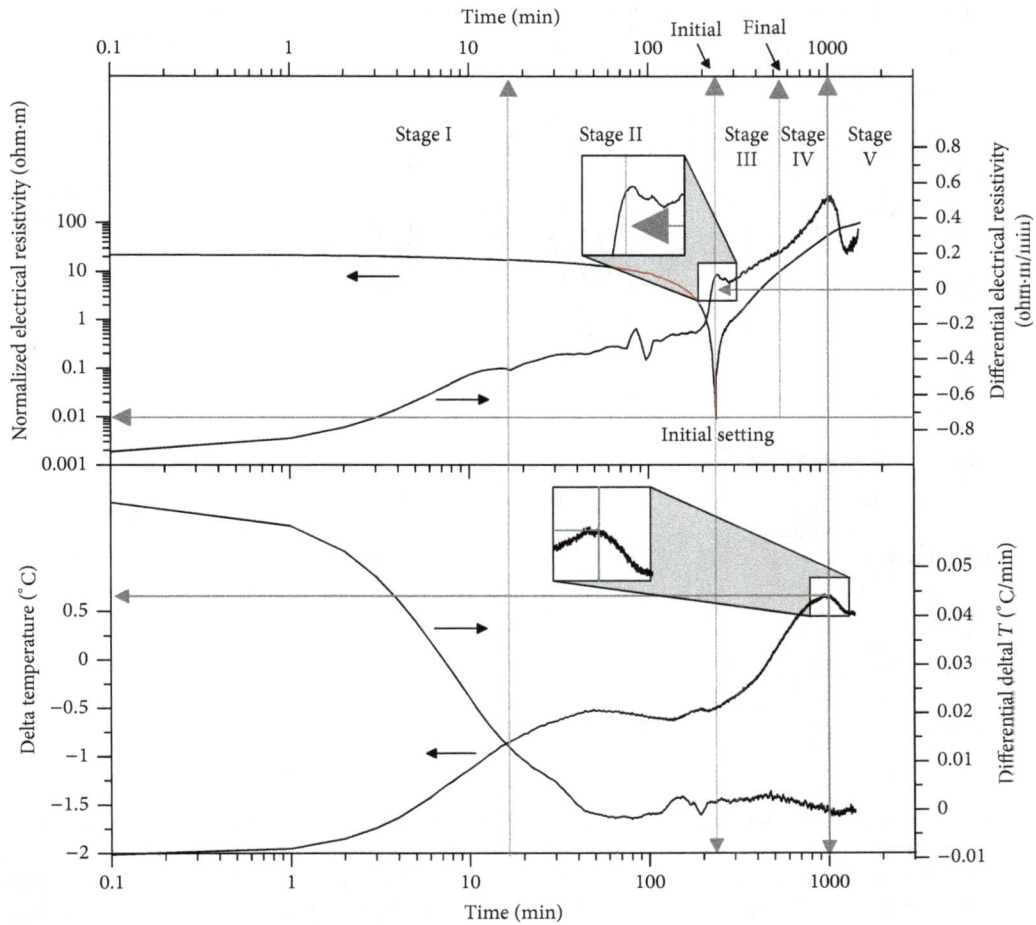

FIGURE 3: *Electrical resistivity of* Portland cement during hydration: five stages of Portland cement hydration according to [19].

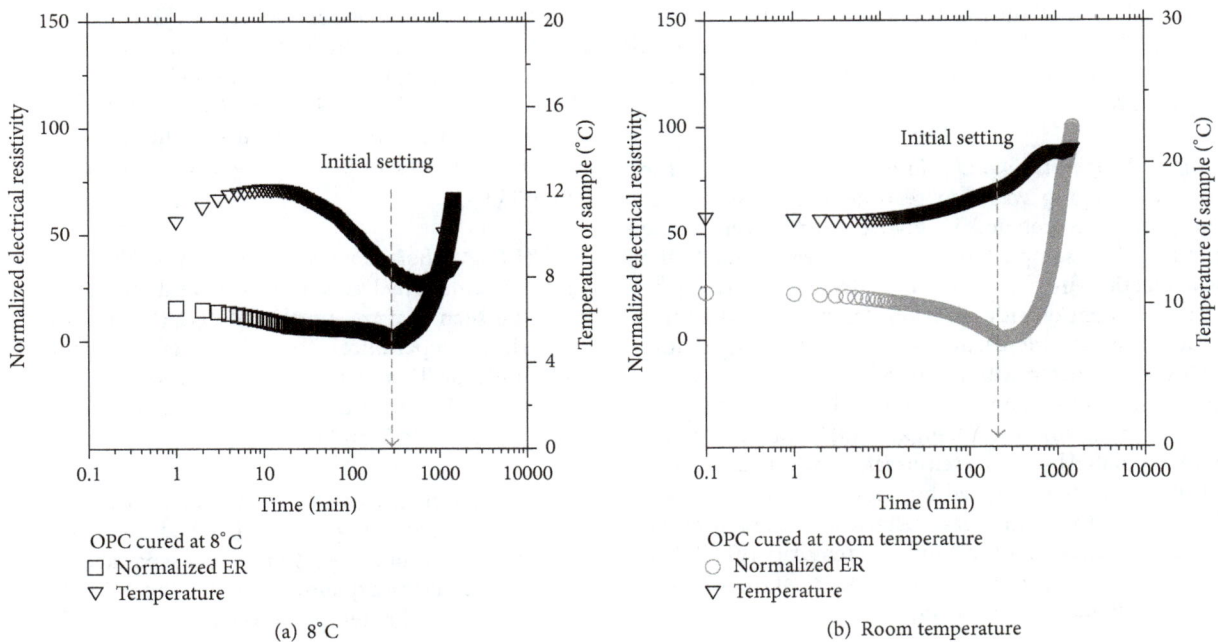

(a) 8°C

(b) Room temperature

FIGURE 4: Sample temperature variation.

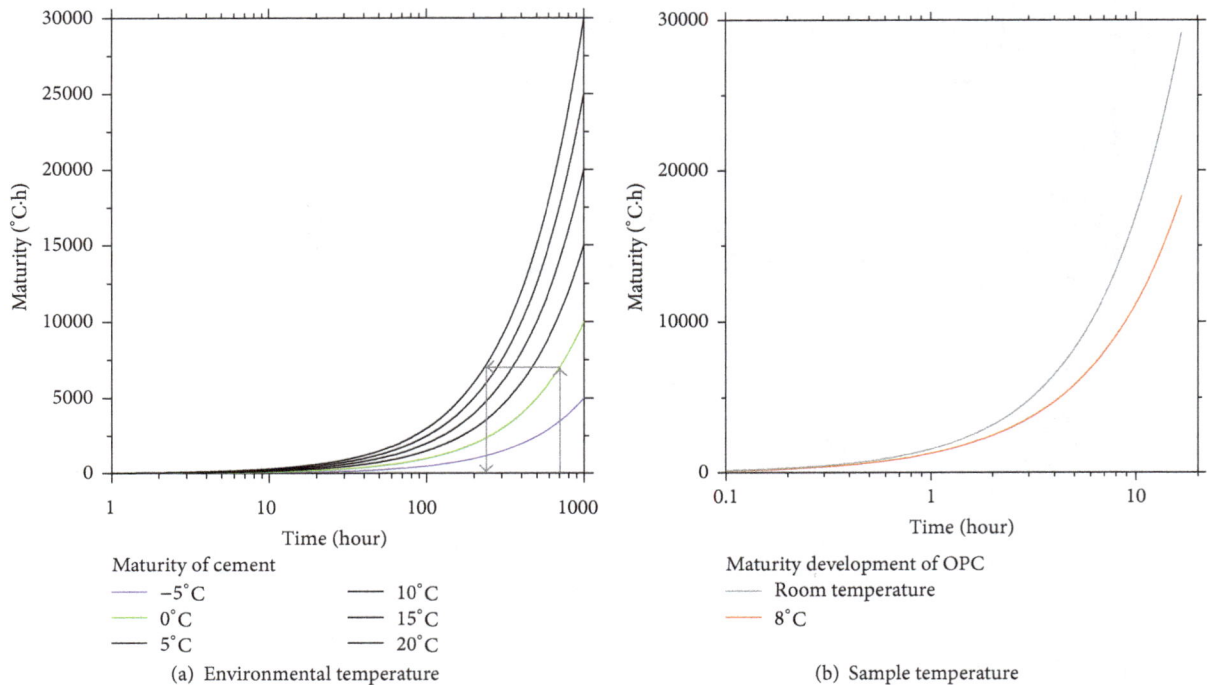

(a) Environmental temperature

(b) Sample temperature

FIGURE 5: Maturity development of cement paste.

conclusion is the basis to adapt the temperature data of sample in maturity calculation below.

Figure 4 presents the normalized resistivity data of Portland cement pastes under 8°C (Figure 4(a)) and 20°C (Figure 4(b)). Two important characteristics could be noticed in the curves: (i) low temperature (8°C) delays the bottom (initial setting time position) of normalized resistivity curve; (ii) the initial plot of normalized resistivity curve at low temperature (8°C) was lower than that at general condition (20°C). They are due to the chemical reaction rate reduced by low temperatures.

3.2. Sample Temperature and Maturity Development. The temperature changing was also recorded by sensors (seeing Figure 2(b)) as detailed in Figure 4. In lower curing temperature (8°C), sample temperature increased due to the chemical reaction in paste and then decreased caused by cooler environment outside. In room temperature the sample temperature kept increasing, where sample temperature showed less relationship with electrical resistivity of paste.

Figure 5 shows the maturity calculated based on the curing/sample temperature via (2). Figure 5(a) is the ideal maturity curve calculated by curing temperature, while Figure 5(b) showed the maturity calculated by sample temperature (see (2)). As believed, the mechanical behaviors of cement-based composites have great relationships with maturity [10]. Taken in this sense, Figure 5 could act as an evidence of the strength delaying in previous studies [17, 18].

3.3. Degree of Hydration. Degree of hydration (DoH) responds to the reaction process of cement paste; hence

the DoH of cement paste cured in lower temperatures was measured based on Powers' model and shown in Figure 6. Beyond the curing time, the DoH of cement with the same curing temperature increased, while higher hydration rate was obtained with higher temperatures. Take −5°C, for instance, its DoH beyond 90 d was 63.2%, much lower than that of a general condition (91.9% at 20°C). This result agreed with the conclusion of FHWA's report [23]. Meanwhile, the experiment indicates the Portland cement can still hydrate at −5°C; for example, the DoHs of OPC at −5°C were 16.7%, 25.5%, 47.4%, 55.3%, 61.9%, and 63.2% beyond 1, 3, 7, 28, 60, and 90 days, respectively. This result explains the slow strength achievement of cement pastes exposed in negative temperatures.

3.4. XRD Analysis. Figure 7 shows the XRD analysis of Portland cement hydrates under different temperatures (1 d). It can be seen that the portlandite ($Ca(OH)_2$) peak differs by curing temperature. To −5°C, there was no obvious portlandite peak, and the mineral phase can be observed clearly (1 d). There were no AFt peaks at 1 day for the pastes cured under −5, 0, 5, and 8°C.

3.5. SEM. The microstructure of hydrated paste was arranged in Figure 8. According to the DoH above, less Portland cement hydrated under −5°C; thus the separated particles in Figure 8(a) could be explained in which the solid did not bind with others. To other temperatures, the pastes hardening was related to the curing temperature. If focused on the hydrates, there were some separated particles in Figure 8(b) (0°C), while there were few unhydrated particles in Figure 8(c)

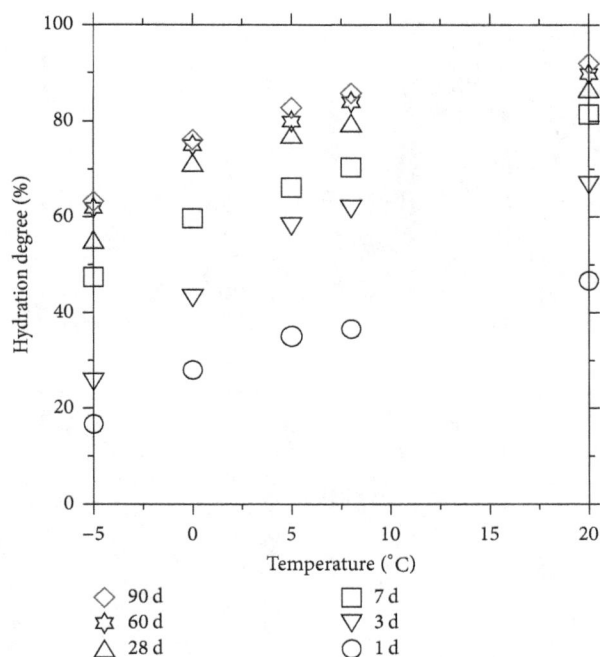

FIGURE 6: Degree of hydration development over curing temperature variation.

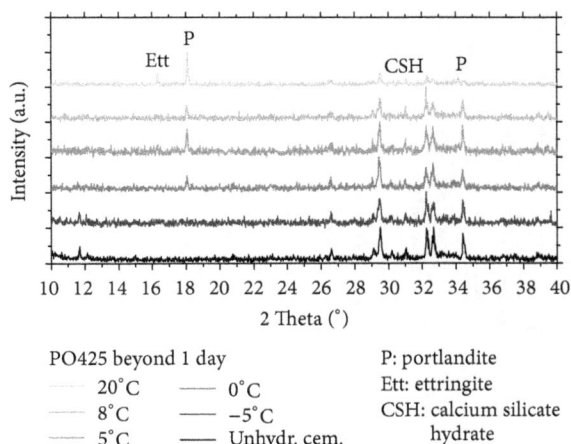

FIGURE 7: XRD result of OPC responded to curing temperature (1 day).

then bind with each other. Here, another reason should not be ignored that the particles in cement paste were moved and separated by ice formed at −5°C (negative temperatures).

4. Thermodynamic Calculation of Portland Cement Hydration at Low Temperature

4.1. Portland Cement Hydrates over Curing Time. Thermodynamic calculation of Portland cement hydration has been proved by a number of studies. Figure 9 shows the hydrates evolution at 20°C based on the thermodynamic simulation. As shown in the figure, the hydrates increased over curing time. The Aft was converted into AFm at 1 d and then disappeared at 2 d, while the C_3AH_6 phase appeared after 3 d. This calculation result was proved by our experimental data as the low temperature effects have been reported previously. The final hydrates will be potentially changed within 0~10°C. Mineral phase variation can be seen in Figure 10.

4.2. Relationships between Strength, Hydrates, and Temperature. The mechanical behavior of cement paste is linked significantly to its hydrates. For instance, the relationship between mechanical behavior and CSH content has been proved in our previous study [17]. Figure 10 shows the relationships between curing temperature, compressive strength, and hydrates volume fraction. The hydrate fractions were collected from thermodynamic calculation (age = 10000 d).

The strength was measured under 0, 5, 8, and 20°C and beyond 3 d, 7 d, and 28 d. It can be seen that (1) the content of CSH was not significantly affected by curing temperature beyond long ages; (2) the volume fractions of AFt, AFm, and portlandite were changed with curing temperature; (3)

(5°C). If the curing temperature was higher than 8°C, less unhydrated particles could be observed in SEM images (see Figures 8(d) and 8(e)). Considering the DoH (47.4%) of Portland cement cured at −5°C in 7 d, the cement should produce some hydrates to bind the particles in paste; however, the DoH of this sample (−5°C, 7 d) was just similar to the one cured at 20°C in 1 d (46.7%, see Figure 4). At this DoH level, particles in paste did not react to bind with others. Actually, the Portland cement just finished phase-boundary reaction and reached diffusion controlled hydration [27] at this DoH level while the spaces between particles were not filled by hydrates. With the DoH increasing (i.e., related with temperature), spaces between hydrates would be filled and

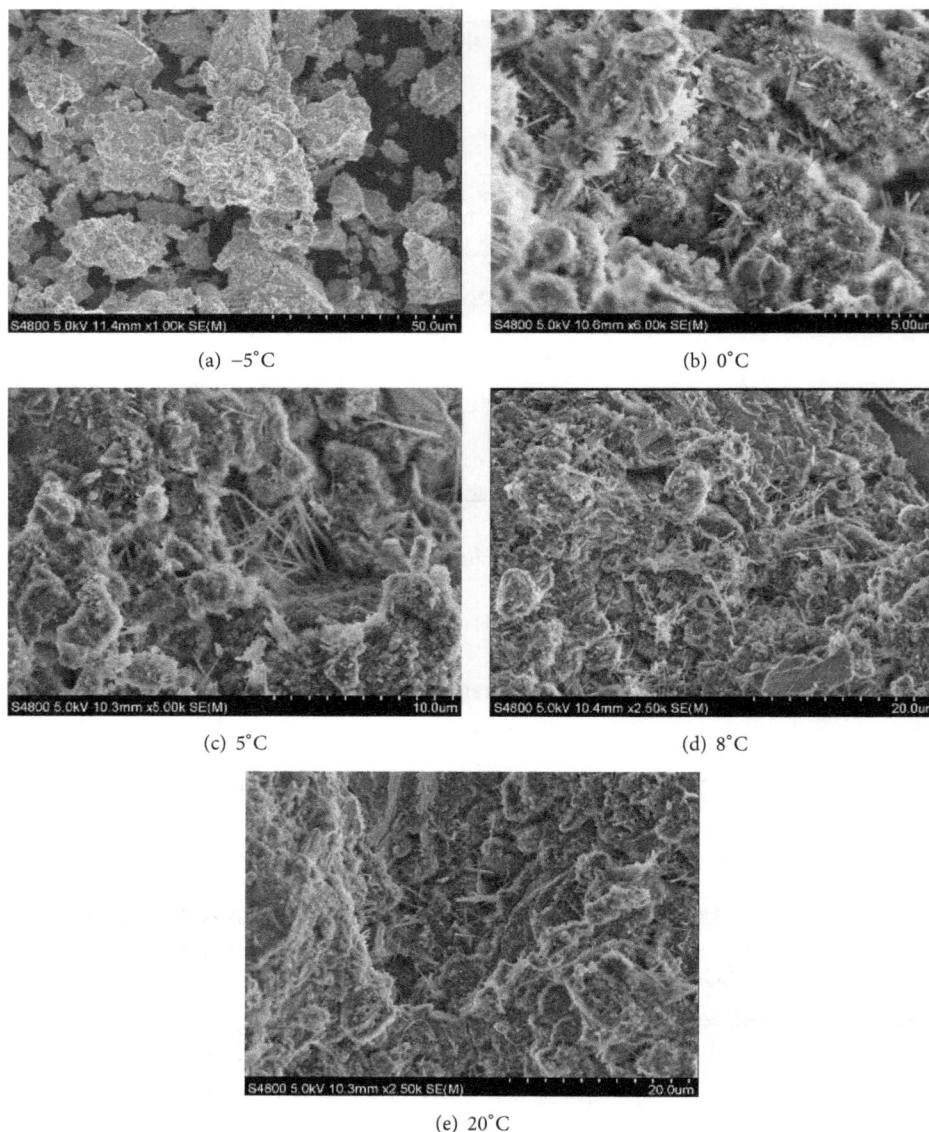

(a) −5°C

(b) 0°C

(c) 5°C

(d) 8°C

(e) 20°C

FIGURE 8: SEM images of solid cement pastes cured under low temperatures (7 d).

the long-term (28 d) strength showed few relationships with CSH content but was linked greatly with AFt, AFm, and portlandite, at the point of 10°C. With the reduction of CH, the compressive strength was increased. Also, the AFm might benefit the mechanical strength in Portland cement at low temperatures.

This finding was very interesting because we always thought that the mechanical behaviors were linked to the CSH volume fraction tightly, but in this study on the premise of different early curing temperatures we found that the mechanical strength changing had no relationship with the CSH content (strength differs with a similar/the same CSH volume fraction) but is linked significantly with the produced aluminate hydrates and portlandite. What should be noted is that the above conclusion may not be 100% right, but we would like to consider that the temperature effects on

early mechanical behavior should have deeper explanations, thermodynamically. However, these deductions need more evidence in the future.

5. Summary and Conclusions

The Portland cement (PO42.5) hydration characteristics at early ages including electrical resistance, temperature evolution, degree of hydration, and phase evolution were observed in laboratory under the curing temperature of −5, 0, 5, 8, and 20°C. Thermodynamic calculation based on GEMS-PSI software was also utilized to explain and verify the experimental results. Conclusions can be drawn as follows:

(1) Low temperatures (−5, 0, 5, and 8°C) reduced the hydration rate but did not terminate the hydrate reaction; also, the hydration process remains as generate

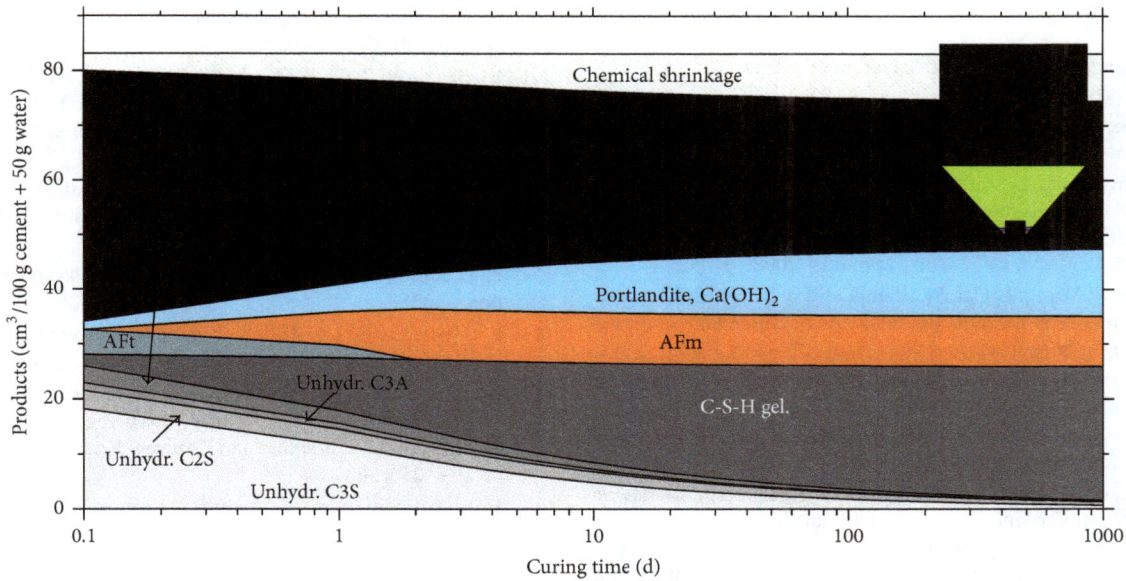

FIGURE 9: Hydrates evolution over curing time.

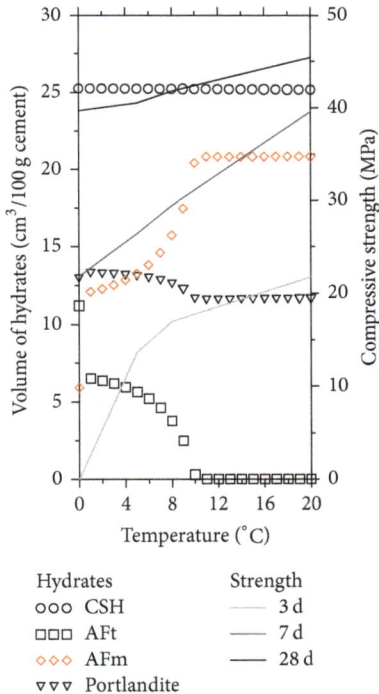

FIGURE 10: Relationships between strength, hydrates, and temperature.

condition. Based on the CCR-II temperature sensors, the maturity might explain the hydration delaying.

(2) There was a linear relationship between DoH and curing temperature to 1–7 d; however, a breakpoint appeared at 7–28 d. In experiment, the breaking point appeared at 5~8°C. This DoH breaking point

is very similar to the phase transformation point in thermodynamics (10°C).

(3) The calculation indicated that early mechanical behavior might have deeper explanations from the view of thermodynamics, where the hydrates differ under low temperatures. On the premise of this study (different early curing temperature), the compressive strength of old pastes (28 d) has less relationship with the CSH content but is linked tightly with aluminate compounds and portlandite.

Conflicts of Interest

The authors declare that they have no conflicts of interest.

Acknowledgments

The authors would like to acknowledge the financial supports from the Natural Science Foundation of China (NSFC, no. 51708045) and the National Key Technology R&D Program, China (2014BAG05B04), as well as address their thanks to graduate students Yongwei Lu and Wenxiu Jiao for their experiment assistant. They also appreciate the discussion and comments from experts in World Transport Convention (WTC, Beijing, 2017), where the data of this research was initially presented, partly. Laboratory experiments of this study were conducted in the Key Laboratory for Special Area Highway Engineering, MOE, China, and the Materials Analysis Center, School of Materials Science and Engineering, Chang'an University, and the Xi'an Mineral Resources Surveillance and Test Center, Ministry of Land and Resource of China. They gratefully acknowledge the support making these laboratories and their operation possible.

References

[1] J. F. Young, S. Mindess, and D. Darwin, *Concrete*, Prentice Hall, New Jersey, NJ, USA, 2002.

[2] R. L. Kozikowski, W. C. McCall, and B. A. Suprenant, *Cold Weather Concreting Strategies*, edit by Rush Jr, W. E., 2014.

[3] B. Lothenbach, T. Matschei, G. Möschner, and F. P. Glasser, "Thermodynamic modelling of the effect of temperature on the hydration and porosity of Portland cement," *Cement and Concrete Research*, vol. 38, no. 1, pp. 1–18, 2008.

[4] Y. Qiao, H. Wang, L. Cai, W. Zhang, and B. Yang, "Influence of low temperature on dynamic behavior of concrete," *Construction and Building Materials*, vol. 115, pp. 214–220, 2016.

[5] L. Xu, P. Wang, and G. Zhang, "Formation of ettringite in Portland cement/calcium aluminate cement/calcium sulfate ternary system hydrates at lower temperatures," *Construction and Building Materials*, vol. 31, pp. 347–352, 2012.

[6] F. Karagol, R. Demirboga, and W. H. Khushefati, "Behavior of fresh and hardened concretes with antifreeze admixtures in deep-freeze low temperatures and exterior winter conditions," *Construction and Building Materials*, vol. 76, pp. 388–395, 2015.

[7] G. C. Lee, T. S. Shih, and K. C. Chang, "Mechanical properties of concrete at low temperature," *Journal of Cold Regions Engineering*, vol. 2, no. 1, pp. 13–24, 1988.

[8] T. Miura, "The properties of concrete at very low temperatures," *Materials and Structures*, vol. 22, no. 4, pp. 243–254, 1989.

[9] R. Demirboğa, F. Karagöl, R. Polat, and M. A. Kaygusuz, "The effects of urea on strength gaining of fresh concrete under the cold weather conditions," *Construction and Building Materials*, vol. 64, pp. 114–120, 2014.

[10] R. W. Nurse, "Steam curing of concrete," *Magazine of Concrete Research*, vol. 1, no. 2, pp. 79–88, 1949.

[11] A. G. Saul, "Principles underlying the steam curing of concrete at atmospheric pressure," *Magazine of Concrete Research*, vol. 2, no. 6, pp. 127–140, 1951.

[12] M. Maslehuddin, C. L. Page, and Rasheeduzzafar, "Temperature effect on the pore solution chemistry in contaminated cements," *Magazine of Concrete Research*, vol. 48, no. 5, pp. 5–14, 1997.

[13] ACI Committee, "Standard Specification for Cold Weather Concreting, (Reapproved 2002)," Tech. Rep. ACI 306.1-90, American Concrete Institute, Michigan, Mich, USA, 2010.

[14] ACI Committee, Guide to Cold Weather Concreting ACI 306R-10, American Concrete Institute, Michigan, Mich, USA, 2010.

[15] J. Liu, Y. Li, P. Ouyang, and Y. Yang, "Hydration of the silica fume-Portland cement binary system at lower temperature," *Construction and Building Materials*, vol. 93, pp. 919–925, 2015.

[16] R. Polat, "The effect of antifreeze additives on fresh concrete subjected to freezing and thawing cycles," *Cold Regions Science and Technology*, vol. 127, pp. 10–17, 2016.

[17] Z. Liu, A. Sha, L. Hu, and X. Zou, "A laboratory study of Portland cement hydration under low temperatures," *Road Materials and Pavement Design*, vol. 18, no. sup3, pp. 12–22, 2017.

[18] Z. Liu, A. Sha, L. Hu et al., "Kinetic and thermodynamic modeling of Portland cement hydration at low temperatures," *Chemical Papers*, vol. 71, no. 4, pp. 741–751, 2017.

[19] L. Xiao and Z. Li, "New understanding of cement hydration mechanism through electrical resistivity measurement and microstructure investigations," *Journal of Materials in Civil Engineering*, vol. 21, no. 8, pp. 368–373, 2009.

[20] ASTM, Standard Practice for Mechanical Mixing of Hydraulic-Cement Pastes and Mortars of Plastic Consistency C305-14, ASTM International West Conshohocken, Philadelphia, Pennsylvania, Pa, USA, 2014.

[21] J. J. Beaudoin, P. Gu, J. Marchand, B. Tamtsia, R. E. Myers, and Z. Liu, "Solvent replacement studies of hydrated portland cement systems: the role of calcium hydroxide," *Advanced Cement Based Materials*, vol. 8, no. 2, pp. 56–65, 1998.

[22] T. C. Powers and T. L. Brownyard, "Studies of the physical properties of hardened Portland cement paste," *Journal of the American Concrete Institute*, 1946.

[23] K. D. Smith, Maturity Testing for Concrete Pavement Applications, 2005.

[24] L. J. Parrot and D. C. Killoh, "Prediction of cement hydration," *Proceedings of the British Ceramic Society*, p. 41, 1984.

[25] F. Lin and C. Meyer, "Hydration kinetics modeling of Portland cement considering the effects of curing temperature and applied pressure," *Cement and Concrete Research*, vol. 39, no. 4, pp. 255–265, 2009.

[26] H. F. W. Taylor, *Cement Chemistry*, Thomas Telford Publishing, London, UK, 2nd edition, 1997.

[27] F. Tomosawa, "Development of a kinetic model for hydration of cement," in *Proceedings of the 10th Interenational Congress on the Chemistry of Cement*, p. 5158, 1997.

6

Surface Corrosion and Microstructure Degradation of Calcium Sulfoaluminate Cement Subjected to Wet-Dry Cycles in Sulfate Solution

Wuman Zhang, Sheng Gong, and Bing Kang

School of Transportation Science and Engineering, Beihang University, Beijing, China

Correspondence should be addressed to Wuman Zhang; wmzhang@buaa.edu.cn

Academic Editor: Charles C. Sorrell

The hydration products of calcium sulfoaluminate (CSA) cement are different from those of Portland cement. The degradation of CSA cement subjected to wet-dry cycles in sulfate solution was studied in this paper. The surface corrosion was recorded and the microstructures were examined by scanning electron microscopy (SEM). The results show that SO_4^{2-}, Na^+, Mg^{2+}, and Cl^- have an effect on the stability of ettringite. In the initial period of sulfate attack, salt crystallization is the main factor leading to the degradation of CSA cement specimens. The decomposition and the carbonation of ettringite will cause long-term degradation of CSA cement specimens under wet-dry cycles in sulfate solution. The surface spalling and microstructure degradation increase significantly with the increase of wet-dry cycles, sulfate concentration, and water to cement ratio. Magnesium sulfate and sodium chloride reduce the degradation when the concentration of sulfate ions is a constant value.

1. Introduction

Calcium sulfoaluminate (CSA) cement was initially developed and produced in China in the late 1960s and was initially designed as an addition to Portland cement (PC) due to its expansive (or shrinkage-compensating) characteristics. Although it has only been in use for less than 50 years, it is one of the most commonly used cements in China and is beginning to be used in all areas of the world [1–3].

Zhu et al. [4] used CSA cement and Silica Fume to design ultrahigh early strength self-compacting mortar. Fu et al. [2], Sahu et al. [5], and He et al. [6] studied the hydration mechanism of CSA cement. Some researchers have also tried to examine the durability of CSA cement concrete. Zhang et al. [7] investigated influences of water/cement ratio and admixtures on carbonation resistance of CSA cement-based high performance concrete. Geng et al. [8] found that carbonation depth of CSA cement concrete was smaller than that of PC concrete regardless of curing time. Zhao et al. [3] concluded that CSA cement was an ideal cementitious material which can significantly improve the resistance to chloride ion erosion of the concrete. Duan et al. [9] indicated that chloride ions migration coefficient of CSA cement concrete was distinctly smaller than that of Portland cement concrete, especially at early stage. Zhao et al. [10] reported that CSA cement concrete had better chloride ion penetration resistance capacity than Portland cement concrete before and after the freeze–thaw cycle.

Compared with Portland cement, the hydration products of calcium sulfoaluminate (CSA) cement contain rich ettringite, C-S-H, and Al gel, but without $Ca(OH)_2$ phase [11] which is regarded as the source of chemical sulfate attack on concrete. Due to the absence of $Ca(OH)_2$, CSA cement concrete showed excellent resistance against chemical sulfate attack. In addition, field experiences [12] indicate that the concrete surface scaling above sulfate environment ground level is more commonly observed than fully submerged concrete damage. Therefore, the damage of CSA cement concrete partially immersed in sulfate solution is worth being studied for not only the application of CSA cement in sulfate environment but also the further understanding of the so-called physical sulfate attack on concrete [13].

TABLE 1: Chemical compositions of cement (%).

	CaO	SiO$_2$	Al$_2$O$_3$	Fe$_2$O$_3$	MgO	SO$_3$	R$_2$O*	Loss
OPC	62.28	21.08	5.47	3.96	1.73	2.63	0.80	1.61
CSA	44.17	11.79	28.74	1.98	2.61	8.71	—	0.16

*R$_2$O = Na$_2$O + 0.658K$_2$O.

TABLE 2: Molar concentrations of ions in sulfate solutions.

Molar concentration (10^{-4} mol/ml)	SO$_4^{2-}$	Na$^+$	Mg^{2+}	Cl$^-$
3% Na$_2$SO$_4$	2.18	4.36	—	—
5% Na$_2$SO$_4$	3.71	7.42	—	—
3% Na$_2$SO$_4$ + 1.80% MgSO$_4$	3.71	4.36	1.53	—
3% Na$_2$SO$_4$ + 1.80% MgSO$_4$ + 3% NaCl	3.71	9.65	1.53	5.29

2. Experimental Program

2.1. Materials.
CSA cement, which is industrially produced in Tangshan Six-Nine Cement Industry Co. Ltd. and set by Chinese national standards GB20472-2006, was used. The chemical compositions of CSA cement and ordinary Portland cement (OPC) are listed in Table 1. Four sulfate solutions, namely, 3% Na$_2$SO$_4$, 5% Na$_2$SO$_4$, 3% Na$_2$SO$_4$ + 1.80% MgSO$_4$, 3% Na$_2$SO$_4$ + 1.80% MgSO$_4$ + 3% NaCl, were used. The molar concentrations of ions in sulfate solutions are shown in Table 2.

2.2. Specimens and Tested Program.
CSA cement pastes were prepared with tap water and the water to cement ratios (w/c) were 0.30, 0.40, and 0.50 by mass of cement. Molded specimens size was made with 20 mm × 20 mm × 20 mm. After demoulding (24 h after casting), the specimens were cured at a constant temperature (20 ± 2°C) and 90% relative humidity for 28 days.

The wet-dry cycles were carried out by exposure to an alternate conditions. CSA cement specimens were subjected to semi-immersion (two-third part of specimen in solution) in the sulfate solution at 20 ± 2°C for 12 hours and drying in laboratory air at 20 ± 2°C for 12 hours (as shown in Figure 1), during which no temperature effects are considered. This regime is used to simulate cyclic outdoor environments such as rainy days and unclouded days. The wet-dry regime of specimens is shown in Figure 1. The sulfate solution was replaced by the original solution each week in order to provide a constant ions concentration. Spalling samples were taken out from the sulfate solutions and washed by the anhydrous ethanol and acetone to stop their hydration. The microstructure of the CSA cement samples was examined by scanning electron microscopy (SEM).

3. Visual Assessment of Cube Degradation

3.1. Effect of Wet-Dry Cycles.
Figure 2 shows the surface change of CSA cement 0.50 w/c specimen subjected to different wet-dry cycles in 5% Na$_2$SO$_4$. It is obvious that the degrees of surface spalling or damage increase rapidly with the increase of wet-dry cycles. Firstly, the sodium sulfate crystals and powder particles are observed on the surface after the specimen is subjected to 18 wet-dry cycles. Then the surface layer is stripped off from the specimen when the wet-dry cycle increases from 18 to 31. And the degrees of spalling increase rapidly with the increase of the wet-dry cycles until the specimen is broken into two parts.

SEM images of SAC 0.50 w/c specimen subjected to different wet-dry cycles in 5% Na$_2$SO$_4$ solution are given in Figure 3. Microcrack is not observed from the SEM image of the specimen subjected to 18 wet-dry cycles, although powder particles are found on the specimen surface. However, microcracks are observed when the wet-dry cycle is raised from 18 to 31, and the number and the connectivity of microcracks obviously increase when the wet-dry cycle is 46. The results indicate that the degrees of microstructures degradation increase with the increasing of wet-dry cycles.

3.2. Effect of W/C.
Figure 4 shows surface change of CSA cement specimens with different w/c subjected to wet-dry cycles in 5% Na$_2$SO$_4$ solution. SAC specimen with 0.50 w/c is broken into two parts when the wet-dry cycle is 46. However, the specimen with 0.40 w/c shows a slight surface spalling at the same wet-dry cycle. Two spallings are observed at the corner of the specimen with 0.30 w/c until the wet-dry cycle increases from 46 to 119. It can be seen that the degrees of surface spalling increase rapidly with the increase of w/c.

SEM images of CSA cement specimens with different w/c subjected to wet-dry cycles in 5% Na$_2$SO$_4$ are given in Figure 5. It is clear that the connected microcracks and a long microcrack are observed from the SEM images of the specimen with 0.50 and 0.40 w/c under 46 wet-dry cycles, respectively. A large number of microcracks are also observed with the specimen with 0.30 w/c under 119 wet-dry cycles. The results indicate that CSA cement specimens are damaged when they are subjected to a certain number of wet-dry cycles in 5% Na$_2$SO$_4$ solution, although the degrees of degradation decrease with the decrease of w/c.

It is well known that w/c is an important factor for mechanical properties and durability of cement paste and concrete. Higher w/c makes the porosity higher and the pore structure coarser, which will reduce the durability of cement paste [15]. CSA cement is no exception to this trend.

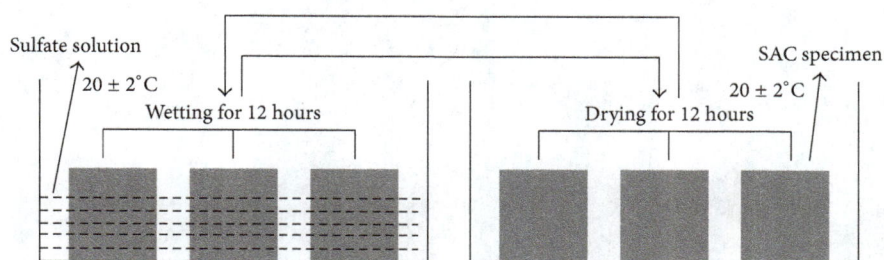

FIGURE 1: The wet-dry regime of specimens.

| (a) 18 cycles | (b) 22 cycles | (c) 25 cycles | (d) 31 cycles |
| (e) 39 cycles | (f) 41 cycles | (g) 43 cycles | (h) 46 cycles |

FIGURE 2: Surface change of CSA cement 0.50 w/c specimen under wet-dry cycles in 5% Na_2SO_4 solution.

3.3. Effect of Ion Concentration. Figure 6 shows the surface change and SEM image of CSA cement 0.50 w/c specimen subjected to 46 wet-dry cycles in 3% Na_2SO_4 solution. It can be seen that Na_2SO_4 solution migrates to the surface of specimen during the wetting process and begins to precipitate Na_2SO_4 crystals during the drying process. Compared to specimen with the same w/c and wet-dry cycles in 5% Na_2SO_4 solution (see Figures 2(h) and 3(h')), a slight surface spalling and a number of microcracks are observed when the specimen is subjected to wet-dry cycles in 3% Na_2SO_4 solution. However, the degrees of damage decrease significantly. The chemical reactions in the pore solution can occur at any sulfate concentration and decrease the possibility of physical attack due to the consumption of sulfates. Moreover, a low concentration solution decreases the possibility of a chemical reaction [12].

In Figure 7, the surface change and SEM image of CSA cement 0.50 w/c specimen subjected to 46 wet-dry cycles in 3% Na_2SO_4 + 1.80% $MgSO_4$ solution are presented. It is clear that the surface spalling is very slight and several microcracks are found in the SEM image. The degree of deterioration of SAC specimen under wet-dry cycles in 3% Na_2SO_4 + 1.80% $MgSO_4$ solution is lesser than that of specimen subjected to wet-dry cycles in 5% Na_2SO_4 solution. In addition, a large number rod-like crystals are observed in the SEM image.

Figure 8 shows the surface change and microstructures of CSA cement 0.50 w/c specimen subjected to 46 wet-dry cycles in 3% Na_2SO_4 + 1.80% $MgSO_4$ + 3% $NaCl$ solution. There is almost no change except a crack on the edge of the surface. It also can be seen that there is a very few microcracks in the SEM image of specimen. Chloride ions have a higher rate of diffusion than that of the sulfate ions. Chloride ions can permeate through the specimen surface much faster than the sulfate ions and block the internal pore, which decreases the penetration of the sodium sulfate solution. However, this result may be different after a longer wet-dry cycle and this work will be carried out in the future.

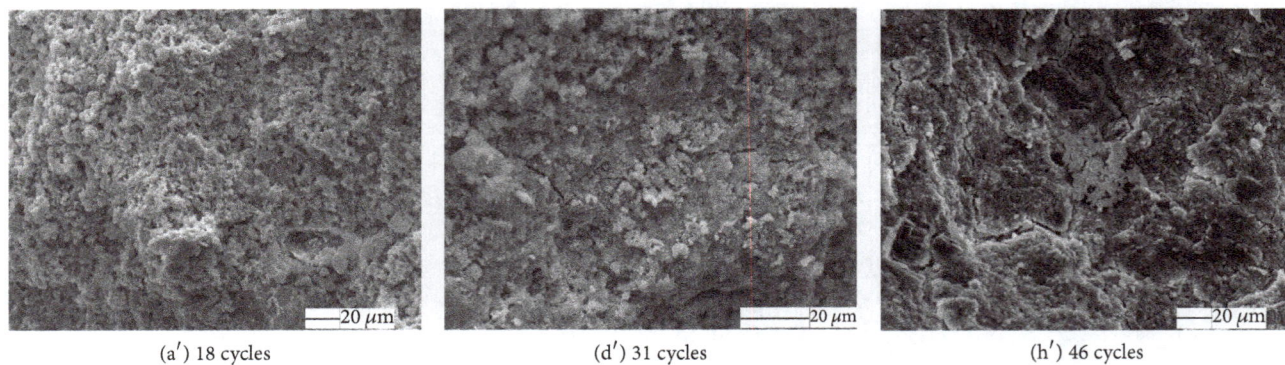

(a′) 18 cycles   (d′) 31 cycles   (h′) 46 cycles

FIGURE 3: SEM images of CSA cement 0.50 *w/c* specimen under wet-dry cycles in 5% Na$_2$SO$_4$ solution.

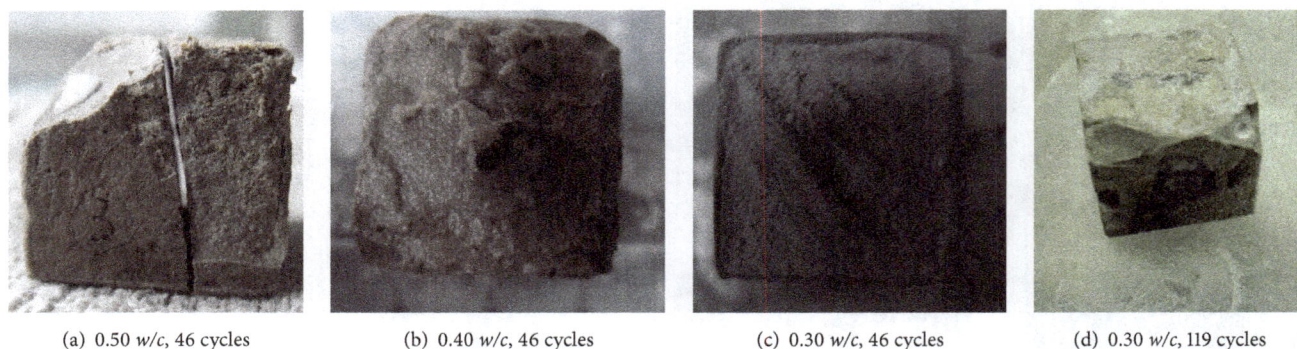

(a) 0.50 *w/c*, 46 cycles   (b) 0.40 *w/c*, 46 cycles   (c) 0.30 *w/c*, 46 cycles   (d) 0.30 *w/c*, 119 cycles

FIGURE 4: Surface change of CSA cement specimens under wet-dry cycles in 5% Na$_2$SO$_4$ solution.

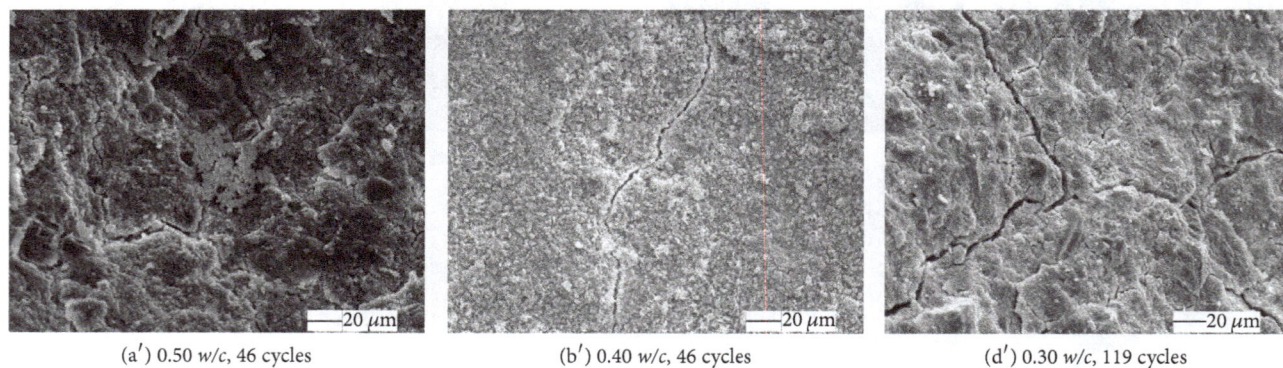

(a′) 0.50 *w/c*, 46 cycles   (b′) 0.40 *w/c*, 46 cycles   (d′) 0.30 *w/c*, 119 cycles

FIGURE 5: SEM images of CSA cement specimens under wet-dry cycles in 5% Na$_2$SO$_4$ solution.

3.4. OPC and CSA Cement. The surface changes of OPC and CSA cement specimens are given in Figures 9 and 10, respectively. It is obvious that there is a significant difference in the surface change between OPC and CSA cement specimens. A serious spalling is observed on each side face and the spalling is mainly above the liquid level in wetting process. There is no significant spalling on the surface of OPC specimen, but netlike cracks are found on each side face. Figure 11 shows the microstructures of OPC and CSA cement specimens. An obvious stripping layer is observed from the SEM image of CSA cement specimens. However, there are no significant microcracks except a certain number of pores in the SEM image of OPC specimen.

This result can be explained as follows. There was a dense process of OPC specimen subjected to sulfate attack because sulfate ions transported into the pore and reacted with calcium hydroxide and calcium aluminate hydrate, to form ettringite and gypsum. The production could block the pore and microcracks in the initial period of sulfate attack [16]. And then the sodium sulfate solution penetrated into OPC specimen would be decreased. However, the chemical reaction may be very slow because of the low Ca(OH)$_2$ content of CSA cement. Much sodium sulfate solution would penetrate into CSA cement specimen during the wetting process, which could rapidly generate a pressure of salt crystallization during the drying process. Therefore, the

(a) Surface change   (b) SEM image

FIGURE 6: CSA cement 0.50 *w/c* specimen subjected to 46 wet-dry cycles in 3% Na_2SO_4 solution.

(a) Surface change   (b) SEM image

FIGURE 7: SAC 0.50 *w/c* specimen subjected to 46 wet-dry cycles in 3% Na_2SO_4 + 1.80% $MgSO_4$ solution.

sulfate attack resistance of CSA cement is not better than that of OPC during the wet-dry cycles.

4. Discussion of Degradation Mechanisms

4.1. Degradation Mechanisms in Sodium Sulfate Solution. Sodium sulfate reacts with calcium aluminate hydrate and calcium hydroxide, leading to the formation of ettringite and gypsum, which is the degradation mechanisms of OPC in sodium sulfate solution. However, it is evident that the hydration products of CSA cement are ettringite, C-S-H, and amorphous AH3. Therefore, the above degradation mechanisms are not suitable for CSA cement because there is very little calcium aluminate hydrate and calcium hydroxide in the hydration products of CSA cement [11].

The degradation has been attributed to the physical sulfate attack which can be regarded as a specific type of salt damage. A common form of physical sulfate attack occurs when sodium sulfate crystallizes in the pore structure of cement or concrete. The typical consequence of physical sulfate attack of concrete is surface erosion. In its early stages, it is manifested primarily by the appearance of white crystallites (efflorescence) on a concrete surface (as shown in Figure 2). The composition of the sulfate salt formed is also

important; sodium sulfates may interconvert between the 10 hydrate and the anhydrous salt. Thenardite has been reported to be capable of producing pore pressures of 400–5000 psi while mirabilite $Na_2SO_4 \cdot 10H_2O$ produces pore pressures of 1000–1200 psi [17].

4.2. Degradation Mechanisms in Magnesium Sulfate Solution. Ettringite is the main hydration product of CSA cement. The Mg^{2+} can influence the stability of ettringite. Figure 12 gives the stability of ettringite as a function of $MgSO_4$ concentration temperature. However the attack reaction can be slow: the kinetics of ettringite decomposition are not rapid even at 85°C in normal seawater concentrations, owing in part to the development of semiprotective layers of reaction products on ettringite [14].

In addition, Mg^{2+} can generate insoluble $Mg(OH)_2$, causing the decomposition of ettringite [18] as the following equation.

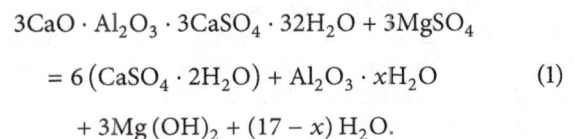

$$3CaO \cdot Al_2O_3 \cdot 3CaSO_4 \cdot 32H_2O + 3MgSO_4$$
$$= 6\left(CaSO_4 \cdot 2H_2O\right) + Al_2O_3 \cdot xH_2O \qquad (1)$$
$$+ 3Mg\left(OH\right)_2 + \left(17 - x\right)H_2O.$$

(a) Surface change   (b) SEM image

FIGURE 8: CSA cement 0.50 *w/c* specimen subjected to 46 wet-dry cycles in 3% Na_2SO_4 + 1.80% $MgSO_4$ + 3% NaCl solution.

FIGURE 9: OPC 0.40 *w/c* specimen subjected to 74 wet-dry cycles in 3% Na_2SO_4 solution.

The damage of the lower part of CSA cement paste in the $MgSO_4$ solution characterized a typical chemical sulfate attack [19]: gypsum starts forming in the region close to the surface due to the solution penetration. The surface zone of the paste, where massive gypsum crystals have been aggregating due to ettringite decomposition, behaves like a skin that is trying to expand. However, the bulk of the paste underneath, which is chemically unaltered, tries to resist this expansion. Thus, a resultant compressive force is generated in the region close to surface causing the outer paste layer detachment. After the outer layer detachment, the same attack process starts causing the second layer formation and detachment and then the third layer and successive layers [13].

4.3. Degradation Mechanisms in Sodium Chloride Solution. Both components of NaCl impact on ettringite stability. Sodium will affect the thermodynamic properties of the aqueous solution coexisting with ettringite and thereby alter its solubility. Chloride has the same action but additionally it stabilises chloride-AFm, Friedel's salt, and this provides another stable host structure which is competitive with ettringite for aluminium. Figure 13 shows experimental data on ettringite stability, based on experiments ranging in

duration between 150 and 400 days [14]. It can been seen that at 25°C the aqueous sulfate concentration approximately doubles between initially distilled water and 0.5 M NaCl. The combined effect of temperature and sodium chloride content is to enhance markedly sulfate solubilities. Ettringite is partially or wholly converted to one or more AFm phases with sulfate and chloride, the chloride phase being either Friedel's salt or a mixed Cl-SO_4 AFm phase.

4.4. Degradation Mechanisms by Carbonation. In addition, the stability of ettringite is influenced by carbonation. Ettringite would decompose to $CaSO_4 \cdot 2H_2O$, $CaCO_3$, and alumina gel as the following equation [20]:

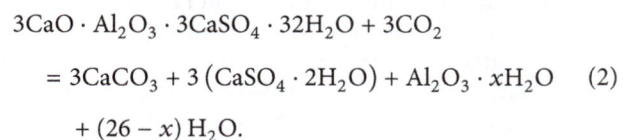

$$3CaO \cdot Al_2O_3 \cdot 3CaSO_4 \cdot 32H_2O + 3CO_2$$
$$= 3CaCO_3 + 3\left(CaSO_4 \cdot 2H_2O\right) + Al_2O_3 \cdot xH_2O \quad (2)$$
$$+ (26 - x)\,H_2O.$$

The carbonation of ettringite caused the formation of $CaCO_3$ and $CaSO_4 \cdot 2H_2O$, but the role of carbonation of ettringite in the paste in the solution should be weak. There was another mechanism causing rich $CaSO_4 \cdot 2H_2O$ formation. According to the basic theory of chemical reaction kinetics the chemical

FIGURE 10: CSA cement 0.40 w/c specimen subjected to 73 wet-dry cycles in 3% Na_2SO_4 solution.

(a) OPC

(b) SAC

FIGURE 11: SEM images of OPC and CSA cement specimens.

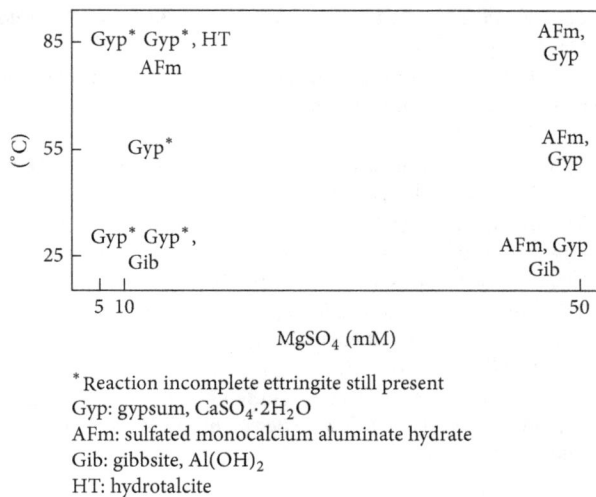

* Reaction incomplete ettringite still present
Gyp: gypsum, $CaSO_4 \cdot 2H_2O$
AFm: sulfated monocalcium aluminate hydrate
Gib: gibbsite, $Al(OH)_2$
HT: hydrotalcite

FIGURE 12: Stability of ettringite as a function of $MgSO_4$ concentration temperature [14].

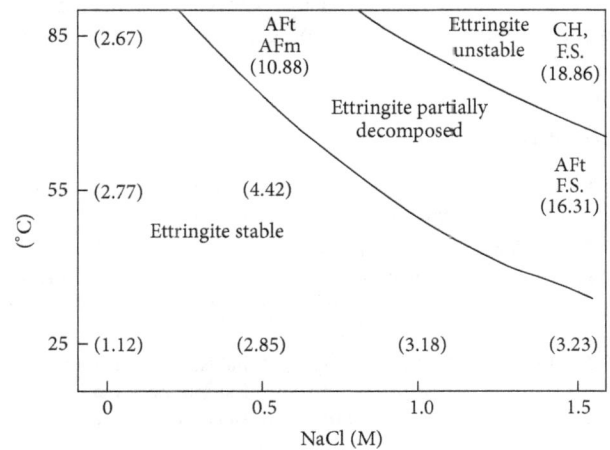

FIGURE 13: Stability of ettringite as a function of NaCl concentration and temperature [14].

5. Conclusions

reaction has to occur if there is water, gas, or insoluble product formation [20, 21].

From the results obtained in this work, the following conclusion can be drawn.

(1) SO_4^{2-}, Na^+, Mg^{2+}, and Cl^- affect the stability of ettringite. In the initial period of sulfate attack, the salt crystallization is the main factor leading to the degradation of CSA cement specimens subjected to wet-dry cycles.

(2) The decomposition and the carbonation of ettringite will cause a long-term degradation of CSA cement specimens under wet-dry cycles.

(3) The surface spalling and microstructure degradation increase significantly with the increase of wet-dry cycles and *w/c*. The degradation increases rapidly when the content of sulfate raises from 3% to 5%.

(4) Magnesium sulfate and sodium chloride can reduce the degradation when the concentration of sulfate ions is a constant value.

(5) The sulfate attack resistance of CSA cement is not better than that of OPC during the wet-dry cycles.

Additional Points

Highlights. (1) SO_4^{2-}, Na^+, Mg^{2+}, and Cl^- have an effect on the stability of CSA cement. (2) Salt crystallization is the main factor of the degradation of CSA cement subjected to wet-dry cycles in sulfate solution. (3) Visual assessment and microstructure analysis of the degradation of CSA cement occurs.

Conflicts of Interest

The authors declare that they have no conflicts of interest.

Acknowledgments

This research was supported by grants from the National Natural Science Foundation of China (no. 51378042, no. 51678022).

References

[1] G. Edward, *Moffatt Durability of Rapid-Set (Ettringite-Based) Concrete*, University of New Brunswick, 2016.

[2] X. Fu, C. Yang, Z. Liu, W. Tao, W. Hou, and X. Wu, "Studies on effects of activators on properties and mechanism of hydration of sulphoaluminate cement," *Cement and Concrete Research*, vol. 33, no. 3, pp. 317–324, 2003.

[3] J. Zhao, G. Cai, and D. Gao, "Analysis of mechanism of resistance to chloride ion erosion of sulphoaluminate cement concrete," *Journal of Building Materials*, vol. 14, pp. 357–361, 2011.

[4] Y. Zhu, B. Ma, X. Li, and D. Hu, "Ultra high early strength self-compacting mortar based on sulfoaluminate cement and silica fume," *Journal Wuhan University of Technology, Materials Science Edition*, vol. 28, no. 5, pp. 973–979, 2013.

[5] S. Sahu, J. Havlica, V. Tomková, and J. Majling, "Hydration behaviour of sulphoaluminate belite cement in the presence op various calcium sulphates," *Thermochimica Acta*, vol. 175, no. 1, pp. 45–52, 1991.

[6] Z. He, H. Yang, and M. Liu, "Hydration mechanism of sulphoaluminate cement," *Journal Wuhan University of Technology, Materials Science Edition*, vol. 29, no. 1, pp. 70–74, 2014.

[7] D. Zhang, D. Xu, X. Cheng, and W. Chen, "Carbonation resistance of sulphoaluminate cement-based high performance concrete," *Journal Wuhan University of Technology, Materials Science Edition*, vol. 24, no. 4, pp. 663–666, 2009.

[8] H. Geng, P. Duan, W. Chen, and Z. Shui, "Carbonation of sulphoaluminate cement with layered double hydroxides," *Journal Wuhan University of Technology, Materials Science Edition*, vol. 29, no. 1, pp. 97–101, 2014.

[9] P. Duan, W. Chen, J. Ma, and Z. Shui, "Influence of layered double hydroxides on microstructure and carbonation resistance of sulphoaluminate cement concrete," *Construction and Building Materials*, vol. 48, pp. 601–609, 2013.

[10] J. Zhao, G. Cai, D. Gao, and S. Zhao, "Influences of freeze-thaw cycle and curing time on chloride ion penetration resistance of Sulphoaluminate cement concrete," *Construction and Building Materials*, vol. 53, pp. 305–311, 2014.

[11] P. Chaunsali, *Early-Age Hydration and Volume Change of Calcium Sulfoaluminate Cement-Based Binders*, University of Illinois at Urbana-Champaign, 2015.

[12] Z. Liu, D. Deng, and G. De Schutter, "Does concrete suffer sulfate salt weathering?" *Construction and Building Materials*, vol. 66, pp. 692–701, 2014.

[13] Z. Liu, X. Li, D. Deng, and G. De Schutter, "The damage of calcium sulfoaluminate cement paste partially immersed in MgSO4 solution," *Materials and Structures*, vol. 49, pp. 719–727, 2016.

[14] F. P. Glasser, *The Stability of Ettringite International RILEM Workshop on Internal Sulfate Attack and Delayed Ettringite Formation*, K. Scrivener and J. Skalny, Eds., 2004.

[15] R. A. Cook and K. C. Hover, "Mercury porosimetry of hardened cement pastes," *Cement and Concrete Research*, vol. 29, pp. 933–934, 1999.

[16] J. Gao, Z. Yu, L. Song, T. Wang, and S. Wei, "Durability of concrete exposed to sulfate attack under flexural loading and drying-wetting cycles," *Construction and Building Materials*, vol. 39, pp. 33–38, 2013.

[17] P. W. Brown, "Thaumasite formation and other forms of sulfate attack," *Cement and Concrete Composites*, vol. 24, no. 3-4, pp. 301–303, 2002.

[18] F. M. Lea, *The Chemistry of Cement and Concrete*, Edward Arnold, London, UK, 3rd edition, 1970.

[19] M. Santhanam, M. D. Cohen, and J. Olek, "Mechanism of sulfate attack: a fresh look part 2 : proposed mechanisms," *Cement and Concrete Research*, vol. 33, no. 3, pp. 341–346, 2003.

[20] T. Nishikawa, K. Suzuki, S. Ito, K. Sato, and T. Takebe, "Decomposition of synthesized ettringite by carbonation," *Cement and Concrete Research*, vol. 22, no. 1, pp. 6–14, 1992.

[21] X. Chen and R. Zou, "Kinetic study of ettringite carbonation reaction," *Cement and Concrete Research*, vol. 24, no. 7, pp. 1383–1389, 1994.

7

Development of a Semirigid Pavement Incorporating Ultrarapid Hardening Cement and Chemical Admixtures for Cement Grouts

Jin Wook Bang,[1] Byung Jae Lee,[2] and Yun Yong Kim[3]

[1]R&D Center, Tongyang Construction Materials Co., Ltd., No. 2822-1, Gimpo-dearo, Wolgot-myeon, Gimpo-si, Gyeonggi-do 10024, Republic of Korea
[2]R&D Center, JNTINC Co., Ltd., No. 9, Hyundaikia-ro, 830 Beon-gil, Bibong-myeon, Hwaseong-City, Gyeonggi-do 18284, Republic of Korea
[3]Department of Civil Engineering, Chungnam National University, Daejeon 34134, Republic of Korea

Correspondence should be addressed to Yun Yong Kim; yunkim@cnu.ac.kr

Academic Editor: Xiao-Yong Wang

Mechanical tests were carried out to evaluate the influence and effects of fluidity and compressive strength of cement grout on semirigid asphalt pavement. An open graded asphalt skeleton was designed in order to achieve target porosity in the range of 18~22%. In addition, four types of cement grout mixtures were produced with varying mix proportions with ultrarapid hardening cement and chemical admixtures, that is, accelerating and retarding agents. For the semirigid pavement specimens, mechanical experiments to measure properties such as porosity, flexural strength, Marshall stability, and wheel tracking resistance were carried out. The test results demonstrated that the flow time (fluidity) of cement grout is the most significant factor that determines the mechanical properties of semirigid asphalt specimens under constant condition of the open graded asphalt skeleton. For the semirigid pavement mixing proportion in the current study, it is recommended that the porosity of the open graded asphalt skeleton and flow time of cement grout should be 20% and within 12 seconds, respectively.

1. Introduction

Asphalt and concrete pavement is the most commonly applied pavement method worldwide and is used for a variety of applications including vehicle roads and airport pavement. Rigid pavements, which mainly use concrete, are highly resistant to traffic load, have a high bearing capacity, offer long-term durability, and resist fatigue and deformation [1, 2]. In asphalt pavement, there is excellent deformation compatibility for expansion and shrinkage with both leading to lower maintenance cost than concrete pavement [3–6]. These two types of pavement methods generally have opposite characteristics and are selectively applied according to environmental conditions such as external load and soil profiles. On the other hand, semirigid pavement, which can complement the advantages and disadvantages of asphalt and concrete pavement, has been studied [7–9] by a number of

researchers since it was first applied to airport pavement in France [10]. In Korea, semirigid pavement has been studied since 2001 and various studies [11, 12] have been conducted to improve its performance. Based on these studies, semirigid pavement has been recently adopted for many construction members such as bridge-decks surfacing pavement, petrol station and heavy distribution centers. Semirigid pavement is a pavement method in which a cement grout having a relatively high stiffness and strength is filled with the skeleton of an asphalt mixture which contains 20~30% voids. The most significant characteristic of semirigid pavement is its ability to combine the ductility of asphalt and the stiffness and strength of concrete to provide excellent running performance and durability. In addition, it can be produced with various properties depending on the characteristics of the cement grouts and the skeleton of the asphalt mixture. In the current study, the physical properties of semirigid pavement were

TABLE 1: Properties of polyacrylate polymer.

Material	Density (g/cm^3)	Solid contents (%)	Viscosity (cps)	Ionicity	pH	Appearance
Polyacrylate polymer	1.0	45.3	360	Anionic	8.5	White liquid

TABLE 2: Mixing proportions according to different types of ultrarapid hardening cement grout.

Mix.	Cement types	Blending proportions (wt.%)			Unit weight (kg/m^3)		
		Cement minerals	Accelerating agent	Retarding agent	Cement	Water	Polyacrylate polymer
PG-1	Proto	46.9	53	0.1	1,150	319.5	255.5
SG-1	Early Strength	39.9	60	0.1	1,150	319.5	255.5
FG-1	Retarding	39.8	60	0.2	1,150	319.5	255.5
FG-2	Strength	39.7	60	0.3	1,150	319.5	255.5

TABLE 3: Characteristics of polymer modified asphalt binder.

Properties	Value	Standard
Penetration at 25°C (1/10 mm)	70	—
Softening point (°C)	72.4	—
Flash point (°C)	362	≥230
Viscosity (cp)	2001	≤3,000
Density at 15°C (g/cm^3)	1.028	—
Storage stability (°C)	0.5	≤2.0

evaluated using a cement grout formulated with ultrarapid hardening characteristics. This paper includes the results of a laboratory investigation into the properties of void content, Marshal stability, flexural strength, and wheel tracking resistance depending on flow time and compressive strength of the cement grout mixed with ultrarapid hardening cement and chemical admixtures including accelerating agent and retarding agent.

2. Materials

2.1. Cement Grout. Four types of cement grout mixture proportions depending on the type of ultrarapid hardening cement were developed and tested. Three types of ultrarapid hardening cement including prototype, early strength type, and two kinds of retarding strength type were designed. The ultrarapid hardening cement was produced by blending cement minerals including alite (C_3S), microsilica (SiO_2), and gypsum and chemical admixtures such as accelerating agent and retarding agent. In addition, all cement grout mixtures incorporated liquid polyacrylate polymer as seen in Table 1 for the purpose of preventing cracks and shrinkage. An overview of the various mixture proportions is presented in Table 2. The fresh cement grouts were produced by using a stand mixer which was able to control mixing speed to 750 RPM. This mixing process was carried out under temperature and relative humidity conditions of 20°C and 60%, respectively.

2.2. Maternal Open Graded Asphalt Mixture

2.2.1. Asphalt Binder. For mixing the open graded asphalt mixture, a polymer modified asphalt (PMA) with 4.0 wt% styrene butadiene styrene (SBS) was used in this study. The molecular structure of SBS is linear. The PMA commercial product type of PG 76–22 was manufactured by a domestic company in Korea and met the quality requirements for open graded asphalt mixture. The characteristics of PMA were measured based on the testing regulations and the results were listed in Table 3.

2.2.2. Aggregates and Filler. The impregnation characteristics of cement grouts into the asphalt skeleton were directly affected by its fluidity as well as by the porosity of the open graded asphalt mixture. A high porosity asphalt mixture enhanced the filling efficiency of cement grouts; however, having the appropriate porosity of asphalt mixture is significant for the strength requirements. In order to design the porosity of the open graded asphalt mixture, it is important that the application of coarse and fine aggregates satisfies the grading requirements of the specification. Fine and coarse aggregates used to accomplish this study were sourced from Gong-ju, Korea. Two types of 0.08–5 mm and 5–13 mm crushed granite gravels were selected as aggregates. All mixtures incorporated limestone powder filler with a range of 0.08–0.15 mm. To achieve the target porosity of open graded asphalt mixture in this investigation, coarse, fine aggregates

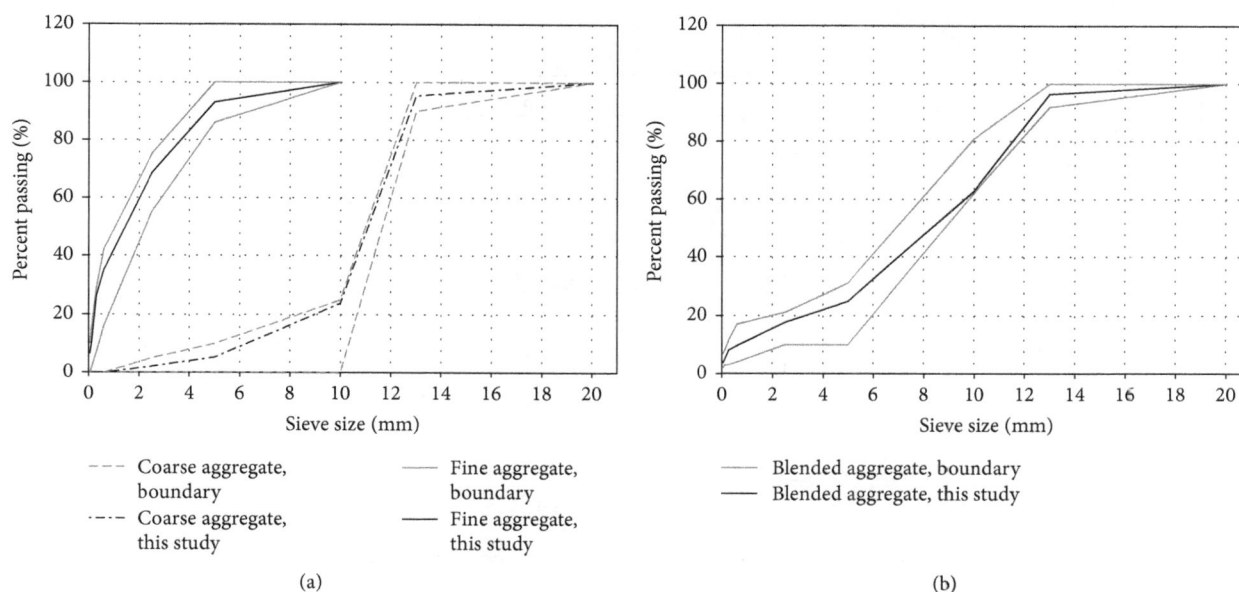

(a)

(b)

FIGURE 1: Size distribution and boundary limits of aggregates according to specification. (a) Fine and coarse aggregate grading curve. (b) Blended aggregates grading curve.

and filler were blended in accordance with the following equation:

$$P(i) = A(i) \times a + B(i) \times b + C(i) \times c + \cdots + N(i) \times n. \tag{1}$$

In this formula, $P(i)$ refers to the percentage of blended aggregates passing through the corresponding (i) sieve, (i) refers to the sieve size, $A(i), B(i), \ldots, N(i)$ refer to the percentage of each aggregate passing through the corresponding (i) sieve, $A, B, \ldots, N$ refer to all kinds of aggregate and filler, and $a, b, \ldots, n$ refer to the contents proportion of each aggregate and filler, and the sum total is 100%. The grading curve of all types of aggregates used has been made to meet the open graded asphalt grading requirements of the Korean standard specification as shown in Figure 1(a). The Korean standard specification suggests the boundary limits for the blended aggregates and filler used for the open graded asphalt mixture as shown in Figure 1(b). The mechanical properties of fine aggregate, including density, water absorption, and fineness modulus, are 2.66 g/cm³, 0.9%, and 3.02, while for coarse aggregate they are 2.67 g/cm³, 0.9%, and 6.08, respectively. Water content and density of limestone filler were measured as 0.2% and 2.72 g/cm³.

2.2.3. Mixing Open Graded Asphalt. Based on the previous studies [10] and the Korean standard of open graded asphalt mixture design [13], the porosity of asphalt mixture is designed with a range of 18~22%. The dosage of asphalt is 3.5%~6.0%. The open graded asphalt mixture proportions are presented in Table 2. In order to mix the open graded asphalt mixtures, all aggregates are cleaned and kept preheated to a temperature of 120°C to 150°C, while asphalt was heated to

a temperature of 150°C. All specimens are cast into the molds and compacted under hot condition according to the experiment standards. All specimens are taken out from the mold when the temperature of specimens reached 50°C.

2.2.4. Semirigid Asphalt Composite. Open graded asphalt mixtures were prepared according to the experimental method of Korea standards and four types of cement grout were injected and compacted. After that, all semirigid asphalt specimens were cured until standardized curing age.

3. Test Program

3.1. Cement Grout

3.1.1. Fluidity. The flow cone test has been carried out to evaluate the fresh fluidity of cement grout. The flow cone used in this study is for the hydraulic cement grout flow test regulated in KS F 4044 method. It has a height of 190 mm, a capacity of 1725 ± 5 mL, and an internal orifice diameter of 12.7 mm. All the grouts were charged to the calibration height and the dropping time was measured. Flow time is an index of the fluidity of the cement grout. Figure 2 shows the geometry of the flow cone used in this study. In the previous study [14] the flow time of the flow cone with a capacity of 1,200 mL and an internal orifice diameter of 10 mm was measured from 11 to 20 seconds. Lei [15] reported that, for the cone with a capacity of 1,000 mL and an internal orifice diameter of 10 mm, flow time was proposed to be in the range of 8 to 10 seconds. Also, because the flow value is slightly different depending on the cone size in the previous studies [16, 17] the flow value of this study can be used for a relative comparison of the grout formulations.

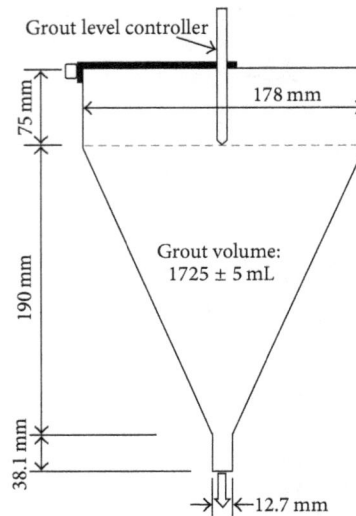

FIGURE 2: Flow cone geometry for hydraulic cement grout flow test.

3.1.2. Compressive Strength. A total of sixty cubic 50 × 50 mm specimens for the compressive strength test were fabricated and cured under standard conditions. The compressive strength test was carried out using a universal test machine with a capacity of 500 kN, and loading rate was controlled to 2.4 kN/sec. The compressive strength was calculated based on the average of three specimens.

3.2. Semirigid Asphalt Composite

3.2.1. Porosity. The open graded asphalt specimens were filled with asphalt mixture of 1,200g in a mold with a diameter of 101.6 mm and a height of 63.5 mm, and the compaction hammer was freely dropped from a height of 450 mm to compact 50 times on each surface, respectively. The specimens were coated with paraffin due to the fact that the moisture content of all specimens exceeded 2%. The porosity test was performed in accordance with KS F 2353 and ASTM D 1188. First of all, specific bulk density (γ_{bulk}) of all the specimens was calculated using (2). On the other hand, the theoretical maximum density ($\gamma_{theoretical}$) of the specimen, which can be compared with the apparent density, was calculated based on (3) using an automatic device of theoretical maximum density in order to quantify the effect of the grain size, the asphalt content, and the temperature condition. The porosity of all specimens including open graded asphalt mixtures and semirigid asphalt mixtures is given in (4).

$$\text{Specific bulk density} \ (\gamma_{bulk})$$
$$= \frac{P_{air}}{P_{pa} - P_{wa} - \left(P_{pa} - P_{air}\right)/r_{pa}}, \tag{2}$$

$$\text{Theoretical maximum density} \ (r_{theoretical})$$
$$= \frac{P_{air}}{P_{air} + P_{bowl} - P_{all}}, \tag{3}$$

$$\text{Porosity} \ (\%) = \left[1 - \frac{\gamma_{bulk}}{\gamma_{theoretical}}\right] \times 100, \tag{4}$$

where, P_{air} is dry weight of specimen in air (g), P_{pa} is paraffin coated dry weight of specimen in air (g), P_{wa} is paraffin coated dry weight of specimen in water (g), r_{pa} is bulk density of paraffin (g/cm^3), P_{bowl} is weight of bowl filled with water at 25°C (g), and P_{all} is weight of bowl filled with water and specimen at 25°C (g).

3.2.2. Marshall Stability. For the Marshall stability test, the specimen was fabricated in the same shape and method as the specimen prepared for the porosity test. The fabricated asphalt mixture specimens were cooled at room temperature for five hours and taken out from the mold using a sample extractor. After that, cement grout was injected and cured at a room temperature of 20°C for 1 day. All the semirigid asphalt specimens were tested using Marshall stability equipment. The test was carried out at a constant loading rate of 50 mm/min until failure in accordance with the KS F 2337 standard. The measured maximum vertical load refers to the Marshall stability value.

3.2.3. Wheel Tracking. For the wheel tracking test regulated by KS F 2374, the asphalt mixture was filled in a square mold having a length × width × height of 300 mm × 300 mm × 50 mm, and a specimen was prepared by using a compaction roller of which surface temperature was kept at 130°C. After demolding, cement grout was injected and cured at a room temperature of 20°C for 1 day. The wheel tracking test was carried out by placing each specimen in a chamber maintained at a temperature of 60°C for 5 hours so that the test can be carried out under the same temperature conditions for all specimens. The chamber was equipped with steel wheel tires with a diameter of 200 mm and a width of 50 mm. The wheel was attached with rubber with a thickness of 15 mm to simulate the wheel of the rear vehicle. The wheel consisted of a vertical loading device and a horizontal moving device. The vertical device applies a load continuously at 686 N and the horizontal device performs a horizontal reciprocating run at a speed of 0.5 Hz (30 times/min.) for a maximum runway distance of 300 mm for 1 hour. The central deflection of the specimen was measured at 5-minute intervals using LVDT with resolution capacity of 1/100 mm attached on the wheel device. In particular, the dynamic stability represents the number of reciprocating runs required for 1.0 mm deflection at the center of the wheel tracking specimen and is calculated using (5) below. The deformation rate (mm/min.) was calculated between 45 and 60 minutes.

$$DS = 42 \times \frac{t_2 - t_1}{d_2 - d_1}, \tag{5}$$

where DS is the dynamic stability, d_1 refers to the central deflection at 45 minutes (t_1), and d_2 refers to the deflection at 60 minutes (t_2).

3.2.4. Flexural Strength. The bending specimen was prepared by dividing the wheel tracking specimen into three equal parts. The shape of the specimen is length (l) × width (b) × height (h) of 300 mm × 100 mm × 50 mm. After allowing all the specimens to air dry, all specimens were put in a constant temperature chamber set at −10°C for 1 day in order to maintain the specimen temperature in the same condition. The test was carried out while maintaining a constant room temperature of 20°C. In order to prevent the abrupt failure of the specimen, the one-point load was continuously applied to the center of the specimen at a speed of 0.5 mm/min using the displacement control method. The applied load (P) and central displacement (d) of the specimen were measured at load cell with capacity of 50 kN and a linear variable displacement transducer (LVDT) with capacity of 10 mm was used to measure the applied load corresponding to the central displacement of the specimen. All the specimens were tested until the load was 0 N, and all data was saved in the TDS-530 data logger. The maximum flexural strength and corresponding flexural strain can be calculated using (6) and (7), respectively.

$$\text{Flexural strength} \left(f_r \right) = \frac{3lP}{2bh^2}, \tag{6}$$

$$\text{Flexural strain} \left(\varepsilon \right) = \frac{6hd}{l^2}. \tag{7}$$

4. Results and Discussion

4.1. Cement Grout

4.1.1. Fluidity. The flow time results in accordance with the ratio of accelerating and retarding agent content are shown in Figure 3. The flow time of PG-1 mixture was measured to be 10.5 s. For SG-1, FG-1, and FG-2, which were mixed with accelerating agent, it was measured as 12.8 s, 11.3 s, and 10.9 s, respectively. The use of accelerating agent increased the flow time of cement grout compared to the PG-1 mixture by 22%, 10%, and 4%, respectively. However, it can be seen that, with the same dosage of the accelerating agent, the flow time decreases as the amount of retarding agent used increases. The flow times of FG-1 and FG-2 mixtures were 91% and 85%,

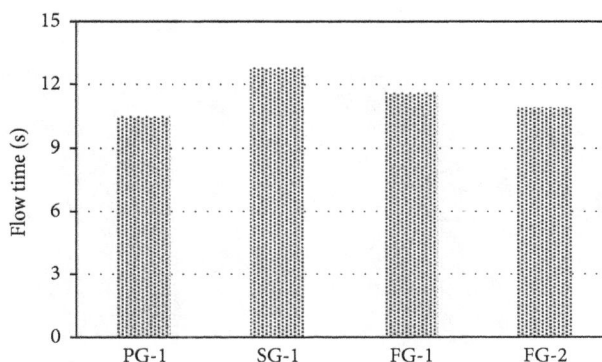

FIGURE 3: Flow time according to the cement grout mixtures.

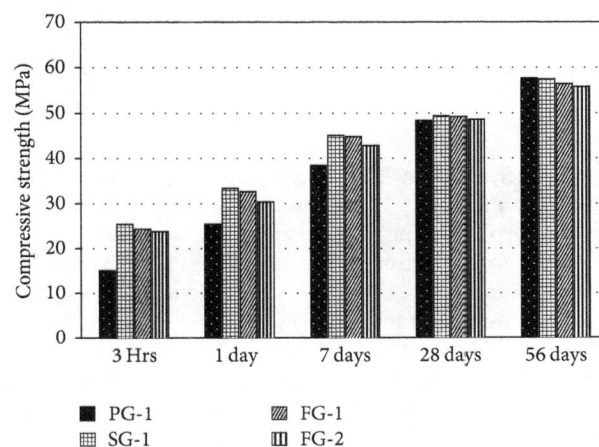

FIGURE 4: Compressive strength results of cement grout mixtures.

respectively, compared with the SG-1 mixture. Therefore, in order to secure the rapid hydration of the cement grout, it was shown that the use of the retarding agent can control the fluidity sufficiently when the accelerating agent was incorporated into the cement grout. Since the cement grout plays an important role in filling the pores of the open graded asphalt mixture to form a semirigid pavement, it should have sufficient fluidity to fill the pores. Afonso et al. [14] reported a flow time of 11 s to 20 s for a cone with a capacity of 1,200 mL and an internal orifice diameter of 10 mm. Lei has proposed a cone with a capacity of 1,000 mL and an internal orifice diameter of 10 mm within a range of 8 s to 10 s. Because the flow time differed according to the cone size in the previous study, the flow value in this current study was able to be used for relative comparisons of cement grout mixtures.

4.1.2. Compressive Strength. The findings related to compressive strength of all cement grout mixtures are presented in Figure 4. For all cement grouts using ultrarapid hardening cement, it was clear that the compressive strength exceeded 15 MPa for 3 hours. Compressive strength of SG-1, FG-1, and FG-2 mixtures that incorporated accelerating agent was 168%, 161%, and 158% higher, respectively, than that of the PG-1 mixture for 3 hours. As the amount of retarding agent

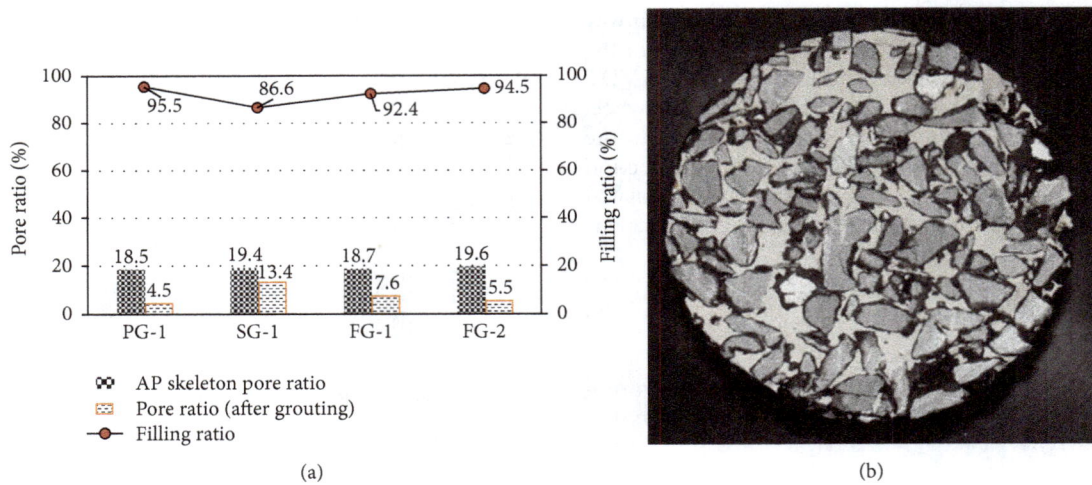

(a)   (b)

FIGURE 5: Pore ratio results of semirigid asphalt specimens. (a) Pore ratio results of semirigid asphalt specimens. (b) Specimen for porosity test.

dosage increased from 0.1% to 0.3%, the compressive strength at early-age including 3 hours and 1 day tended to decrease when the mixing content of accelerating agent was fixed in the cement grout mixture. It can be seen in Figure 4, however, that the compressive strength of all cement grout mixtures can be maintained at a similar level in long-term ages. The use of retarding agents is expected to have a positive effect on the increase of the fluidity of the initial cement grout and is not significantly affected by the reduction of compressive strength in all ages.

4.2. Semirigid Asphalt Composite

4.2.1. Porosity. The pore ratio and filling ratio of the open graded asphalt skeleton and the semirigid asphalt specimens filled with four types of cement grout mixture are shown in Figure 5(a), and the FG-1 specimen for porosity test is represented in Figure 5(b). From the porosity test, the average porosity of four specimens for each open graded asphalt skeleton was evaluated as 19.1%, with porosity being in a range of 18.5% to 19.6%. After filling cement grout into the open graded asphalt skeleton, the porosity of the PG-1, SG-1, FG-1, and FG-2 specimens was measured as 4.5%, 13.4%, 7.6%, and 5.5%, and, using (2), the filling ratios of the cement grout mixtures were calculated as 95.5%, 86.6%, 92.4%, and 94.5%, respectively. As can be seen from Figures 3 and 5, a flow time of cement grout mixtures within 12 s enabled an excellent filling ratio over 90% to be secured. These results indicated that the flow time, which is a typical index of fluidity, has a strong influence on the filling rate of the semirigid asphalt pavement given similar porosity condition of the open graded asphalt skeleton. In this current study, PG-1, FG-1, and FG-2 are suitable cement grout proportions for semirigid pavement that could secure outstanding filling property.

4.2.2. Marshall Stability. The results of the Marshall stability test were shown in Figure 6. In the open graded asphalt skeleton, the Marshall stability of the specimens was measured as

5.43~5.82 kN and the average value was calculated as 5.68 kN. Semirigid pavement mixtures such as PG-1, SG-1, FG-1, and FG-2 filled with cement grout exhibited outstanding Marshall stability. The average value is 430%, 374%, 404%, and 413% higher than that of skeleton mixture, respectively, and the semirigid pavement specimens showed very stable results after grouting. On the other hand, although the compressive strength of the SG-1 for 1 day was the highest among all cement grout mixtures, the Marshall stability value of the SG-1 was 86.9%, 92.5%, and 90.5% that of the PG-1, FG-1, and FG-2 mixtures, respectively. From the test results, it can be inferred that the Marshall stability of the semirigid pavement was strongly dependent on the fluidity of the cement grout rather than the strength of the cement grout. Meanwhile, the relationship between Marshall stability and filling ratio of all semirigid pavement specimens is shown in Figure 7. Marshall stability was found to increase linearly as the filling ratio increased, and the correlation coefficient was calculated as 92%. This result indirectly shows that the filling degree of the cement grout in the pores of asphalt skeleton determines the level of Marshall stability performance.

4.2.3. Flexural Strength. Figure 8 presents the flexural strength and midspan deflection relation curve of the asphalt skeleton and semirigid pavement specimens. The flexural strength of open graded asphalt skeleton and correlating flexural strain were calculated as 0.64 MPa and 0.524%, respectively. The flexural strength of PG-1, SG-1, FG-1, and FG-2 was calculated as 3.21 MPa, 2.92 MPa, 3.18 MPa, and 3.10 MPa, and corresponding flexural strain was 0.212%, 0.210%, 0.237%, and 0.247%, respectively. The test results were listed in Table 4. It was found that the semirigid pavement specimens were able to improve the flexural strength of asphalt skeleton by at least 453% and by up to 498%, and the flexural strain corresponding to maximum flexural strength could be decreased by a range of 40.1% to 47.0%. It was clearly found that the flexural strength of semirigid pavement specimens was enhanced by filling cement grout into the asphalt

TABLE 4: Flexural test results of all specimens.

	Flexural strength (MPa)	Midspan deflection (mm)	Flexural strain (%)
Asphalt skeleton	0.64	1.57	0.524
PG-1	3.21	0.64	0.212
SG-1	2.92	0.63	0.211
FG-1	3.18	0.71	0.237
FG-2	3.10	0.74	0.247

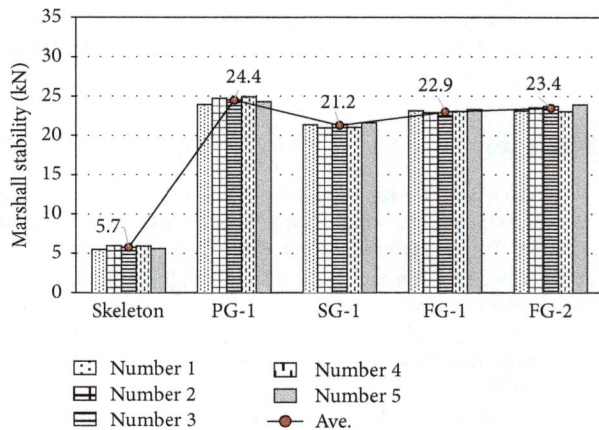

FIGURE 6: Results of Marshall stability test.

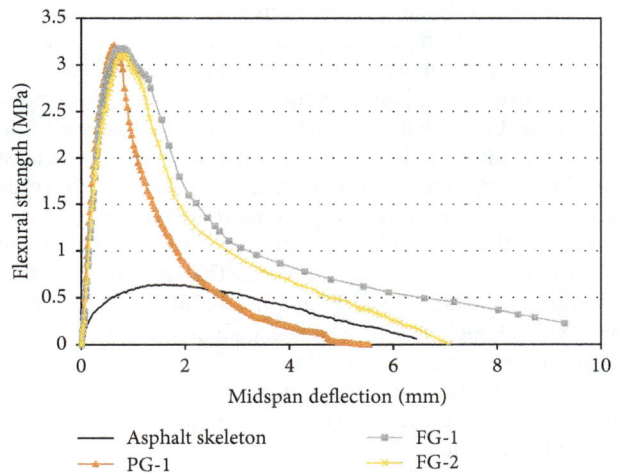

FIGURE 8: Results of flexural test of asphalt skeleton and semirigid asphalt specimens.

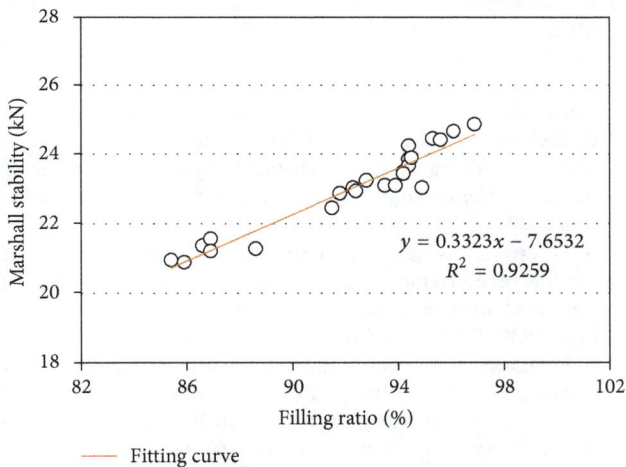

FIGURE 7: Relationship between Marshall stability and filling ratio of all semirigid pavement specimens.

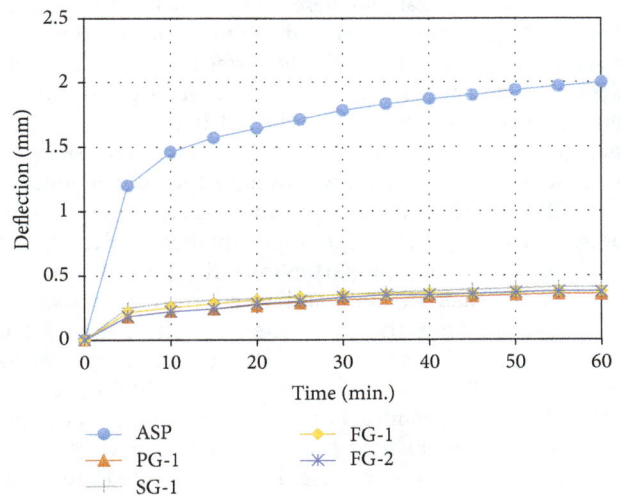

FIGURE 9: Results of wheel tracking test of asphalt skeleton and semirigid asphalt specimens.

skeleton. However, as expected, the lowest fluidity of the SG-1 cement grout mixture was measured as 2.92 MPa, and the other three types of semirigid pavement specimens with a similar level of fluidity showed no significant difference according to the type of cement grout. Similarly to the results of Marshall stability, the flexural strength was also significantly affected by the fluidity of the cement grout filled in the asphalt skeleton pores. From the flexural test results, it was found that the semirigid pavement could strongly

enhance the low stiffness of the open graded asphalt skeleton with flexural strength.

4.2.4. Wheel Tracking Resistance. The relationship between the vertical loading and correlating central deflection measured by the wheel tracking test is shown in Figure 9. In the open graded asphalt skeleton, the initial center deflection rapidly increased to 1.2 mm by 5 min. For semirigid pavement specimens such as PG-1, SG-1, FG-1, and FG-2 mixtures, however, it was found that the initial deflection at the center was 0.18 mm, 0.25 mm, 0.21 mm, and 0.18 mm. These values range from 4.8 times to 6.7 times lower than that of the open graded asphalt skeleton. This result shows that cement grouting is effective in controlling the initial deformation for the open graded asphalt skeleton. As can be seen from the graph slope in Figure 9, the deflection of the asphalt skeleton was continuously increased until the end of the experiment. For the semirigid pavement specimens, however, the center deflection gradually increased and then converged after 40 minutes. It was found that the final deflection of the open graded asphalt skeleton specimen was the largest at 2.0 mm, but it was 0.36 mm to 0.41 mm for semirigid pavement, which could reduce the final deflection by at least 4.88 times compared to the asphalt skeleton. The dynamic stability of all the semirigid pavement specimens was calculated as 31,500 times/mm, while for the asphalt skeleton it was 6,300 times/mm. In addition, the deformation rate of semirigid pavement specimens was 0.0013 mm/min compared with 0.0067 mm/min of the asphalt skeleton. The test results show that it was possible to effectively enhance the dynamic stability and deformation rate for deformation, thereby improving the durability of asphalt pavement under repeated vehicle load.

5. Conclusions

This current study investigated the effect of cement grout controlled with ultrarapid hardening cement, accelerating and retarding chemical agent on the mechanical properties of semirigid pavement. In the first stage of this study, the effect of chemical admixtures on the cement grout fluidity and compressive strength was evaluated. In the next stage, the open graded asphalt pavement mixture, which is the skeleton of the semirigid pavement, was designed to have a porosity of 18~20%. Finally, semirigid pavement specimens were prepared by injecting cement grout into an open graded asphalt pavement. Mechanical properties including porosity, flexural strength, Marshall stability, and wheel tracking resistance were evaluated. From the test results it was clearly found that the porosity and filling rate of semirigid pavement specimens were greatly influenced by the flow time (fluidity) of cement grout. The use of chemical admixtures was found to be able to control the initial flow time; the compressive strength property, however, was not significantly affected. Through evaluating the strength characteristics through Marshall stability, flexural strength, and wheel tracking test, it was found that the fluidity favorable to filling of the cement grout in the voids of asphalt skeleton greatly influences the strength of semirigid pavement rather than the cement grout compressive strength property. Semirigid pavement could improve the strength and stiffness of open graded asphalt mixture, and the semirigid pavement's performance could be adjusted to the target performance by controlling the flow and strength characteristics of the cement grout. Based on the test results, it was recommended that the porosity of open graded asphalt be 20% and the flow time of cement grout be within 12 seconds.

Conflicts of Interest

The authors declare that there are no conflicts of interest regarding the publication of this article.

Acknowledgments

This research was supported by Basic Science Research Program through the National Research Foundation of Korea (NRF) funded by the Ministry of Science, ICT & Future Planning (no. 2016R1A2B4011810).

References

[1] P. Saxena and L. Khazanovich, "Determination of critical bending stresses in portland cement concrete layer with asphalt overlay," *Transportation Research Record*, vol. 2306, 2013.

[2] Y.-H. Cho, H. Chiu, and B. F. McCullough, "Asphalt overlay design methods for rigid pavements considering rutting, reflection cracking, and fatigue cracking," Tech. Rep., Center for Transportation Research, The University of Texas, Austin, Tex, USA, 1998.

[3] M. Dinis-Almeida, J. Castro-Gomes, M. L. Antunes, and L. Vieira, "Mix design and performance of warm-mix recycled asphalt," *Proceedings of Institution of Civil Engineers—Construction Materials*, vol. 167, no. 4, pp. 173–181, 2013.

[4] N. Kawamura, Y. Morikawa, M. Murayama, T. Hirato, and R. Maekawa, "Durability of high-stability asphalt mixture under aircraft loading," in *Proceedings of the 2014 FAA Worldwide Airport Technology Transfer Conference*, pp. 1–14, Galloway, NJ, USA, 2014.

[5] C.-T. Chiu, T.-H. Hsu, and W.-F. Yang, "Life cycle assessment on using recycled materials for rehabilitating asphalt pavements," *Resources, Conservation and Recycling*, vol. 52, no. 3, pp. 545–556, 2008.

[6] A. Setyawan, "Asessing the compressive strength properties of semi-flexible pavements," in *Proceedings of the 2nd International Conference on Rehabilitation and Maintenance in Civil Engineering, ICRMCE 2012*, vol. 54, pp. 863–874, Indonesia, 2013.

[7] K. E. Hassan, A. Setyawan, and S. E. Zoorob, "Effect of cementitious grouts on the properties of semi-flexible bituminous pavement," in *Proceedings of the Fourth European Symposium on Performance of Bituminous and Hydraulic Materials in Pavement*, pp. 113–120, Nottingham, UK, 2002.

[8] S. Koting, M. R. Karim, H. B. Mahmud, and N. A. A. Hamid, "Mechanical properties of cement-bitumen composites for semi-flexible pavement surfacing," *The Baltic Journal of Road and Bridge Engineering*, vol. 9, no. 3, pp. 191–199, 2014.

[9] S. Li, X. Liu, and Z. Liu, "Interlaminar shear fatigue and damage characteristics of asphalt layer for asphalt overlay on rigid pavement," *Construction and Building Materials*, vol. 68, pp. 341–347, 2014.

[10] B. Yang and X. Weng, "The influence on the durability of semi-flexible airport pavement materials to cyclic wheel load test," *Construction and Building Materials*, vol. 98, pp. 171–175, 2015.

[11] T. S. Park, "Evaluation of the performance and moisture retaining ability in semi-rigid pavement," *Journal of Korean Society of Road Engineers*, vol. 10, no. 1, pp. 69–79, 2008.

[12] J. W. Kim and K. W. Kim, "A review of long-term serviceability of semi-rigid pavement," *Magazine of Korean Society of Road Engineers*, vol. 12, no. 2, pp. 35–40, 2010.

[13] "Provisional guidelines for porous asphalt pavement," Provisional guidelines for Porous Asphalt Pavement, No.11-1611000-001712-01, 2011.

[14] M. L. Afonso, M. Dinis-Almeida, L. A. Pereira-De-Oliveira, J. Castro-Gomes, and S. E. Zoorob, "Development of a semi-flexible heavy duty pavement surfacing incorporating recycled and waste aggregates—preliminary study," *Construction and Building Materials*, vol. 98, pp. 171–175, 2015.

[15] W. Lei, "The design and construction of semi-rigid pavement materials in perfusion type," in *Proceedings of the International Conference on Computer Distributed Control and Intelligent Environmental Monitoring*, pp. 711–715, IEEE, Hunan, China, March 2012.

[16] H. Zhang, "Semi-flexible pavement with cement mortar mix design," *Jilin Highway and Transportation Research*, vol. 2, pp. 8–23, 1999.

[17] D. Q. Wu, Daud, and Y. Zhang, "The semi-rigid pavement with higher performances for roads and parking aprons," in *Proceeding of the 29th Conference of ASEAN Federation of Engineering Organizations-Sustainable Urbanization—Engineering Challenges and Opportunities*, pp. 1–7, The Rizqun International Hotel, Brunei Darussalam, November 2011.

8

An Analysis of the Mechanical Characteristics and Constitutive Relation of Cemented Mercury Slag

Xinwei Li,[1,2] Sui Zhang,[1] E-chuan Yan,[2] Duoyou Shu,[1] Yangbing Cao,[3] Hui Li,[4] Siyang Wang,[1] and You He[1]

[1]Geological Party 103 Guizhou Bureau of Geology and Minerals, Tongren 554300, China
[2]China University of Geosciences, Wuhan 430074, China
[3]Fuzhou University, Fuzhou 350108, China
[4]Environmental Geological Prospecting Institute of Hebei Province, Shijiazhuang 050011, China

Correspondence should be addressed to Xinwei Li; lixinwei0510@163.com

Academic Editor: Enzo Martinelli

This study focuses on mercury slag in the Tongren area of Guizhou Province, China. Computed tomography (CT) is used with uniaxial and triaxial compression tests to examine the mechanical changes in cemented mercury slag and its formation. The CT results for the uniaxial compression test reveal the overall failure process of the mercury slag structure. Based on the coarse-grained soil triaxial test, a modified Duncan-Chang model is compared with the actual monitoring results and is found to be suitable for the analysis of the slag constitutive model.

1. Introduction

China was one of the first countries to discover and exploit mercury resources. According to data collected in 2011, China has proven mercury reserves of 14.38×10^4 t and total reserves of 8.14×10^4 t. Guizhou Province accounts for more than 70% of the total reserves. China's largest mercury mine is within 30 km of Tongren City, Guizhou Province, and this mine's mercury production in the last century has ranked first worldwide. Thus, this resource is known as "China's mercury" [1–4].

Mercury is found in nature in relatively low quantities. The mercury content in ore must be greater than 0.8% to qualify as industrial grade. Thus, large amounts of mining waste and smelting tailings are produced during mercury mining. In early mining efforts, large quantities of slag waste were left untreated in the mining area near the free stack without awareness of the potential implications [5–8]. Over time, however, the accumulated formation of cemented mercury slag has led to concerns over its stability.

Mercury slag is similar to other slag materials in that it consists of slag or granular waste residue. The structure is loose, the water permeability is high, and the stability is generally poor. Under the action of an external disturbance or long-term environmental forces, slag debris flow occurs easily, which can lead to geological disasters (e.g., collapse and landslides) and mercury pollution of the environment. For example, the Donglin coal gangue hill of the Chongqing Nantong Mining Bureau collapsed on 25 October 2005, resulting in one person being buried. In another case, the gangue hill for an old house-based coal preparation plant of the Guizhou Panjiang Coal and Electricity Group Company experienced a major landslide on 16 April 2006 that buried seven people. The accumulation of large amounts of mercury slag adversely influences the mining landscape and environmental safety (leading to, e.g., rocky desertification, geological disasters, and environmental pollution). Therefore, the mechanical properties and constitutive relation of cemented mercury slag aggregates must be studied, and the stability of the slag accumulation body must be analyzed. Such studies are important for ensuring the geological safety of a mining area and ensuring a relatively unpolluted living environment.

2. Research Status in China and Elsewhere

Because of the complexity and variability of the structure of cemented mercury slag, there have been few studies on the stability of slag accumulation. The mechanical properties and constitutive parameters of cemented mercury slag in the stacking state continue to be a source of controversy.

In the study of the crystallisation mechanism and cementation mode of cemented mercury slag, researchers have relied on methods developed to analyze other types of materials. For example, some researchers [9–14] have analyzed the basic mechanical properties and influence factors for natural-grade gravel material mixed with a small amount of cemented rock and found that cemented rock-fill has both the properties and characteristics of concrete. Chen et al. [15] analyzed the effect of soil structure, soil erosion, and moisture retention; the result indicated that the type of backfill soil affects the quality of the artificial soil significantly. Kara and Baykara [16] investigated the characteristics of three neighboring soils from the NE of Turkey and elucidated the effect of different land-use management on the soil aggregate stability and microbial biomass in Galyan-Atasu dam watershed. All these results showed that the supply and demand of nutrients to the soil in the samples are largely dependent on the type, quantity, and presence of the cemented substances. However, the lack of basic research on the structure of cemented mercury slag has limited the scientific management of such slag.

The lack of experimental research on the mechanical properties of cemented mercury slag has resulted in the absence of a constitutive model. The triaxial test is commonly used to study the mechanical properties of cemented soil materials [17], such as cemented carbide slag. Asghari et al. [18] simulated hydrated lime cement and performed a triaxial shear test on cemented coarse-grained soils from the Tehran suburbs. That group found that the cement strength plays a decisive role in the peak shear strength of cemented coarse-grained soils. Kongsukprasert et al. [19] conducted a series of triaxial shear tests on cemented gravel and found that the moisture content, strike strength, and mass ratio of cement ash to gravel have a significant effect on the stress-strain relationship.

Because of the lack of a constitutive model for cemented mercury slag, we are left to draw on the current classical constitutive relation based on the engineering characteristics and special requirements of cement residue and to perform repeated experiments to obtain a more practical cemented mercury slag constitutive model. Candidate classical constitutive models include the following: the Cambridge model, the Duncan-Chang hyperbolic nonlinear elastic model, the Lade-Duncan elastic-plastic model [20], Matsuoka and Nakai's spatial mobilised plane [21], the multiple yield surface model [22–25], and the nonlinear K-G model [26]. Because each constitutive model has its own scope of application and each material has its own complexity, corrections must be made to the models, depending on the

conditions. For example, Jia et al. [27] proposed a modified Mohr-Coulomb model that considers the maximum tensile stress criterion to compensate for the limitations of the Mohr-Coulomb criterion when describing the tensile properties of a rock mass over a certain range. In Liu et al. [28], based on the "elastic modulus method" derived from the hypothesis of strain equivalence and test data of complete stress-strain curves of marble, granite, and sandstone under uniaxial compression, a damage model in the form of logistic equation is proposed to simulate the stress-strain relation of rocks. This model provides a method for estimating the elastic modulus of undamaged rocks using uniaxial compression test results and it presents a reasonable explanation of the chaos phenomenon occurring in uniaxial compression test. Saberi et al. [29] considered the particle breakage and proposed a constitutive modeling of gravelly soil-structure interface. Ma et al. [30] present a new approach for the development of an elastoplastic constitutive model to predict the strength and deformation behavior of soils under general stress conditions. Stutz and Mašín [31] proposed a new hypoplastic interface model and pointed out that this model can describe a number of important phenomena of clay-structure interfaces. Xiao and Liu [32] proposed an elastoplastic model considering state dependence and particle breakage and pointed out that the model considering particle breakage can well predict the stress-strain and particle-breakage behaviors of rock-fill materials at various confining pressures. Their model can simultaneously reflect the mechanisms of compressive and shear deformations. Because the physical properties of cemented mercury slag are similar to those of this soil and the stone materials, the research described above can be used to develop a constitutive relationship for cemented mercury slag.

Our literature review indicates that there has been no basic research on the structure, physical properties, mechanical properties, or constitutive relation of cemented slag in China or elsewhere. By referring to the research of other materials, however, the physical properties and constitutive relationship of the slag can be determined to analyze the stability of mercury slag landfills.

3. Methodology

Mercury slag in Tongren, Guizhou Province, was selected as the research object. First, computed tomography (CT) and uniaxial and triaxial compression tests were performed to study the mechanical changes to cemented mercury slag and its formation [33]. Second, based on a constitutive model of coarse-grained soil, a constitutive model of cemented mercury slag under typical conditions was developed. Finally, the constitutive model and Duncan-Chang model were applied to a stability simulation analysis of a representative cement-type mercury slag accumulation body and compared with the actual monitoring results of the mine to verify the rationality of the cemented slag constitutive model. The conclusions of this study and the technical methods used during the research process can provide basic theoretical support for managing mercury slag stability.

TABLE 1: Relative density test results (700 samples).

Years	Maximum dry density (g/cm^3)	Relative density of 0.75 dry density (g/cm^3)	Relative density of 0.80 dry density (g/cm^3)	Relative density of 0.85 dry density (g/cm^3)	Sample site density (g/cm^3)
40	2.275	1.706	1.820	1.933	1.721
70	2.358	1.769	1.886	2.004	1.794

TABLE 2: Particle size distributions (700 samples).

Particle size (mm)	60–40	40–20	20–10	10–5	5–2	2–1	1–0.5	0.5–0.25	0.25–0.1	<0.1	Total
40-year group											
Content (%)	10.8	23.6	31.6	14.5	4.0	2.7	2.0	3.5	3.0	4.3	100
70-year group											
Content (%)	5.4	27.1	30.9	15.6	4.0	2.9	2.3	3.6	3.5	4.7	100

3.1. Materials. The mercury slag in Tongren is composed primarily of waste slag, tailings, and soot. The physical and chemical properties of the slag are stable in its natural state, and the physical and mechanical properties of the slag are stable under natural conditions. The physical and mechanical properties of the slag increase with its mechanical strength. Tailings due to calcination can cause the material composition to change. This process can destabilise the chemical properties and cause a chemical reaction to occur. Tongren has a humid climate with frequent rain; the tailings can slag through the leachate to crystallise the cement. The cement flows with the leachate and fills the skeleton of the waste slag, tailings, and ashes to form a cement paste.

To consider the time effect on the cementation of mercury slag, 40- and 70-year aggregates of 700 mercury slag heaps were selected for field sampling. Table 1 presents the relative densities of the cemented mercury slag, and Table 2 presents the particle size distribution.

Due to cementation, the cohesion of the cemented mercury slag was enhanced, and a hardened "shell" appeared outside the accumulation body. An investigation on the instability of cemented mercury slag accumulation in Tongren found that the main failure mode was a failure at the slope toe, as shown in Figure 1. In fact, because of the protective effect of the hardened shell, the other three types of damage were also directly related to the slope. Therefore, it is of practical importance to study the common occurrence of cemented mercury slag, which plays a controlling role in the destruction of traceability.

Based on the similarity between cemented dross and gravel, cemented mercury slag with residual strength in the saturated state plays a decisive role in the stability of the deposited body and was found to be the most common cemented mercury slag in Tongren. In this study, this composition was taken to be the representative state of cemented mercury slag.

3.2. Description of the Experiment

3.2.1. Uniaxial Compression CT Test. Figure 2 shows the CT visualisation system of the Key Laboratory of Ministry of Water Resources of Chang'an Academy that was used for the indoor uniaxial compression CT test. Samples were collected from the initial state, preloading, loading, and failure stages of glial gravel grit mercury slag. The stress-strain relationship of the sample and the structural meso failure response process were analyzed. Table 3 presents the main technical parameters of the CT testing system (SOMATOM Sensation 40, Siemens).

The procedure for the uniaxial compression CT test was as follows.

(1) Sample Preparation. A handheld gasoline drill and standard sampler were used to take samples of 40- and 70-year cementation particles from a <30 mm layer of glial gravel grit mercury slag. The samples were 100 mm in diameter and 200 mm in length.

(2) Scanning before Experiment. Before the test, the samples of glial gypsum gelled mercury slag were measured, as shown in Figure 3(a). The test samples were installed in the triaxial instrument, as shown in Figure 3(b). A CT scan was performed to determine the initial state of the glial gravel grit mercury residue slag samples, as shown in Figure 3(c).

(3) Uniaxial Compression to Sample Failure. The samples were subjected to uniaxial compression until destruction while being monitored throughout the process, as shown in Figure 3(d). CT images of the same position at different time points were selected, and the mechanical structure of the damaged sample structure was analyzed.

3.2.2. Triaxial Compression Test of Coarse-Grained Soil. A large-scale triaxial test was performed on coarse-grained soil to simulate the failure of the specimen under multidimensional dynamic and static conditions and obtain the stress-strain relationship of cemented mercury slag. Table 4 presents the main technical parameters of the 2000 kN microcomputer-controlled electrohydraulic servo triaxial testing machine (Model TAJ-2000) of China's Geological University Zigui Practice Base Laboratory. This type

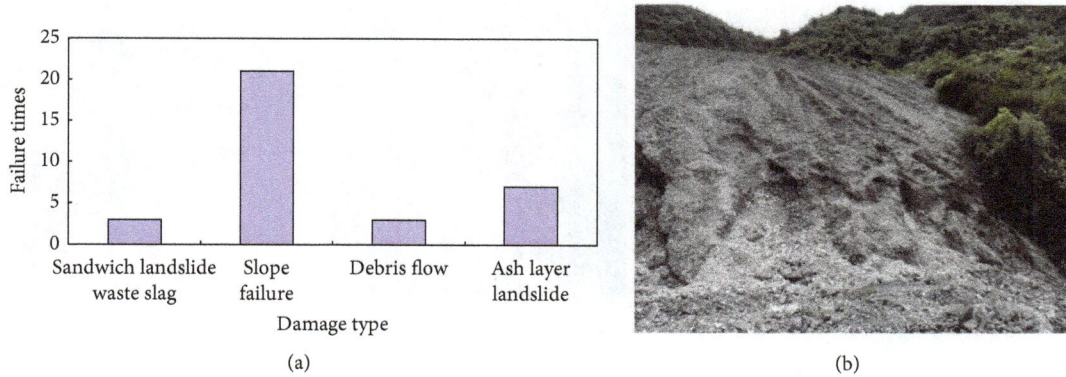

FIGURE 1: Investigation results for the main failure mode of mercury slag accumulation in the Tongren area: (a) types and (b) slope foot source damage.

TABLE 3: Main technical parameters of SOMATOM Sensation 40.

Project	Content
Aperture	70 cm
Tilt angle	±30°
Rack rotation speed	162 r/min
Detector	40 rows of rare earth ceramic (UFC) variable array detector (AAD) acquisition system
Number of detectors	26,880
Data collection system	40 layers; 1344 channels per layer, 4640 times per revolution projection
Tube	Electron beam control Straton tube
Tube heat capacity	0.6 MHU
Anode heat dissipation rate	5 MHU/min
High-voltage generator	70 kV
Voltage range	80, 100, 120, 140 kV
Tube current	28–580 mA

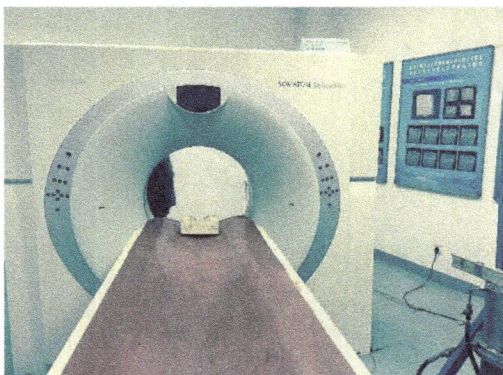

FIGURE 2: CT visualisation system.

of instrument is used primarily for coarse particles: axial pressure and lateral pressure strength tests of sands and soils and soil dynamics tests.

The test procedure was as follows.

(1) Sampling. 700 samples of 40- and 70-year accumulated mercury slag were produced using a portable drill and standard samplers with a length of 200 mm and a diameter of 100 mm. Each sample was formed by partitioning slag into eight pieces.

(2) Sample Preparation. The specimen height was set to 600 mm, and the diameter was set to 300 mm. The control dry density was 1.8 g/cm^3. The stratified compaction vibration method was used, as shown in Figures 4(a)–4(c). A vacuum was applied between the sample and rubber film to remove bubbles and ensure that the sample would not be damaged and that the gap would be reasonably filled, as shown in Figure 4(d). After a sample was installed, the head saturation method was used to bring the sample to the saturation point, as shown in Figure 4(e).

(3) Consolidation Drainage Test. After a sample was installed, gas was exhausted through the exhaust pipe. The initial water head was recorded, and the sample was examined to determine if it was fully saturated; if not, it was resaturated. Shear tests were performed on samples with confining pressures of 0.2, 0.4, 0.6, and 0.8 MPa. During the shear process, the strain rate was controlled, and the changes to the axial load

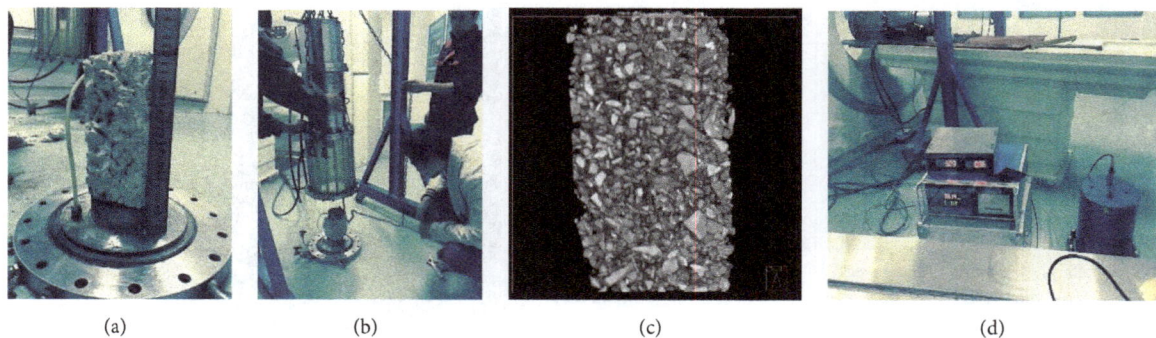

FIGURE 3: Steps of the uniaxial compression CT test: (a) sample, (b) sample installation, (c) test sample facade structure, and (d) data reading in the trial.

TABLE 4: TAJ large moving and static triaxial instrument.

Project	Main indicators
Test size	Φ300 mm × 600 mm
Axial maximum test force	Static: 2000 kN, dynamic test range: 0–1000 kN
Axial displacement measurement range	0–300 mm
Confining system	Maximum confining pressure: 10 MPa; dynamic confining pressure: 0–3 MPa
Pore water pressure	0–10 MPa (accuracy: ±1% RO)
Vibration frequency	$F \geq 0.1$ Hz (full load: ±700 kN, displacement amplitude: ±70 mm) $F \leq 10$ Hz (full load: ±200 kN, displacement amplitude: ±1 mm)
Load form	Sine, triangular, square, and random waves; the impact load speed is greater than 10 mm/s
Control system	Can use stress and strain control, vertical and circumferential load synchronisation, or phase control

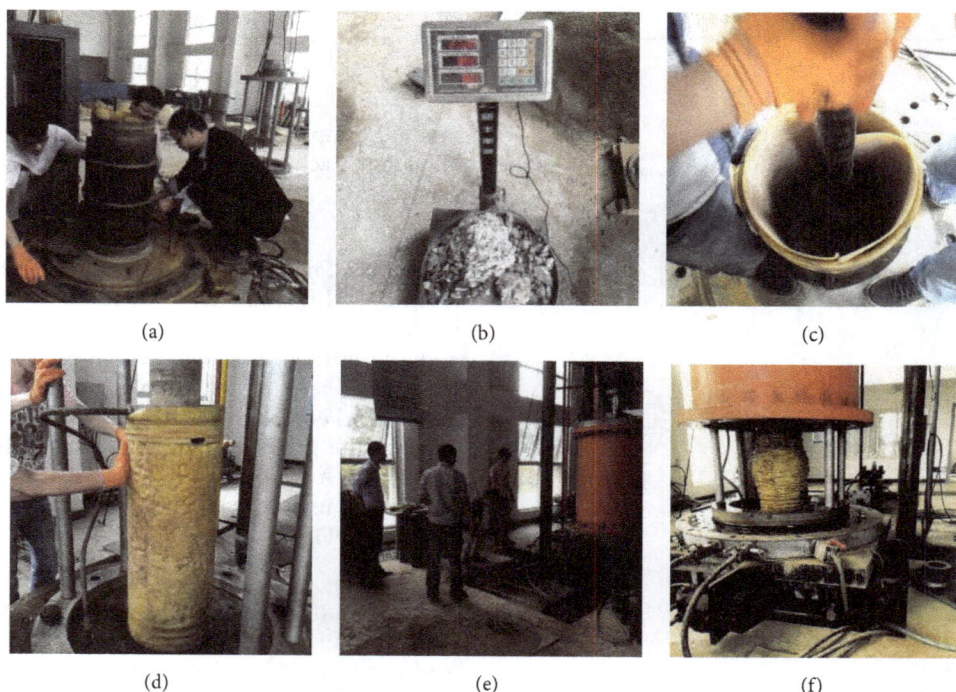

FIGURE 4: Triaxial compression test procedure: (a) mould tube installation, (b) sample weighing, (c) sample layer vibrating chamber, (d) sample vacuum forming, (e) sample saturation, and (f) sample shear failure.

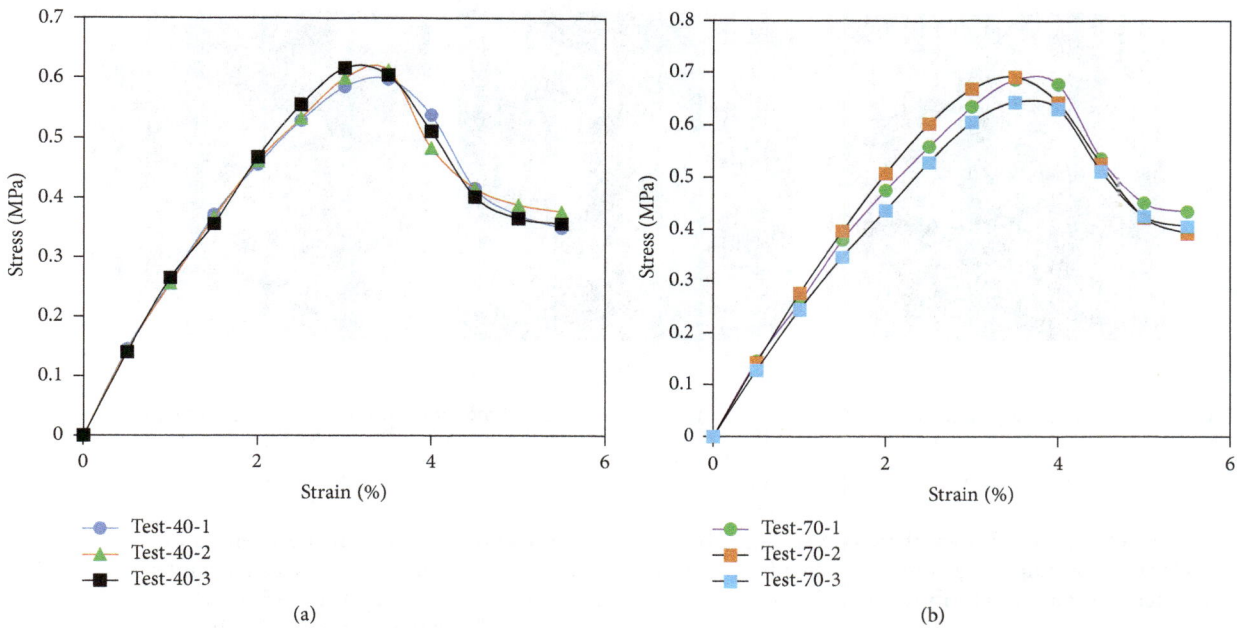

FIGURE 5: Stress-strain curves for cemented mercury slag: (a) 40 years and (b) 70 years.

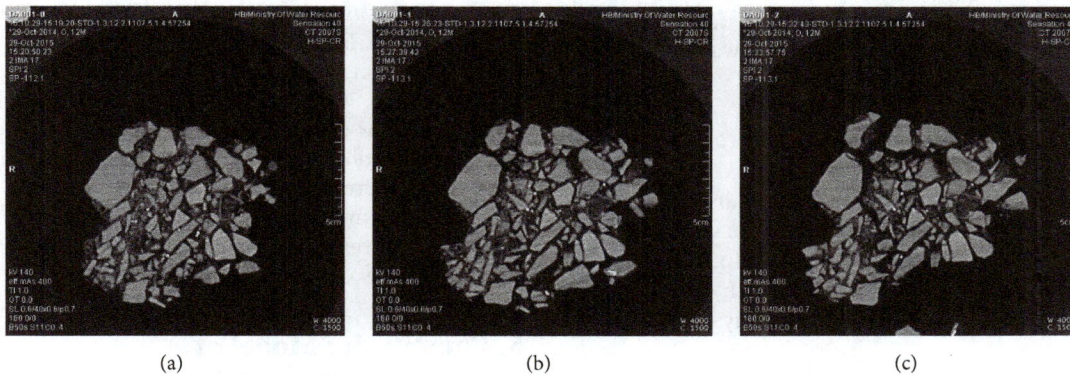

FIGURE 6: CT scan results for a cracked cross section of the 40-year sample at the same location for different compression times.

and volume were recorded at strain increments of 1 mm. The results are shown in Figure 4(f).

4. Results

4.1. Analysis of Uniaxial Compression Failure of Cemented Structures. Figure 5 shows the macroscopic stress-strain curves that were obtained from the uniaxial compression test on 40- and 70-year glial gravel gelled mercury slag raw materials. By analyzing the corresponding cross-sectional structure of each sample in the same position, as shown in Figures 6 and 7, the stress-strain response of the typical mercury slag structure was obtained.

(1) The stress-strain curve and CT scan showed that the internal cementation structure tended to be closed before the stress was relatively high. The strength increased with the stress, and the cemented mercury

slag was elastic in the initial stress-strain process. The stress-strain curve can be used to characterise the linear elastic behavior.

(2) When the stress concentration reached its peak, the cemented mercury slag structure presented a corresponding response. When the stress increased with the initial crack, the strain did not show any significant increase. The tangent stress curve was relatively flat, and the slope was small. At this point, the cemented mercury slag structure maintained the overall peak strain.

(3) As the stress increased, the structure of the cemented mercury slag was rapidly damaged. One overall and two local failure modes were observed. The overall failure and stress-strain curve corresponded to rapid expansion of the crack after unloading. When the force increased slightly, the crack expanded, as shown

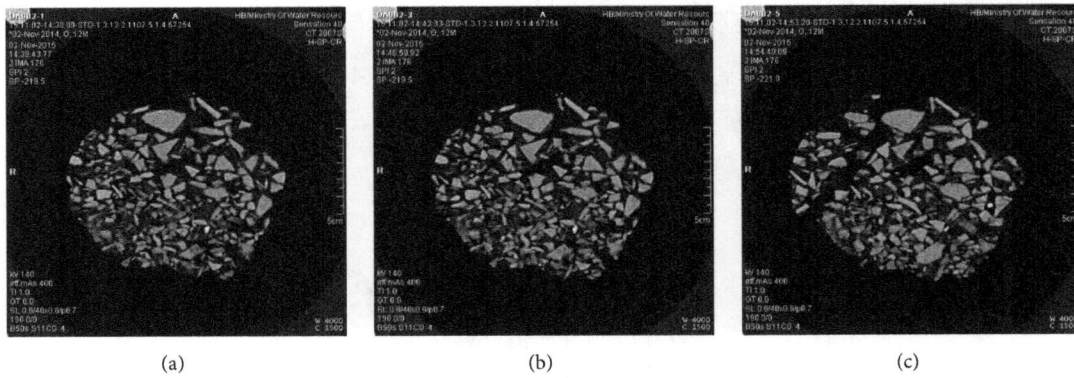

<table>
<tr><td>(a)</td><td>(b)</td><td>(c)</td></tr>
</table>

FIGURE 7: CT scan results for a cracked cross section of the 70-year sample at the same location for different compression times.

in Figures 6 and 7. The local damage was caused by the collapse of blocks at the edge or surface from cracks that led to separation from the main body.

(4) When the strain did not change with increasing stress, the mercury slag structure lost its response mechanism and was completely destroyed. At this time, the structure of the specimen was thoroughly penetrated, and the specimen was completely destroyed. The mechanical strength of the structure was terminated, as shown in Figure 7.

4.2. Analysis of the Strength Parameters of Cemented Mercury Slag under Typical Conditions. Cemented mercury slag typically has a granular structure that generally cannot bear tensile stress. More research is required on the shear strength characteristics of this slag. The relationship between the normal stress, shear stress, and shear strength on the failure surface under shear conditions is generally described as follows:

$$t_f = c + \sigma \tan \varphi. \tag{1}$$

Based on the maximum axial stress and confining pressure derived from the triaxial experiment, a Mohr circle can be drawn for the two sets of samples with $\sigma_1 + \sigma_3/2$ as the centre and $\sigma_1 - \sigma_3/2$ as the radius. The values of c and φ for the cemented mercury slag in the saturated state are obtained as shown in Figures 8(a) and 9(a). In the case of cemented mercury slag, which is similar to coarse-grained soil, the particle distribution with crushing occurs at a low confining pressure. At this point, the stress between particles is redistributed, and the connecting forces between the particles are weakened, resulting in lower friction. The strength of the envelope is not a straight line and presents a downward bending curve, as shown in Figures 8(b) and 9(b). The following equation is typically used to obtain φ_0, $\Delta\varphi$, and the relationship with the shear strength for the sample:

$$\varphi = \varphi_0 - \Delta\varphi \lg \frac{\sigma_3}{P_a}, \tag{2}$$

where p_a is the atmospheric pressure and φ_0 is the value of φ when $\sigma_3 = p_a$.

The contrast between the strength envelopes of the 40- and 70-year cemented slags can be interpreted using the Mohr-Coulomb strength criterion. The cemented mercury slag with 40 years of accumulation had lower bond strength than the slag with 70 years of accumulation and a lower friction angle. This result indicates that the consolidation stress of cemented mercury slag increases over a long duration, resulting in greater cohesion of the sample with longer stacking times. In addition, because the internal particles of the slag did not change, the internal friction angle did not greatly change. When plotted, (2) clearly shows the increase in the curve as accumulation occurs over a longer period. The consolidation stress further increases, and φ_0 shows a substantial increase. As the consolidation stress increases with the accumulation time, the particles inside the slag harden at a lower confining pressure. Thus, the slag becomes stronger and less prone to deformation.

5. Constitutive Model of Cemented Mercury Slag

Three notable coarse-grained soil constitutive models are the Rowe dilatancy model, the K-G model, and the Duncan-Chang model. The Rowe dilatancy model is based on the principle that the minimum interaction can be compared using energy principles; the interaction between pairs of particle is used to simulate the global interaction between all particles. This model can capture the deformation of coarse grains; however, the impact of particle crushing is not considered, and limitations remain. For the K-G model, the expression is relatively simple: the stress and strain are decomposed into a dilatational (or hydrostatic) tensor and a deviatoric tensor, where the dilatational stress tensor corresponds to the dilatational strain tensor and the deviatoric stress tensor corresponds to the deviatoric strain tensor. The Duncan-Chang model, which is also called the hyperbolic model, is a type of subelastic model that employs a nonlinear elastic constitutive relation. It is straightforward to determine the parameter values through experiment. The formula is simple, the concept is clear, and much engineering application experience has been accumulated using this formula.

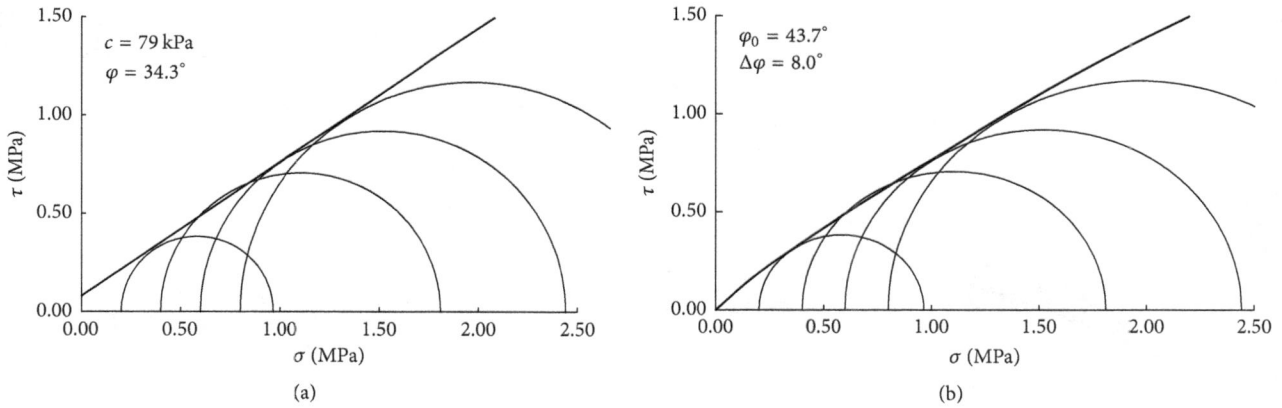

FIGURE 8: Shear strength envelope of the 40-year cemented mercury slag.

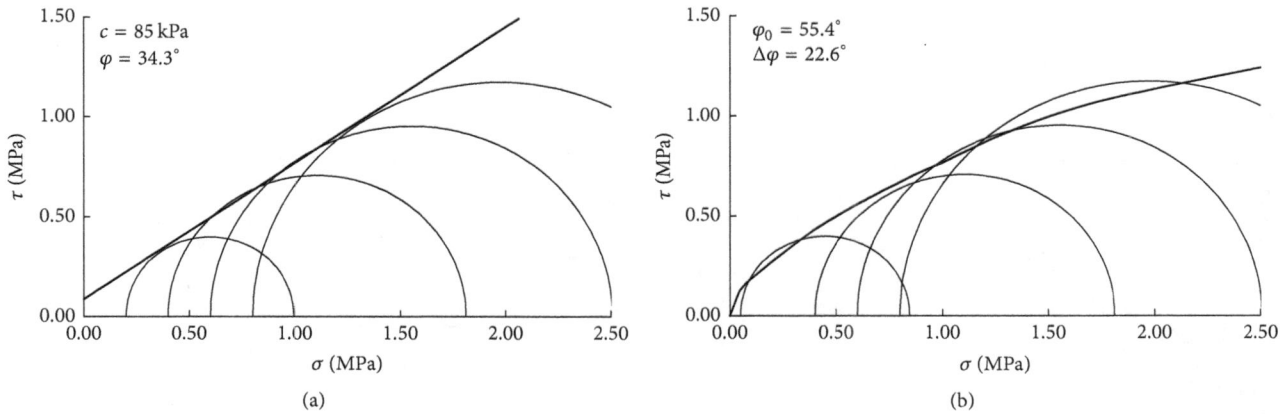

FIGURE 9: Strength envelope of the 70-year cemented mercury slag.

Cement-type slag is predominantly coarse-grained, and the stress-strain relationship is similar to that of coarse-grained soil. The Duncan-Chang constitutive model was used in this study. Based on an analysis of the experimental data, the ε_1-ε_3 relationship was improved, and an expression for Poisson's ratio ν_t was derived to obtain an improved Duncan-Chang (E-ν) constitutive model.

A triaxial compression test was conducted on the 40- and 70-year mercury slag samples. The test results were used to improve the ε_1-ε_3 relationship and establish the constitutive model.

5.1. Improved Characterisation of the ε_1-ε_3 Relationship.
According to the conventional triaxial test results, as shown in Figures 10(a) and 10(b), the main stress difference ($\sigma_1 - \sigma_3$) increased with the axial strain ε_1 before finally stabilising, but the volume strain continued to increase. If the axial strain ε_3 is defined as positive, then the ε_1-ε_3 relationships with pressure are as shown in Figure 11.

Data analysis indicated a parabolic relationship between ε_1 and ε_3 set at the origin. Based on this finding, the relationship between the two was set to

$$\varepsilon_1{}^2 = A\varepsilon_3{}^2 + B\varepsilon_3, \qquad (3)$$

TABLE 5: A and B values under different confining pressures.

Confining pressure (MPa)	0.2	0.4	0.6	0.8
A	3.7474	3.2756	3.4282	3.1515
B	0.2662	0.4427	0.4948	0.7257

where A and B are test parameters for which a fitting curve is available through data analysis, ε_1 is the axial strain, ε_3 is the lateral strain, and $\varepsilon_1{}^2$ and ε_3 are fitting results (specified in Table 5) for a parabolic relationship.

According to Table 5, the value of A for a sample under different confining pressures changes only minimally. Thus, the average value of A under all tested confining pressures can be used. The value of B increases with the confining pressure, and the growth rate gradually decreases. Through data analysis, B and $\lg(\sigma_3/p_a)$ are established to have the following linear relationship:

$$B = t_1 \lg\left(\frac{\sigma_3}{p_a}\right) + t_0, \qquad (4)$$

where t_0 and t_1 are test constants obtained by linear fitting, p_a is the standard atmospheric pressure, and σ_3 is the test confining pressure.

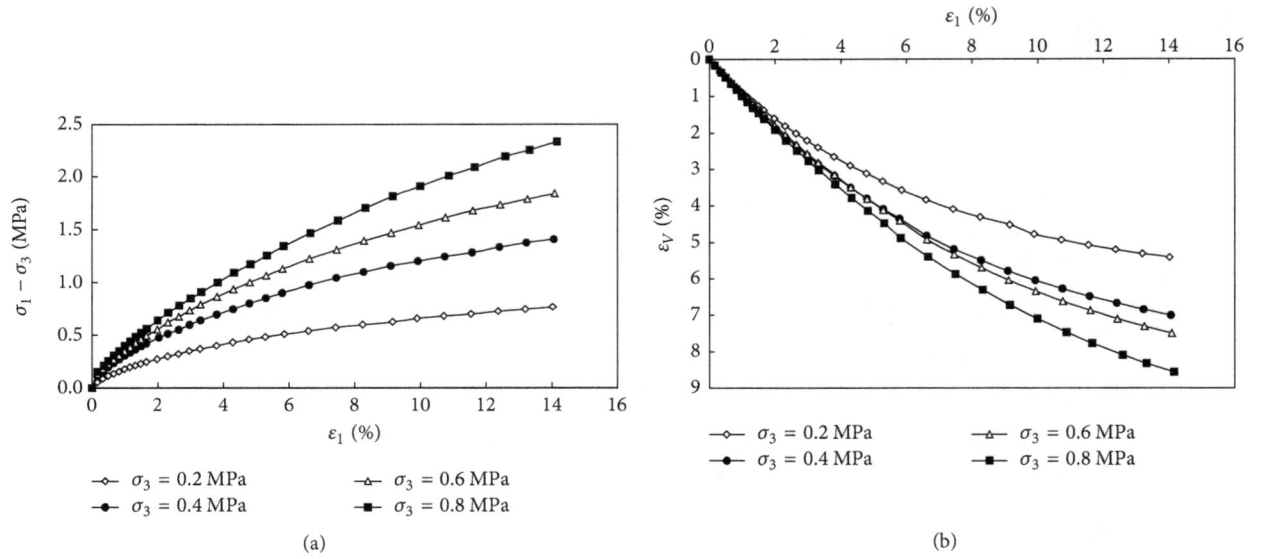

FIGURE 10: The conventional triaxial test results.

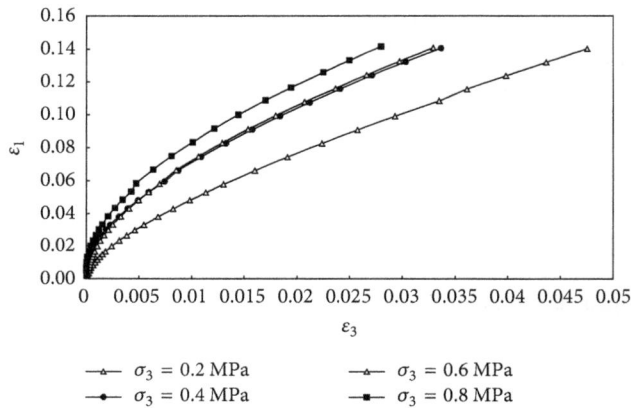

FIGURE 11: Sample relationship curves of ε_1-ε_3.

5.2. Discussion of Volume Strain ε_v.

ε_v is defined as follows:

$$\varepsilon_v = \varepsilon_1 + \varepsilon_2 + \varepsilon_3 = \varepsilon_1 + 2\varepsilon_3. \tag{5}$$

By combining (3) and (5), the following relationship is derived:

$$\varepsilon_v = \varepsilon_1 + \frac{\left(B - \sqrt{B^2 + 4A\varepsilon_1{}^2}\right)}{A}. \tag{6}$$

Figure 12 compares the calculated (scatter) and measured (solid line) values for ε_v. The improved model matches the measured data well and reflects the volume strain of the cemented mercury slag under typical conditions.

5.3. Discussion of Poisson's Ratio ν_t.

According to (3), ν_t can be expressed as follows:

$$\nu_t = \frac{d\varepsilon_3}{d\varepsilon_1} = \frac{\varepsilon_1}{\sqrt{A\varepsilon_1{}^2 + 0.25B^2}}$$

TABLE 6: Measured results for the fitted parameters.

Parameter	A	t_1	t_0
Value	3.4007	0.6873	0.0384

$$= \frac{\varepsilon_1}{\sqrt{A\varepsilon_1{}^2 + 0.25\left[t_1 \lg\left(\sigma_3/p_a\right) + t_0\right]^2}}. \tag{7}$$

Table 6 presents the fitting parameter values based on the sample results.

The value of ν_t predicted using (7) agrees with the measured value, as shown in Figure 13. This finding indicates that the fitted parameter values are reasonable.

Relative to the original Poisson ratio in the original E-K model, the initial Poisson ratio was 0, primarily because the sample started to compress. As the pores of the specimen became closed, the deformation of the pores after lateral deformation was the main source of the initial axial deformation. Therefore, the initial Poisson ratio is more consistent with the test phenomenon. At different confining pressures, the tangential Poisson ratio tended to stabilise with increasing axial strain. For the experimental specimen, the final sample deformation and tangential Poisson's ratio stabilised with increasing axial stress. For the test sample, as the axial stress increased, the final sample deformation and tangential Poisson's ratio stabilised. The improved Poisson ratio was consistent with the experimental results. Therefore, the improved Duncan-Chang model better reflects the stress-strain relationship of cemented mercury slag under typical conditions.

6. Simulation for the Stability of Cemented Mercury Slag

The Duncan-Chang E-ν model was transformed into the incremental expression required for secondary development

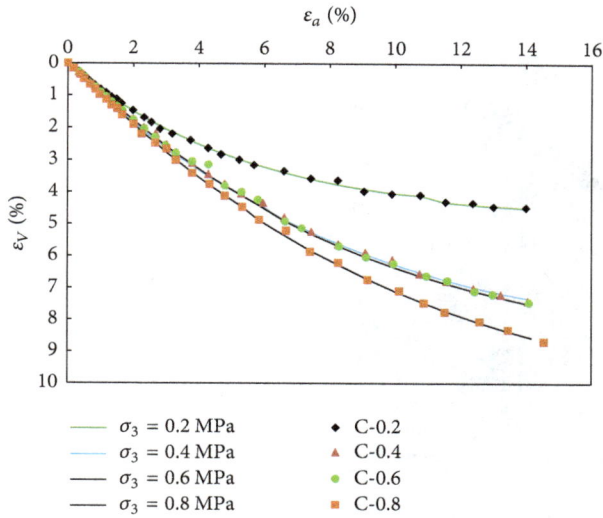

FIGURE 12: Measured and calculated ε_v values.

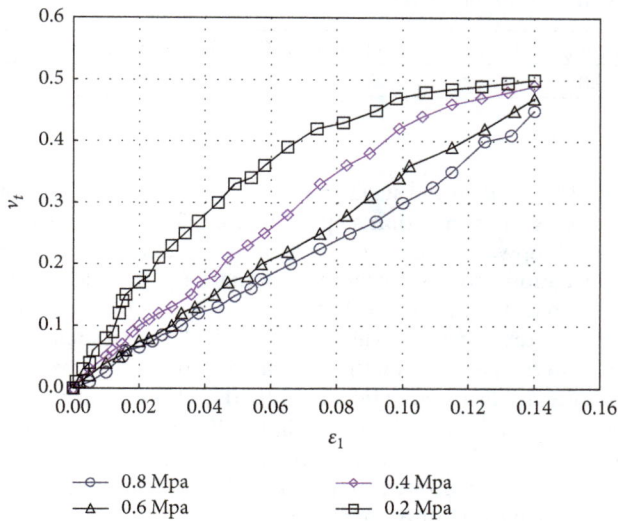

FIGURE 13: Chart comparing the measured and predicted v_t values.

of the FLAC^{3D} constitutive relation. The constitutive model library of FLAC^{3D} was used in this process. The improved Duncan-Chang constitutive model was applied to numerical analysis of the slope of a mercury slag accumulation body and compared with the actual deformation and failure process of the slope to further verify the correctness and applicability of the model.

6.1. Overview of Project Examples.
The type of cemented mercury slag of interest was formed over approximately 30

years. The maximum elevation of the top of the slag was 873.00 m. The length of the deposited body was approximately 50.0 m, the width was approximately 35.0 m, the measured plane area was approximately 1476.3 m^2, and the volume was approximately 2657.3 m^3. Thus, the body was considered a small slag accumulation body.

The Duncan-Chang model can be used to describe the stress-strain characteristics of cemented mercury slag in the accumulation body, which is characterised by a loose slag structure, high porosity, and low mechanical strength. The thickness of the bulk layer was 0.2–2.5 m with an average thickness of 1.8 m. The thickness decreased from the slope top to the foot. The bedrock was 106° ∠ 20°, and the thickness was 216–386 m. Figure 14 shows the section structure.

6.2. Mechanical Parameters of Bedrock.
The indoor test results show that the upper part of the accumulated slag had a density of 1560 kg/m^3, an apparent density (basket method) of 2828 kg/m^3, and a porosity of ≤4.8%. Based on the high porosity of the deposit, the natural gravity of the slag accumulation body is expected to be 16.70–18.5 kN/m^3. Based on the engineering geology analogy method and local rock and soil surveying experience, the physical and mechanical parameters of the bedrock in the field are as given in Table 7 according to the *Basic Code for Building Foundation Design* (GB500007-2002).

6.3. Incremental Expression for the Constitutive Relation.
FLAC^{3D} uses an incremental constitutive relation; the stress increment is calculated from the strain increment, and the total stress is obtained by superimposing the stress calculated in the previous period. According to FLAC^{3D}'s own UDM custom constitutive documents in the stensor.h header file, the stress tensor structure StnS export function EXPORT bool *Resoltopris* () can be the current unit stress field along the three principal axes.

In this study, the FLAC^{3D} constitutive model was developed using the E_t-v_t model; the modified Duncan-Chang E_t-v_t replaces the stress sign:

$$E_t = K p_a \left(\frac{-\sigma_1}{p_a} \right)^n \left[1 - R_f \frac{(1 - \sin\varphi)(\sigma_3 - \sigma_1)}{2c\cos\varphi - 2\sigma_1 \sin\varphi} \right]^2 ,$$

$$v_t = \frac{\varepsilon_1}{\sqrt{A\varepsilon_1{}^2 + 0.25 \left[t_1 \lg(-\sigma_1/p_a) + t_0 \right]^2}} . \tag{8}$$

ε_1 can be used to derive the tangent elastic modulus E_t:

$$\varepsilon_1 = \frac{(\sigma_3 - \sigma_1)}{K p_a (-\sigma_1/p_a)^n \left[1 - R_f ((1 - \sin\varphi)(\sigma_3 - \sigma_1) / (2c\cos\varphi - 2\sigma_1 \sin\varphi)) \right]} . \tag{9}$$

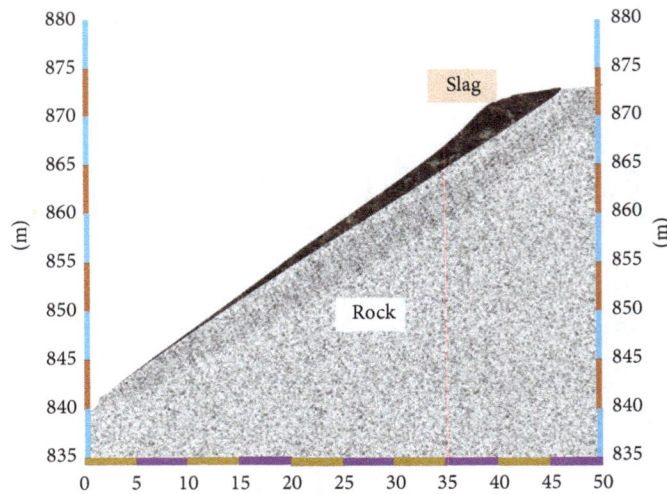

FIGURE 14: Stratigraphic profile.

TABLE 7: Physical and mechanical parameters of the bedrock.

Natural gravity (kN/m³)	Poisson's ratio	Elastic modulus (GPa)	Cohesion (MPa)	Internal friction angle (°)	Tensile strength (MPa)
26.50	0.35	5.0	9.0	27	1.2

6.4. FLAC^{3D} Modeling. According to the slope shown in Figure 14 to establish the FLAC^{3D} calculation model, the vertical scale for the change in the slope of the terrain was not large. Thus, a plane model was used. The left and right boundaries were fixed to have an x-direction displacement of 0, and the bottom boundary had a z-direction displacement of 0. Because a plane model was used, the entire model had a y-direction displacement of 0. In terms of regional tectonics, the field is located at the junction of the Yangtze platform and South China fold belt. The regional folds are the Kyrgyzstan–Yuzhan oblique and Feng–Akira anticline, and the twofold axes are parallel to the northeast direction and lie 12 km apart. The impact on the construction of this area is small; thus, the model considered only the gravity field.

For the comparative study, the improved and unmodified Duncan-Chang E-v models were used to improve the constitutive model parameters presented in Tables 8 and 9 under the boundary conditions for the bulk body model and gravity field. Table 8 presents the constitutive model parameters. Considering that the stacking age of this sample was approximately 30 years, Table 8 presents the selected model parameters. Table 9 presents the corresponding model parameters for a sample with a stacking age of 40 years. The simulation results were saved every 10,000 h. In total, 30,000 h were simulated.

6.5. Analysis of the Results. Figures 14 and 16 show the maximum horizontal displacement cloud and maximum shear strain cloud, respectively. As shown in Figures 15(a), 15(c), and 15(e), when the improved constitutive model was used on the slag, the maximum horizontal displacement of accumulation body was concentrated predominantly near the

foot. The slope of the surface had a relatively small range of change as the simulation steps increased. The body showed gradual upward expansion, which is consistent with the deformation process. Because no actual displacement monitoring data were available, the results could not be compared with the displacement values. Because of the influence of errors in the parameters, the simulated displacement value is not necessarily accurate but can fully reflect the deformation and evolution law. As shown in Figures 15(b), 15(d), and 15(f), the maximum horizontal displacement of the accumulation slope was concentrated at the top of the slope when the unmodified constitutive model was used.

As shown in Figures 16(a), 16(c), and 16(e), the maximum shear strain of the deposited body was concentrated primarily near the foot of the slope, and the initial small range of the slope surface increased with the simulation step. Figures 16(b), 16(d), and 16(f) show that the maximum shear strain was largest at the slope position of the deposited body. As the simulation time steps increased, the maximum shear strain was distributed predominantly in the rock near the contact surface rather than the slope surface. To further quantify the data, the modified model was used to simulate the deformation process in comparison with the actual monitoring results, as shown in Figure 17.

As shown in Figure 17, the horizontal displacements of the monitoring points at the same time step were in the order of A > B > C > D > E. Point A was located at the foot and had the largest displacement when the model simulation began, followed by points B, C, D, and E. The gradual increase in slope surface displacement reflects the deformation process.

As described above, the Duncan-Chang E-v model can be used to simulate the plastic deformation of a deposited body.

TABLE 8: Et-vt model parameters.

Years	Shear strength index				Et-vt model parameters							
	C' (kPa)	φ (°)	Φ_0 (°)	$\Delta\varphi$ (°)	K	n	K_b	m	R_f	G	F	D
40	78.6	34.28	43.75	7.99	150	0.538	29.9	0.51	0.744	0.088	0.063	6.88
70	85.1	34.33	55.36	22.55	332	0.141	55.5	0.13	0.774	0.089	0.074	6.93

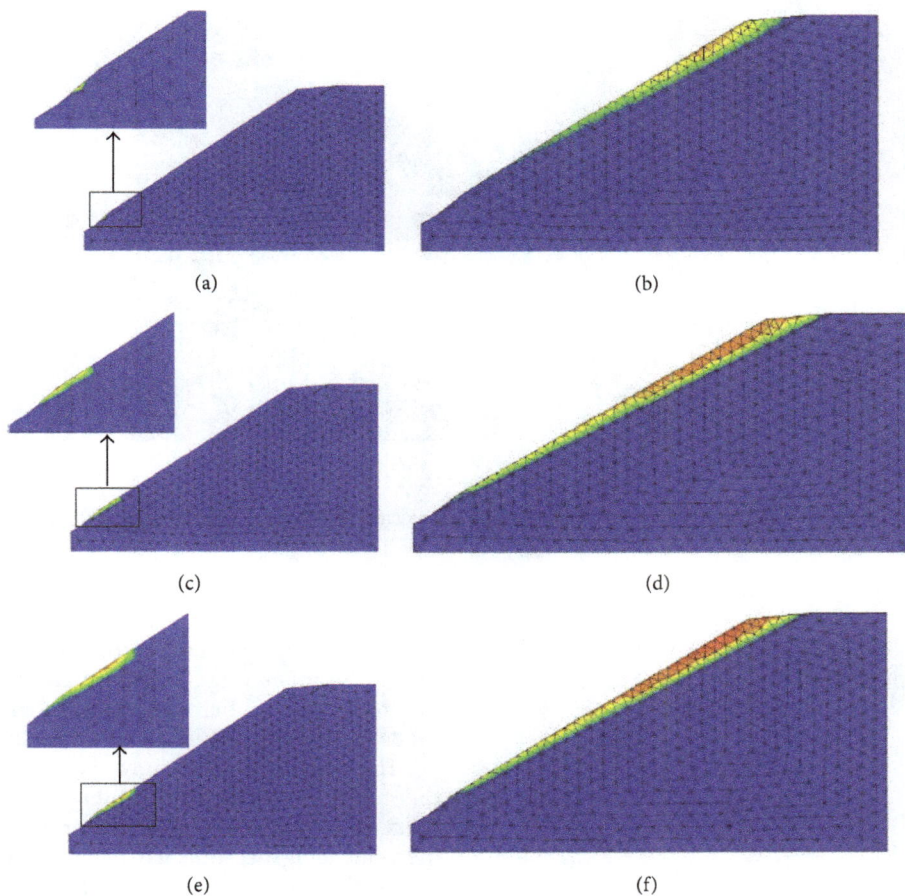

FIGURE 15: Maximum horizontal displacements of the deposited body: improved model after (a) 10,000, (c) 20,000, and (e) 30,000 time steps; unmodified model after (b) 10,000, (d) 20,000, and (f) 30,000 time steps.

TABLE 9: Parameters for improving tangential Poisson's ratio in the Et-vt model.

A	t_1	t_0
3.4007	0.6873	0.0384

Moreover, the modified Duncan-Chang E-v constitutive model can be used to study the displacement and shear strain to analyze the evolutionary process of deformation. When the unmodified Duncan-Chang E-v constitutive model was used, the simulated cumulative slope deformation and evolution process did not match the actual slope deformation and failure phenomenon. This finding establishes that the modified Duncan-Chang E-v constitutive model can reflect the actual situation better than the unmodified Duncan-Chang

E-v constitutive model and that the improved constitutive model is not only correct but also meaningful.

7. Conclusion

This study focused on the stability of the slag accumulation body in the Tongren area of Guizhou Province, China. CT was used with uniaxial and triaxial compression tests to examine the mechanical properties and constitutive relation of cemented mercury slag aggregates. The main findings of this study are as follows.

(a) The stress-strain structural response of cemented mercury slag can be divided into four aspects. The internal cementation structure tends to be closed before stress exposure. When the stress concentration reaches its peak, the

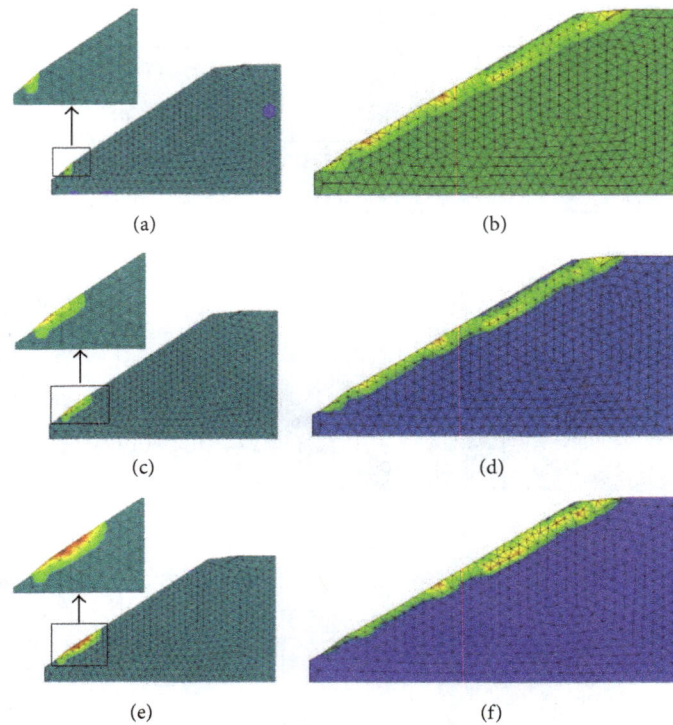

FIGURE 16: Maximum shear strain clouds of the deposited body: improved model after (a) 10,000, (c) 20,000, and (e) 30,000 time steps; unmodified model after (b) 10,000, (d) 20,000, and (f) 30,000 time steps.

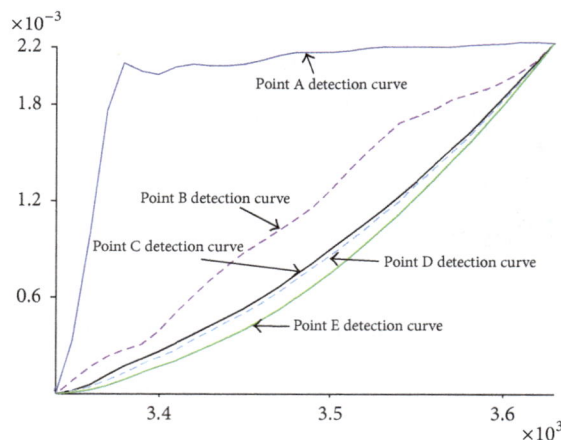

FIGURE 17: Measured horizontal shift of monitoring points.

structure of the cemented mercury slag presents the corresponding initial crack response. As the stress increases, the structure of the cemented mercury slag is rapidly damaged and exhibits one global and two local failure modes. When the strain does not change with increasing stress, the structure is completely destroyed, and the response of the mercury slag structure terminates.

(b) As the consolidation stress increases with the accumulation time, the particles inside the slag exhibit a hardening characteristic at lower confining pressures. This process makes the slag stronger and less prone to deformation.

(c) The modified Duncan-Chang E-ν constitutive model is more likely to reflect the actual situation than the unmodified classic Duncan-Chang E-ν constitutive model. This finding establishes that the improvement to the constitutive model is not only correct but also meaningful.

The results of this study are important for ensuring the geological safety of a mining area and establishing a nontoxic living environment. Future research may focus on the influencing mechanism of the grain roughness of mercury slag on the degree of consolidation.

Conflicts of Interest

The authors declare that there are no conflicts of interest regarding the publication of this paper.

Acknowledgments

The authors acknowledge support received for the study of surface seepage prevention technology and rehabilitation scheme optimisation of mercury slag accumulation (Bureau of Geology and Minerals: 2016-4) and the manganese, aluminium, and barite ore resources in deep mining and deep processing technology and demonstration research (2902).

References

[1] X. Liu, "Experimental Study on Mechanical Properties of Slag," *Yangtze River Sci. Res. Inst*, vol. 11, no. 4, pp. 46–51, 1994 (Chinese).

[2] Z. Wang, H. Lan, and J. Wu, "Present situation of Chinas raw mercury industry and suggestions for future shutdown policy," *Earth Environ*, vol. 42, no. 5, pp. 659–662, 2014 (Chinese).

[3] H. Yang, L. Ping I, and G. Qiu L, "Mercury pollution in mercury mining areas throughout the world: an overview," *Earth & Environment*, 2009 (Chinese).

[4] X. Feng, G. Li, and G. Qiu, "A preliminary study on mercury contamination to the environment from artisanal zinc smelting using indigenous methods in Hezhang County, Guizhou, China: Part 2. Mercury contaminations to soil and crop," *Science of the Total Environment*, vol. 368, no. 1, pp. 47–55, 2006.

[5] M. Horvat, J. Kotnik, M. Logar, V. Fajon, T. Zvonarić, and N. Pirrone, "Speciation of mercury in surface and deep-sea waters in the Mediterranean Sea," *Atmospheric Environment*, vol. 37, no. 1, pp. S93–S108, 2003.

[6] G. Qiu, X. Feng, S. Wang, X. Fu, and L. Shang, "Mercury distribution and speciation in water and fish from abandoned Hg mines in Wanshan, Guizhou province, China," *Science of the Total Environment*, vol. 407, no. 18, pp. 5162–5168, 2009.

[7] M. Fillion, D. Mergler, C. J. Sousa Passos, F. Larribe, M. Lemire, and J. R. D. Guimarães, "A preliminary study of mercury exposure and blood pressure in the Brazilian Amazon," *Environmental Health: A Global Access Science Source*, vol. 5, article no. 29, 2006.

[8] S. Wang, X. Feng, and G. Qiu, "The study of mercury exchange rate between air and soil surface in Hongfeng reservoir region, Guizhou, PR China," *Journal De Physique IV*, vol. 107, no. 1, pp. 1357–1360, 2003.

[9] M. E. Hines, M. Horvat, J. Faganeli et al., "Mercury biogeochemistry in the Idrija River, Slovenia, from above the mine into the Gulf of Trieste," *Environmental Research*, vol. 83, no. 2, pp. 129–139, 2000.

[10] Y.-N. Zhang and Y.-L. Chen, "Mechanical characteristics of cemented rockfill materials with high rock block proportions under uniaxial compression," *Journal of Northeastern University*, vol. 37, no. 11, pp. 1630–1634, 2016.

[11] Q. Zheng, C. Wang, and J. Zhang, "The strength and stress-strain behavior of the cemented rockfill," *Advanced Materials Research*, vol. 598, pp. 565–568, 2012.

[12] A. Fakhimi and B. Hemami, "Axial splitting of rocks under uniaxial compression," *International Journal of Rock Mechanics and Mining Sciences*, vol. 79, pp. 124–134, 2015.

[13] S. Zhang, H. Tang, H. Zhan, G. Lei, and H. Cheng, "Investigation of scale effect of numerical unconfined compression strengths of virtual colluvial-deluvial soil-rock mixture," *International Journal of Rock Mechanics and Mining Sciences*, vol. 77, pp. 208–219, 2015.

[14] A. Kesimal, E. Yilmaz, B. Ercikdi, I. Alp, and H. Deveci, "Effect of properties of tailings and binder on the short-and long-term strength and stability of cemented paste backfill," *Materials Letters*, vol. 59, no. 28, pp. 3703–3709, 2005.

[15] Z. Chen, R. Luo, Z. Huang et al., "Effects of different backfill soils on artificial soil quality for cut slope revegetation: Soil structure, soil erosion, moisture retention and soil C stock," *Ecological Engineering*, vol. 83, pp. 5–12, 2015.

[16] O. Kara and M. Baykara, "Changes in soil microbial biomass and aggregate stability under different land uses in the northeastern Turkey," *Environmental Modeling & Assessment*, vol. 186, no. 6, pp. 3801–3808, 2014.

[17] X.-W. Fang, Z.-H. Chen, C.-N. Shen, S.-G. Sun, and Q.-M. Wang, "Test study on residual soil under special stress paths," *Rock and Soil Mechanics*, vol. 26, no. 6, pp. 932–936, 2005.

[18] E. Asghari, D. G. Toll, and S. M. Haeri, "Triaxial behaviour of a cemented gravely sand, Tehran alluvium," *Geotechnical and Geological Engineering*, vol. 21, no. 1, pp. 1–28, 2003.

[19] L. Kongsukprasert, F. Tatsuoka, and M. Tateyama, "Several factors affecting the strength and deformation characteristics of cement-mixed gravel," *Soils and Foundations*, vol. 45, no. 3, pp. 107–124, 2005.

[20] J. M. Duncan and C. Y. Chang, "Nonlinear analysis of stress and strain in soils," *ASCE Soil Mech. Found. Div. J*, vol. 96, no. 5, pp. 1629–1653, 1970.

[21] H. Matsuoka and T. Nakai, "Stressstrain relationship of soil based on the SMP in," in *Proceedings of the 9th International Conference on Soil Mechanics and Foundation Engineering, Special Session*, vol. 9, pp. 153–162, London, UK, 1977.

[22] K. S. Choi and J. Pan, "A generalized anisotropic hardening rule based on the Mroz multi-yield-surface model for pressure insensitive and sensitive materials," *International Journal of Plasticity*, vol. 25, no. 7, pp. 1325–1358, 2009.

[23] W. Hu, "A novel quadratic yield model to describe the feature of multi-yield-surface of rolled sheet metals," *International Journal of Plasticity*, vol. 23, no. 12, pp. 2004–2028, 2007.

[24] H. L. Fang, H. Zheng, and J. Zheng, "Micromechanics-based multimechanism bounding surface model for sands," *International Journal of Plasticity*, vol. 90, pp. 242–256, 2017.

[25] R. L. Jackson and J. L. Streator, "A multi-scale model for contact between rough surfaces," *Wear*, vol. 261, no. 11-12, pp. 1337–1347, 2006.

[26] G. Li, *Discussion on three-dimensional constitutive relation of soil and model verification*, Tsinghua University, Beijing, China, 1985 (Chinese).

[27] S.-P. Jia, W.-Z. Chen, J.-P. Yang, and P.-S. Chen, "An elastoplastic constitutive model based on modified Mohr-Coulomb criterion and its numerical implementation," *Rock and Soil Mechanics*, vol. 31, no. 7, pp. 2051–2058, 2010.

[28] D. Liu, M. He, and M. Cai, "A damage model for modeling the complete stress–strain relations of brittle rocks under uniaxial compression," *International Journal of Damage Mechanics*, 2017, 105678951772080.

[29] M. Saberi, C. Annan, and J. Konrad, "Constitutive Modeling of Gravelly Soil–Structure Interface Considering Particle Breakage," *Journal of Engineering Mechanics*, vol. 143, no. 8, p. 04017044, 2017.

[30] C. Ma, D. Lu, X. Du, and A. Zhou, "Developing a 3D elastoplastic constitutive model for soils: A new approach based on characteristic stress," *Computers & Geosciences*, vol. 86, pp. 129–140, 2017.

[31] H. Stutz and D. Mašín, "Hypoplastic interface models for fine-grained soils," *International Journal for Numerical and Analytical Methods in Geomechanics*, vol. 41, no. 2, pp. 284–303, 2017.

[32] Y. Xiao and H. Liu, "Elastoplastic constitutive model for rockfill materials considering particle breakage," *International Journal of Geomechanics*, vol. 17, no. 1, pp. 1–13, 2017.

[33] X. W. Li, *Study on Mechanical Properties and Constitutive Relation of Cemented Mercury Slag*, China University of Geosciences, 2016.

9

Review on Cement Stabilization/Solidification of Municipal Solid Waste Incineration Fly Ash

Chengcheng Fan, Baomin Wang, and Tingting Zhang ⓘ

Institute of Building Materials, School of Civil Engineering, Dalian University of Technology, Dalian 116024, China

Correspondence should be addressed to Tingting Zhang; tingtingzhang@dlut.edu.cn

Academic Editor: Ling B. Kong

Municipal solid waste incineration (MSWI) fly ash must be treated properly prior to being disposed in the security landfill due to its serious pollution toxicity. Nowadays, lots of studies have demonstrated that cement-based stabilization/solidification could reduce the toxicity pollution effectively by encapsulating the heavy metals into cement matrix, which leads to greater capacity and weight. This paper compares and discusses the MSWI fly ash treatment with the mostly used matrix materials such as Portland cement, phosphate cement, aluminate cement, and alkaline activated cement. Moreover, immobilization mechanism introduced by the interaction between the MSWI fly ash and hydrated cement matrix materials, such as the physical cementing effect, adsorption, isomorphous replacement, and complex precipitation, was explored in depth. The paper also pointed out some reasonable development directions for cement-based stabilization/solidification technology to improve the effectiveness and application of cement-based stabilization/solidification technology.

1. Introduction

Waste incineration technology has been widely used globally because it can achieve the goal of harmlessness, waste minimization, and resource utilization of municipal solid waste incineration. Nonetheless, there will be a large amount of MSWI fly ash coupled with various types of toxic heavy metal materials with high concentration after the incineration, accounting about 3 to 5 percent of the total waste, which not only has severely restricted the enterprise's survival and development, but also causes a great threat to the human survival environment and health. Therefore, disposing MSWI fly ash and solving the problem of heavy metal pollution properly is a task which brooks no delay.

Nowadays, cement-based stabilization/solidification technology has been widely used all over the world for 60 years, it and has become an economic efficiency approach to immobilize toxic and harmful heavy metal in the MSWI fly ash by being fixed in the hydration products, such as calcium silicate hydrate gel (C-S-H) and ettringite (AFt) which were formed by chemical reaction between cement and water [1–4]. Besides, the advantages of cement-based stabilization/solidification not only have simple operation and low processing cost, but also

significant reduction of percolation of heavy metal ions in the hazardous waste landfill, which could meet the Standard for Pollution Control on the Security Landfill. Thus, the United States Environmental Protection Agency (EPA) has considered the method as the best way to deal with poisonous and harmful waste in the world so far [5].

On the basis of existing research, this paper made a comparative analysis on the effect of MSWI fly ash immobilized with different types of cementitious materials, such as Portland cement, phosphate cement, aluminate cement, and alkaline activated cement, and then further explored the influence factors and immobilization mechanism through the interactions between cement hydration products and heavy metals. Moreover, some suggestions were also put forward on the application and development of cement-based stabilization/solidification technology in the next studies.

2. Characteristics of MSWI Fly Ash

Comparative analysis of MSWI fly ash around the world showed that the composition of MSWI fly ash was relatively complicated. It not only consisted of silicate, chlorine salt, calcium hydroxide, and sulfate, but also absorbed a lot of the poisonous and harmful substances in the flue gas, including dioxins, furans, mercury (Hg), copper (Cu), lead (Pb), manganese (Mn), cadmium (Cd), chromium (Cr), nickel (Ni), and so on [6, 7], which had already become a serious threat to the ecological environment, food safety, and human health. However, the organic matters would decompose and evaporate during the incineration process, and its content in the MSWI fly ash was generally below 2%, thus it meant that the treatment and remediation of heavy metals had become the main target in the stabilization/solidification process of MSWI fly ash. Moreover, once heavy metals caused harm to human health and natural environment, it would be serious, long-term, latent, cumulated, and irreversible [8–10]. Therefore, finding the solutions to solve the heavy metal pollution of MSWI fly ash has become a top priority. Tables 1 and 2 summarize the sources and chemical speciation of heavy metals in the MSWI fly ash, which could provide some theory guidance for the MSWI fly ash immobilization study in the following.

3. Effect of Different Cementitious Matrix Materials on MSWI Fly Ash Immobilization

3.1. Portland Cement.
Lombardi et al. and Polettini et al. [11, 12] reported that cement could effectively immobilize contaminants, but the compressive strength will be reduced and leaching concentration will be increased when the MSWI fly ash addition is increased. Also, Dermatas and Meng [13] pointed out that the most obvious shortcoming of cement stabilization/solidification was high capacity, weight increased rate, and poor long-term stability. The capacity and weight increased rate was up to 2.0. However, when the MSWI fly ash was prewashed with water, phosphoric acid, or other solutions before cement-based stabilization/solidification, it would drastically reduce the toxicity of heavy metals and reduce the consumption of cement. However, there was still tricalcium aluminate (C_3A) in the cement solidified body after the 28-day curing period, but C_3A would greatly reduce its compressive strength [14], and it was difficult for Portland cement solidified body to achieve high mechanical strength whether the MSWI fly ash was pretreated or not [15].

What is more, some literature studies proposed that some heavy metals in the MSWI fly ash such as Cr^{6+}, Cd^{2+}, Pb^{2+}, and Cu^{2+} could dramatically accelerate C_3S hydration whether it was in the late or initial stage of hydration reaction, but the Zn^{2+} only promoted the C_3S hydration during the later period of hydration [16, 17]. But some studies reported that heavy metals would delay the cement hydration to some extent. In detail, Cu^{2+} and Pb^{2+} usually bound on the surface of cement clinker particles in the form of insoluble salt, such as carbonates and sulfates, which inhibited the cement hydration. Ma et al. [18] proposed that Cu, Zn, and Pb could not obviously change the cement hydration process according to the analysis of XRD spectrum, and Cu, Zn, and Pb were immobilized within cement hydration products through a series of effect mechanism, such as physical fixation and adsorption mechanisms. Vogg et al. [19] pointed out that heavy metals would leak out little by little in an acidic environment, but there would be no

detection of Cd, Cu, Pb, and Zn under the condition at pH value of 10.0.

3.2. Phosphate Cement.
Magnesium phosphate cement was also used to immobilize MSWI fly ash, the reason of which was not only the alkalescence of potassium phosphate cement with the characteristic of dense structure and high strength, but also the widespread use of phosphate as curing agents. It was found that the compressive strength and leaching toxicity for magnesium phosphate cement on Cr^{3+}, Pb^{2+}, Ni^{2+}, Cd^{2+}, and Zn^{2+} could meet the national corresponding standards [20–22]. On the one hand, there were some macropores and gassy surface in the light burned magnesia powder crystal with a high activity, which brought the strong adsorption for heavy metals. On the other hand, magnesia powder with higher activity could give impetus to the violent reaction with magnesium salt solution and chemical precipitation of heavy metals [23]. Therefore, it was easy for magnesium hydroxide to absorb and remove heavy metals from industrial waste water under the help of its high specific surface area and adsorption capacity.

Besides, the solidification effect on Pb^{2+} of magnesium phosphate cement would be more prominent in contrast with other heavy metals. It was the results from both chemical reaction and physical encapsulation. On the one hand, Pb^{2+} was more likely to react with PO_4^{3-} and HPO_4^{2-}, and then generate insoluble phosphate precipitation containing Pb. Moreover, the heavy metals could be encapsulated securely because of compact reticular structure of magnesium phosphate cement, in which the radioactive elements can be solidified effectively as well [24–26]. And the heavy metals such as Pb, Cd, and Cr almost could not be detected when the addition of magnesium potassium phosphate cement was up to 40%. Besides, it was noted that the leaching concentration of heavy metals always followed a descending order of Cu > Zn > Ni > Pb > Cd > Cr when eluviated in different leaching environments.

3.3. Aluminate Cement.
There were very few studies on the heavy metals immobilization in aluminate cement and sulphoaluminate cement. But Sahu et al. and Luo et al. [27–29] found that sulphoaluminate cement could also reduce the leaching toxicity of heavy metals, and it might be prone to fix Pb^{2+} than Zn^{2+}, Cd^{2+}, and Cu^{2+} due to the selectivity of its hydration products. Besides, the heavy metals mainly deposited in the complex chemical speciation. For example, Pb^{2+} is mainly in the form of $2PbCO_3 \cdot Pb(OH)_2$; Cd^{2+} mainly exists in the form of $CdPbO_3$. Overall, the adsorption stability of Pb^{2+}, Cu^{2+}, Cd^{2+}, and Zn^{2+} immobilized by sulphoaluminate cement is good, especially for Pb^{2+}.

3.4. Alkaline Activated Cement.
Alkaline activated cement is a kind of cementitious material formed by reaction of the alkaline activator and materials with volcanic activity or latent hydraulicity. The chemical reactions including the precipitation of calcium silicoaluminate gel (CASH) and sodium

TABLE 1: Source of heavy metals in MSWI fly ash.

Heavy metal	Source
Pb	Gasoline additives, ammunition, solder, paint, pesticide
Hg	Electrical manufacturing, pharmaceuticals, paints, plastics, paper, batteries, coal, mercury smelters, and mercury preparation plants
Cd	Plastic, enamels with pigment, electroplating, metal covering
Cr	Mine, smelting plant, electroplating factory, chrome tanning system
As	Atmospheric dust, tailings and pesticide, leather, chemical pharmaceutical, metallurgy, decolorant
Cu	Mine, metal processing, machinery manufacturing, steel production
Zn	Galvanizing, smelting, machinery manufacturing, organic synthesis, mine

TABLE 2: Chemical speciation of heavy metals in MSWI fly ash.

Heavy metal	Speciation
Pb	$PbCl_2$, $PbCO_3$, PbO, Pb_3SiO_3, Pb_3SiO_4, $Pb_3O_2SO_4$, $Pb_3Sb_2O_7$
Cd	$Cd(OH)_2$, CdO, $CdSO_4$, $CdCl_2$, $CdSiO_4$
Zn	$ZnCl_2$, ZnO, $Zn(OH)_2$, $ZnCO_3$, $ZnSO_4$, K_2ZnCl_4
Fe	Fe_2O_3, Fe_3O_4
Cu	$Cu(OH)_2$, CuO, $CuCO_3$

silicoaluminate gel (NASH) may occur during the process of setting and hardening, then forming a three-dimensional aluminum silicate network structure which consists of AlO_4 and SiO_4 tetrahedral structure units, which could be used to seal the contaminants tightly in the cavity [30, 31].

Zak and Deja [32] used chemical reagents to synthesize C-S-H, CASH, and NASH, and then found that Zn^{2+}, Cd^{2+}, Pb^{2+}, Cr^{3+}, and Cr^{6+} inhibited the synthesis of C-S-H, CASH, and NASH structures, and C-S-H and NASH could immobilize heavy metal at the degree of 99.9% and 93.1%, respectively. It could be deduced that the alkaline activated cement could be used to seal heavy metals, which was in line with the conclusion from Davidovits [33]; meanwhile, he also stated that the network skeleton of alkaline activated cement was of perfect stability even though under nuclear radiation.

From the above, cement-based stabilization/solidification could effectively fix the heavy metal in the cement with the advantages of simple operation, low cost, and high strength of cement solidified body, but it also gave rise to poor stability, high capacity, and weight increased rate. Furthermore, it should be noted that the cement-based stabilization/solidification of MSWI fly ash just belonged to temporary concentration control, which could not guarantee its long-term security and stability, especially under the acid environment. Moreover, it was indeed feasible to fix the MSWI fly ash with cement when mixed with little MSWI fly ash, but it was hard to ensure that the performances of cement solidified body meet the national landfill standards with increased MSWI fly ash. If there were special salts in the MSWI fly ash, the salts would hinder the cement-hardening process and weaken its mechanical strength, leading to high leaching rates of heavy metals.

Comparative analysis of the effect of different cementitious matrix materials on MSWI fly ash immobilization was summarized in Table 3. There are lots of smaller capillary pores in the magnesium phosphate cement,

TABLE 3: MSWI fly ash immobilized with different cementitious matrix materials.

Cement	Advantages	Disadvantages
Portland cement	(1) Simple operation (2) Low cost (3) Moderate compressive strength	(1) High capacity and weight increased rates (2) Poor long-term security and stability
Phosphate cement	(1) High compressive strength (2) Good dry shrinkage (3) Low cost	(1) High capacity and weight increased rates (2) Poor long-term security and stability
Aluminate cement	(1) Simple operation (2) Low cost	(1) High capacity and weight increased rates (2) Poor stability (3) Poor dry shrinkage
Alkali activated cement	(1) High compressive strength (2) Low cost (3) Good long-term security and stability	(1) High capacity and weight increased rates (3) Poor dry shrinkage

resulting in a more compact microstructure, thus it could immobilize the heavy metals more effectively in theory. However, what calls for special attention is that magnesium phosphate cement would lose the strength and burst into fragments when exposed to the humid environment for a long time due to the poor water resistance, which would put the disposal of MSWI fly ash at risk. Alkaline activated cement was of the ideal performances, including excellent mechanical properties and durability, whereas shrinkage cracking and weak anticarbonization were the two fatal flaws of this material, which would result in secondary pollution to the environment, ultimately affecting the health of human beings. It can be concluded from these literature studies that the solidification effect of these cementitious materials on MSWI fly ash is alkali activated cement > magnesium phosphate cement > Portland cement > aluminate cement.

4. Immobilization Mechanisms

During the process of cement hydration, heavy metals eventually settled in the cement hydration products in the form of hydroxide or complex by reacting with cement by means of the physical cementing effect, chemical absorption, isomorphous replacement, and complex precipitation. Meanwhile, cement also provides the alkaline environment for solidified body so as to effectively inhibit the infiltration of heavy metals. Besides, these hydration products including C-S-H, AFt, single sulfur hydrated calcium thioaluminate (AFm), and CASH, rather than calcium hydroxide, were difficult to dissolve, and these insoluble hydrated products would play important roles during heavy metal immobilization. Thus, the immobilization of MSWI fly ash in cement-based stabilization/solidification was investigated in depth mainly from these aspects such as hydration products and immobilization effect.

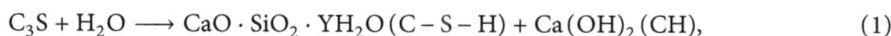

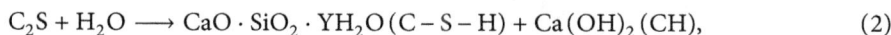

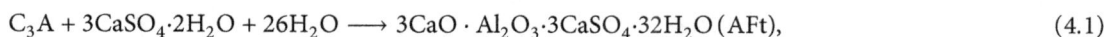

$$C_3S + H_2O \longrightarrow CaO \cdot SiO_2 \cdot YH_2O\,(C-S-H) + Ca(OH)_2\,(CH), \tag{1}$$

$$C_2S + H_2O \longrightarrow CaO \cdot SiO_2 \cdot YH_2O\,(C-S-H) + Ca(OH)_2\,(CH), \tag{2}$$

$$C_3A + 6H_2O \longrightarrow 3CaO \cdot Al_2O_3 \cdot 6H_2O\,(CAH), \tag{3}$$

$$C_3A + 3CaSO_4 \cdot 2H_2O + 26H_2O \longrightarrow 3CaO \cdot Al_2O_3 \cdot 3CaSO_4 \cdot 32H_2O\,(AFt), \tag{4.1}$$

$$3CaO \cdot Al_2O_3 \cdot 3CaSO_4 \cdot 32H_2O + 2C_3A + 4H_2O \longrightarrow 3\,[3CaO \cdot Al_2O_3 \cdot CaSO_4 \cdot 12H_2O]\,(AFm), \tag{4.2}$$

$$C_4AF + 7H_2O \longrightarrow 3CaO \cdot Al_2O_3 \cdot 6H_2O\,(CAH) + CaO \cdot Fe_2O_3 \cdot H_2O\,(CFH). \tag{5}$$

4.1. C-S-H. C-S-H is a type of amorphous microporous phase with high specific surface area and high-density hydrogen bonding, which could tightly bind heavy metals with strong chemical adsorption. In general, the C-S-H colloid of rich calcium with positive charges on the surface has preferential adsorption of anions including OH^-, Cl^-, SO_4^{2-}, and so on, but the C-S-H colloids of rich silicon are preferentially adsorbed cations [34]. Furthermore, it was confirmed that the adsorption of heavy metals is always significantly influenced by pH value, and it would decrease obviously when the pH value surpasses 9.0 [35].

In addition to this, C-S-H had significant cation exchange capacity, and some heavy metals such as Cr^{6+}, Cd^{2+}, Zn^{2+}, and Pb^{2+} often had stronger adsorption affinity in contrast with alkali metal ions. The heavy metals were thus prone to ions exchange, and they would then stabilize in the lattice. Ca^{2+}, Al^{3+}, and Si^{4+} in the cement solidified body are easily replaced by heavy metal cation, and SO_4^{2-} and OH^- can be easily replaced by heavy metal anions [36]. Similarly, Zn^{2+} could replace the Ca^{2+} in C-S-H or react with Ca to produce the oxides or hydroxides containing Ca and Zn [37]. Moreover, Cr^{3+} could be combined into C-S-H by replacing Si^{2+} in C-S-H [38], and CrO_4^{2-} could be fixed in the cement solidified body by replacing SO_4^{2-}.

However, Pb^{2+} in C-S-H cannot replace Ca^{2+}, but it can be fixed in C-S-H in the form of hydroxide precipitation and carbonate precipitation, while Pb-C-S-H was observed by means of X-ray photoelectron spectroscopy (XPS) and SEM/EDS in the research of Cocke [39]. It should be noted that weak alkalescent environment could further retard leaching toxicity of heavy metals in cement solidified body in most cases. It should be noted that hydroxide precipitation occurs when the pH value of leaching solution dissolved heavy metals is raised to some optimum level for a specific metal. Table 4 exhibits that the optimum pH value is always different for each metal and for different valence states of a single metal. For example, the leaching rates for Cd^{2+} decreased when the pH value is more than 8.0, and Pb^{2+} leached to the least when the pH value is 10.3 [40, 41].

4.2. AFt and AFm. AFt phase, $3CaO \cdot Al_2O_3 \cdot 3CaSO_4 \cdot 32H_2O$, is one of the main hydration products of cement, but it can be easily changed into AFm ($Ca_2Al(OH)_6 \cdot 0.5X \cdot H_2O$) when chloride or sulfate is insufficient. As shown in Figure 1, AFt phases are chemically and mineralogically complex, in which columns of $Ca_6Al(OH)_6 \cdot 24H_2O$ are lines of $Al(OH)_6^{3-}$ octahedrally bonded with three calcium polyhedrons, and each calcium polyhedron is with OH^- and four water molecules, which presents an orientational column structure on the whole.

In most cases, AFt can absorb and immobilize the heavy metals through the isomorphous replacement and chemical adsorption; a plenty of studies demonstrated that AFt have a very strong lattice binding effect on heavy metal ions. For example, Cs, Sn, and other metal ions could also be adsorbed on the cylinder surface of AFt [42, 43]. On the one hand, Cd^{2+}, Ni^{2+}, Zn^{2+}, Pb^{2+}, and Co^{2+} could replace Ca^{2+}, and Cr^{3+}, Ti^{3+}, Mn^{3+}, Si^{4+}, and Fe^{3+} can replace Al^{3+} in AFt; on the other hand, Cl^-, CO_3^{2-}, SeO_4^{2-}, BrO_3^-, AsO_4^{2-}, and ClO_3^- can replace SO_4^{2-} [44–47]. Besides, Sabine and Lan et al. found that Cr^{3+}, CrO_4^{2-}, and $Cr_2O_7^{2-}$ can get inside the AFt lattice and then stabilize in the AFt phases. Tashiro et al. [48–50] found that heavy metal oxides and hydroxides could promote the formation and crystal growth of ettringite and produce some changes in microstructure and Cr_2O_3 and $Cu(OH)_2$ exert a considerable influence. Similar to AFt, Ca^{2+}, Al^{3+}, CO_3^{2-}, and SO_4^{3-} in AFm phases are easy to be replaced by other ions, and AFm have a more stable structure than AFt. Several studies have been put forward to explain that AFm could effectively reduce the content of heavy metal ions in the water, such as Cr^{3+}, CrO_4^{2-}, Cd^{2+}, B^{3+}, Se^{6+}, and so on [44, 51, 52].

TABLE 4: Optimum pH value of heavy metal precipitates.

Heavy metals	$Cd(OH)_2$	$Pb(OH)_2$	$Fe(OH)_2$	$Zn(OH)_2$	$Cu(OH)_2$
pH value	11.0	9.7	7.2	6.7	5.5

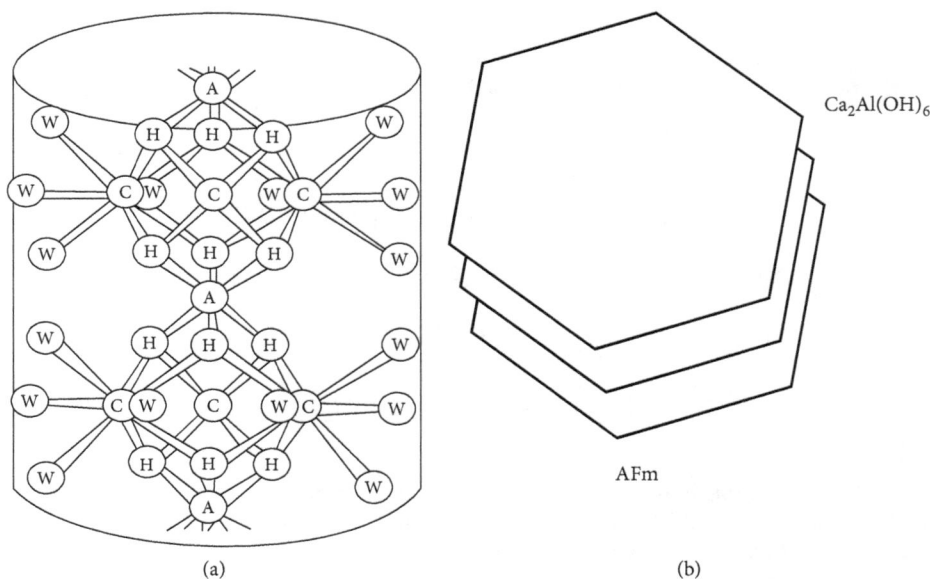

(a)   (b)

FIGURE 1: Cellular structure of AFt and AFm.

Isomorphous replacement is also known as isomorphous substitution. In mineral crystallization, the position occupied by some ions or atoms in the crystal structure is partly occupied by other ions or atoms of similar nature and size, but the crystal structure is basically unchanged. The electronegativity of Cr^{3+} and Al^{3+} is close, and both ions' electricity price is the same, so it can replace the calcium alum stone Al^{3+} form relatively stable substitutional solid solution. On the contrary, Cr^{6+} would change the molecular symmetry and cause a great influence on the lattice structure of AFt when entered into the lattice structure. Therefore, AFt is prone to solidly immobilizing Cr^{3+} than Cr^{6+}.

4.3. CASH. CASH is a kind of zeolite minerals composed of oxygen-silicon tetrahedron, which has strong ability of adsorption and ion exchange capacity due to its higher specific surface area with $500\sim1000\ m^2/g$. However, Si^{4+} in the oxygen-silicon tetrahedron always be substituted by Al^{3+} and form a alumina tetrahedron with some extra negative charge, which needs alkali metal cations to achieve the charge balance. There is poor binding force between alkali metal cations and the CASH crystal, because most of the alkali metal cations are located in a cavity or pore structures. Besides, these heavy metal ions including Cr^{3+}, Cd^{2+}, and Pb^{2+} have stronger adsorption ability than alkali metal cations in CASH, so the heavy metal ions are more likely to be effectively immobilized in CASH.

5. Summary and Prospect

The comparative analysis of the solidification effect and leaching toxicity of MSWI fly ash treated by various types of cement has been investigated and discussed in depth in this paper and found that it was more effective to use the phosphate cement and alkali activated cement to immobilize the heavy metals. And this paper also looks at the interaction of cement hydration products and heavy metals in particular. However, there are also lots of controversies about immobilization mechanism of heavy metals with cement-based stabilization/solidification technology. For example, these immobilization mechanisms always change with external environment and matrix materials structure, and meanwhile, the heavy metal hydrolyses always give rise to variation of pH values and cement hydration process, which would make the study on immobilization mechanisms and hydration products fixing effects of heavy metal more complex. But beyond all that, there are few in-depth studies on the long-term safety and stability after the landfill of MSWI fly ash solidification and no universal practical method to realize the reutilization of MSWI fly ash. Therefore, it is to be hoped that the proper treatment and disposal of MSWI fly ash would be realized with the efforts in the next few years.

Conflicts of Interest

The authors declare no conflicts of interest.

Acknowledgments

The financial support of this work was supported by the National Science Foundation of China (51578108 and 51778101), the Fundamental Research Funds for the Central Universities DUT18ZD211, National Key R&D Program of China (2017YFE0107000), and Dalian High-level Talent Innovation Program (2017RQ051). Thanks are due to all authors of those cited references for the data and achievements. They make it well-founded for this study.

References

[1] U. Richers and L. Birnbaum, "Detailed investigations of filter ashes from municipal solid waste incineration," *Waste Management and Research*, vol. 16, no. 2, pp. 190–194, 1998.

[2] A. Polettini, R. Pomi, L. Trinci, A. Muntoni, and S. L. Mastro, "Engineering and environmental properties of thermally treated mixtures containing MSWI fly ash and low-cost additives," *Chemosphere*, vol. 56, no. 10, pp. 901–910, 2004.

[3] J. Tang, R. Ylmén, M. Petranikova, C. Ekberg, and B. M. Steenari, "Comparative study of the application of traditional and novel extractants for the separation of metals from MSWI fly ash leachates," *Journal of Cleaner Production*, vol. 172, pp. 143–154, 2018.

[4] G. Weibel, U. Eggenberger, D. A. Kulik et al., "Extraction of heavy metals from MSWI fly ash using hydrochloric acid and sodium chloride solution," *Waste Management*, vol. 76, pp. 457–471, 2018.

[5] C. H. K. Lam, J. P. Barford, and G. McKay, "Utilization of municipal solid waste incineration ash in portland cement clinker," *Clean Technologies and Environmental Policy*, vol. 13, no. 4, pp. 607–615, 2011.

[6] L. L. Forestier and G. Libourel, "Characterization of flue gas residues from municipal solid waste combustors," *Environmental Science and Technology*, vol. 32, no. 15, pp. 2250–2256, 1998.

[7] F. Lombardi, T. Mangialardi, and L. Piga, "Mechanical and leaching properties of cement solidified hospital solid waste incinerator fly ash," *Waste Management*, vol. 18, no. 2, pp. 99–106, 1998.

[8] Z. Yang, S. Tian, L. Liu, X. Wang, and Z. Zhang, "Application of washed MSWI fly ash in cement composites: long-term environmental impacts," *Environmental Science and Pollution Research*, vol. 25, no. 12, pp. 12127–12138, 2018.

[9] J. Seniunaitė and S. Vasarevičius, "Heavy metals leaching of MSWI bottom ash: effect of short-term natural weathering," in *Proceedings of 10th International Conference "Environmental Engineering"*, Vilnius, Lithuania, August 2017.

[10] M. D. Dimitrijević, M. M. Nujkić, S. Č. Alagić, S. M. Milić, and S. B. Tošić, "Heavy metal contamination of topsoil and parts of peach-tree growing at different distances from a smelting complex," *International Journal of Environmental Science and Technology*, vol. 13, no. 2, pp. 615–630, 2016.

[11] F. Lombardi, T. Mangialardi, L. Piga, and P. Sirini, "Mechanical and leaching properties of cement solidified hospital solid waste incinerator fly ash," *Waste Management*, vol. 18, no. 2, pp. 99–106, 1998.

[12] A. Polettini, R. Pomi, P. Sirini, and F. Testa, "Properties of Portland cement-stabilized MSWI fly ashes," *Journal of Hazardous Materials*, vol. 88, no. 1, pp. 123–138, 2001.

[13] D. Dermatas and X. Meng, "Utilization of fly ash for stabilization/solidification of heavy metal contaminated soils," *Engineering Geology*, vol. 70, no. 3-4, pp. 377–394, 2003.

[14] N. Alba, E. Vázquez, S. Gassó, and J. M. Baldasano, "Stabilization/solidification of MSW incineration residues from facilities with different air pollution control systems. Durability of matrices versus carbonation," *Waste Management*, vol. 21, no. 4, pp. 313–323, 2001.

[15] X. Gao, W. Wang, T. Ye, F. Wang, and Y. Lan, "Utilization of washed mswi fly ash as partial cement substitute with the addition of dithiocarbamic chelate," *Journal of Environmental Management*, vol. 88, no. 2, pp. 293–299, 2008.

[16] Y. Yao, X. Wang, B. L. Yan, L. Wang, and C. Liu, "The research on heavy metal ions curing and its influence on the cement hydration process," *Bulletin of the Chinese Ceramic Society*, vol. 7, no. 2, p. e31494, 2012.

[17] M. A. Trezza, "Hydration study of ordinary portland cement in the presence of zinc ions," *Materials Research*, vol. 10, no. 4, pp. 331–334, 2007.

[18] B. G. Ma, J. R. Wang, and X. G. Li, "Effect of heavy metals and leaching toxicity of magnesium potassium phosphate cement," *Applied Mechanics and Materials*, vol. 117–119, pp. 1080–1083, 2012.

[19] H. Vogg, H. Braun, M. Metzger, and J. Schneider, "The specific role of cadmium and mercury in municipal solid waste incineration," *Waste Management Research*, vol. 4, no. 1, pp. 65–73, 1986.

[20] Y. Su, J. Yang, D. Liu, S. Zhen, N. Lin, and Y. Zhou, "Effects of municipal solid waste incineration fly ash on solidification/stabilization of cd and pb by magnesium potassium phosphate cement," *Journal of Environmental Chemical Engineering*, vol. 4, no. 1, pp. 259–265, 2016.

[21] I. Buj, J. Torras, M. Rovira, and J. d. Pablo, "Leaching behaviour of magnesium phosphate cements containing high quantities of heavy metals," *Journal of Hazardous Materials*, vol. 175, no. 1-3, pp. 789–794, 2010.

[22] S. C. Zhen, Y. Xun, and B. Q. Miao, "Solidification/stabilization of heavy metals by magnesium potassium phosphate cement," *Advanced Materials Research*, vol. 664, pp. 683–689, 2013.

[23] Y. Li and B. Chen, "Factors that affect the properties of magnesium phosphate cement," *Construction and Building Materials*, vol. 47, pp. 977–983, 2013.

[24] J. H. Yang, M. S. Jin, H. L. Chang, C. M. Heo, M. K. Jeon, and K. H. Kang, "Stabilization of Cs/Re trapping filters using magnesium phosphate ceramics," *Journal of Radioanalytical and Nuclear Chemistry*, vol. 295, no. 1, pp. 211–219, 2013.

[25] A. Covill, N. C. Hyatt, J. Hill, and N. C. Collier, "Development of magnesium phosphate cements for encapsulation of radioactive waste," *Advances in Applied Ceramics*, vol. 110, no. 3, pp. 151–156, 2011.

[26] X. Xu, J. Yang, and Y. Gu, "Properties of magnesium potassium phosphate cement containing heavy metal Pb," *Journal of Building Materials*, vol. 19, no. 1, pp. 29–34, 2016.

[27] S. Sahu, J. Havlica, V. Tomková, and J. Majling, "Hydration behaviour of sulphoaluminate belite cement in the presence of various calcium sulphates," *Thermochimica Acta*, vol. 175, no. 1, pp. 45–52, 1991.

[28] Z. T. Luo, B. G. Ma, Z. Q. Yu et al., "Influence of heavy metal Pb on hydration and leaching toxicity of sulphoaluminate cement," *Journal of Qingdao Technological University*, vol. 30, no. 4, pp. 130–133, 2009.

[29] A. Gabrisová, J. Havlica, and S. Sahu, "Stability of calcium sulphoaluminate hydrates in water solutions with various pH

values," *Cement and Concrete Research*, vol. 21, no. 6, pp. 1023–1027, 1991.

[30] G. Huang, Y. Ji, J. Li, Z. Hou, and C. Jin, "Use of slaked lime and portland cement to improve the resistance of MSWI bottom ash-gbfs geopolymer concrete against carbonation," *Construction and Building Materials*, vol. 166, pp. 290–300, 2018.

[31] A. Wongsa, K. Boonserm, C. Waisurasingha, V. Sata, and P. Chindaprasirt, "Use of municipal solid waste incinerator (MSWI) bottom ash in high calcium fly ash geopolymer matrix," *Journal of Cleaner Production*, vol. 148, pp. 49–59, 2017.

[32] R. Zak and J. Deja, "C-S-H, C-A-S-H and N-A-S-H phases as a matrices for the immobilization of heavy metals," in *Proceedings of 14th International Congress on the Chemistry of Cement*, Beijing, China, October 2015.

[33] J. Davidovits, "Properties of geopolymer cements alkaline cements," in *Proceedings of the First International Conference on Alkali Cements and Concretes*, pp. 131–149, Ukraine, October 1994.

[34] D. L. Cocke, M. Y. A. Mollah, T. R. Hess, and R. K. Vempati, "Advanced concepts in cement solidification and stabilization technology," *International Organization*, vol. 18, no. 1, pp. 207–209, 2010.

[35] E. Fourest and J. C. Roux, "Heavy metal biosorption by fungal mycelial by-products: mechanisms and influence of pH," *Applied Microbiology and Biotechnology*, vol. 37, no. 3, pp. 399–403, 1992.

[36] N. J. Coleman, Q. Li, and A. Raza, "Synthesis, structure and performance of calcium silicate ion exchangers from recycled container glass," *Physicochemical Problems of Mineral Processing*, vol. 50, no. 1, pp. 5–16, 2014.

[37] A. Mollha, M. Yousuf, J. R. Pargat, and D. L. Cocke, "An infrared spectroscopic examination of cement-based solidification/stabilization systems-Portland types V and IP with zinc," *Environmental Letters*, vol. 27, no. 6, pp. 1503–1519, 1992.

[38] L. Rozumová, O. Motyka, K. Čabanová, and J. Seidlerová, "Stabilization of waste bottom ash generated from hazardous waste incinerators," *Journal of Environmental Chemical Engineering*, vol. 3, no. 1, pp. 1–9, 2015.

[39] D. L. Coeke, "The binding chemistry and leaching mechanisms of hazardous substances in cementitious solidification/stabilization systems," *Journal of Hazardous Materials*, vol. 24, no. 2-3, pp. 231–253, 1990.

[40] X. D. Li, C. S. Poon, H. Sun, I. M. C. Lob, and D. W. Kirkc, "Heavy metal speciation and leaching behaviors in cement based solidified/stabilized waste materials," *Journal of Hazardous Materials*, vol. 82, no. 3, pp. 215–230, 2001.

[41] P. Billen, B. Verbinnen, M. D. Smet et al., "Comparison of solidification/stabilization of fly ash and air pollution control residues from municipal solid waste incinerators with and without cement addition," *Journal of Material Cycles and Waste Management*, vol. 17, no. 2, pp. 229–236, 2015.

[42] J. W. Ahn, K. S. You, G. C. Han, and K. H. Cho, "Stabilization behavior of heavy metals derived from wastes on cementitious minerals and hydrates," *Materials Science Forum*, vol. 510-511, pp. 630–633, 2006.

[43] K. S. You, J. W. Ahn, H. C. Cho, G. C. Han, D. Y. Han, and K. H. Cho, "Competing ion effect of stabilization by Cr (III) and Cr(VI) in ettringite crystal structure," *Solid State Phenomena*, vol. 124-126, no. 2, pp. 1629–1632, 2007.

[44] M. L. D. Gougar, B. E. Scheetz, and D. M. Roy, "Ettringite and C-S-H Portland cement phases for waste ion immobilization:

a review," *Waste Management*, vol. 16, no. 4, pp. 295–303, 1996.

[45] H. Y. Na and T. W. Song, "Heavy metal ion immobilization properties of microporous ettringite body," *Journal of the Korean Ceramic Society*, vol. 46, no. 6, 2009.

[46] M. Chrysochoou and D. Dermatas, "Evaluation of ettringite and hydrocalumite formation for heavy metal immobilization: literature review and experimental study," *Journal of Hazardous Materials*, vol. 136, no. 1, pp. 20–33, 2006.

[47] M. Chrysochoou and D. Dermatas, "Application of ettringite on heavy-metal immobilization: a literature reviewand experimental study," *Journal of Hazardous Materials*, vol. 136, no. 3, pp. 20–33, 2004.

[48] K. A. Saeed, K. A. Kassim, H. Nur, and N. Z. M. Yunus, "Strength of lime-cement stabilized tropical lateritic clay contaminated by heavy metals," *KSCE Journal of Civil Engineering*, vol. 19, no. 4, pp. 887–892, 2014.

[49] C. Tashiro and J. Oba, "The effects of Cr_2O_3, $Cu(OH)_2$, ZnO and PbO on the compressive strength and the hydrates of the hardened C3A paste," *Cement and Concrete Research*, vol. 9, no. 2, pp. 253–258, 1979.

[50] J. N. Diet, P. Moszkowicz, and D. Sorrentino, "Behaviour of ordinary portland cement during the stabilization/solidification of synthetic heavy metal sludge: macroscopic and microscopic aspects," *Waste Management*, vol. 18, no. 1, pp. 17–24, 1998.

[51] M. A. Trezza and M. F. Ferraiuelo, "Hydration study of limestone blended cement in the presence of hazardous wastes containing Cr(VI)," *Cement and Concrete Research*, vol. 33, no. 7, pp. 1039–1045, 2003.

[52] I. Baur, P. Keller, D. Mavrocordatos, B. Wehrli, and C. A. Johnson, "Dissolution-precipitation behaviour of ettringite, monosulfate, and calcium silicate hydrate," *Cement and Concrete Research*, vol. 34, no. 2, pp. 341–348, 2004.

10

Effect of Particle Size of Periclase on the Periclase Hydration and Expansion of Low-Heat Portland Cement Pastes

Man Yan ⓘ, Min Deng ⓘ, Chen Wang, and Zhiyang Chen

State Key Laboratory of Materials-Oriented Chemical Engineering, Nanjing Tech University, Nanjing, China

Correspondence should be addressed to Min Deng; dengmin@njtech.edu.cn

Academic Editor: Giorgio Pia

In this paper, low-heat Portland cement (LHC) clinkers were prepared by calcining raw materials at 1350°C for 2.0 hours, 1400°C for 1.0 hour, 1400°C for 1.5 hours, 1400°C for 2.0 hours, 1450°C for 1.0 hour, and 1450°C for 2.0 hours. The clinkers were ground with gypsum to produce LHC. The particle size of periclase was analysed by BSEM. Expansion of LHC pastes due to hydration of periclase was measured. The hydration degree of periclase in LHC pastes was quantitatively determined by XRD internal standard method and BSEM. The results showed that the particle size of periclase was larger when clinkers were calcined at higher temperatures or for longer time. Smaller periclase (2.60 μm) in LHC pastes tended to hydrate faster. As a result, expansion of LHC pastes develops relatively faster. Smaller particle of periclase in clinker tends to result in higher hydration degree of periclase in pastes cured at 20°C for 240 days, and there is a small amount of brucite appearing around periclase. The hydration rate of 4.00 μm periclase particle in cement paste cured at 80°C is obviously faster than that in paste cured at 20°C and 40°C. When cement paste was cured at 80°C for 7 days, the periclase was hydrated for 32.56%. The smaller size periclase (1–3 μm) had fully hydrated when the curing age was 240 days, and a large amount of brucite was produced around the larger periclase particle.

1. Introduction

Low-heat Portland cement (LHC), namely, high belite cement [1], is currently attracting a great deal of interest worldwide. This is largely due to its lower energy consumption, CO_2 emission, and heat release than ordinary Portland cements (OPC) and moderate-heat Portland cements (MHC) [2]. LHC has also many advantages, such as higher long-term strength and smaller dry shrinkage [3, 4]. The MHC is not only one of the largest amount of special cement but also the main binder used in hydraulic concrete in China, amounting for approximately 30% of hydraulic engineering cement so far [5]. Yang et al. [4] indicated that LHC concrete has a better anticrack behavior than MHC concrete. In recent years, LHC has been used in several large hydraulic structures, for example, the Three Gorges Dam, because of its low hydration heat and excellent durability [6, 7].

Like conventional Portland cement, MgO as a minor component exists in LHC. MgO as an expansive additive is often used in dam construction to compensate for the slight natural shrinkage of PC during hydration, which can continue for months or years in service [8, 9]. As we all know, the free MgO in the form of the periclase crystal in clinker convert to $Mg(OH)_2$ with water, and the solid volume expansion increases 118% [10]. However, the presence of excessive amounts of MgO in the hydraulic concrete will lead to the volume expansion [4] and cause the crack formation of the dam. It is an effective and economic method to prevent the crack of dam by using the delayed expansion of periclase to compensate the thermal shrinkage of the concrete [11].

Chen et al. [12] and Lou et al. [13] found the characteristics of delayed expansion of the periclase, which indicated the expansion occurred mainly in the later period. The results of Mo et al. [14] showed that higher calcination temperature and longer residence time will cause grain growth of MgO, thus decrease in inner pore volume and specific surface area, thereby decreasing the hydration activity of MEA. Mehta et al. [15] thought that the expansion rate of the MgO in cement can be artificially adjusted by controlling the calcination temperature and particle size of the periclase. Song et al. [16] reported that periclase content increased with the increase of MgO content, periclase hydrated at a higher rate at early age, and the increase of curing temperature accelerated the hydration rate. There were results showing that increasing the calcination temperature and prolonging holding time had little effect on the whole content of f-MgO in high magnesium clinker, but the number of the larger size particles of the periclase crystal had increased [17, 18].

Periclase in Portland cement clinkers may cause unsound issues to concrete, and its content is often limited in many specifications of cements [19–21]. The expansion due to hydration of periclase in MHC clinkers [22] and especially in Mg-based expansive agents has been successfully used to compensate shrinkage due to decrease of temperature in dam concretes [23–25]. Then, there are many researchers to investigate the expansion control of MgO expansive agent [14, 26]. However, the expansion control of periclase in LHC has not been deeply studied.

In order to achieve the compensation effect on the temperature shrinkage of hydraulic concrete by using the delayed expansion of periclase in the LHC cement, LHC clinker was prepared under different calcination conditions, and the influence of crystal size of periclase on the hydration and expansion of the self-prepared low-heat cement was studied in this paper.

2. Materials and Methods

2.1. Materials. The low-heat Portland cement (LHC) clinkers used were prepared in the laboratory by calcining raw materials of limestone, silica, dolomite, aluminum ore, and copper slag that came from Sichuan Jiahua Special Cement Company. The gypsum used was from Huaneng Power Plant in Nanjing. Table 1 shows the chemical compositions of raw materials. XRD analysis of gypsum is shown in Figure 1. The raw materials were, respectively, calcined at 1350°C for 2.0 hours, 1400°C for 1.0 hour, 1400°C for 1.5 hours, 1400°C for 2.0 hours, 1450°C for 1.0 hour, and 1450°C for 2.0 hours. The theoretic mineral compositions of LHC clinker are shown in Table 2.

The obtained LHC clinkers were ground into powders with less than 10% sieve residue. 95% clinker powders and 5% gypsum were put into the mixing bucket, and the mixture was blended with 12 hours to obtain LHC.

2.2. Test Methods

2.2.1. Expansion Test of Cement Pastes. The cement pastes were prepared according to the JC/T 313-2009 (Chinese Standard). The fresh pastes were cast into a mould of 20 mm × 20 mm × 80 mm. The w/c ratio was set at 0.27. The hydration samples were made by casting the pastes into the mould of 20 mm × 20 mm × 20 mm while molding expansion specimens. The pastes with the mould were cured in a moist environment (98% relative humidity) at 20 ± 1°C for

24 ± 2 h, and then demoulded to measure the initial length L_0. The demoulded specimens were cured in water at 20°C, 30°C, 40°C, and 80°C, respectively. The length L_i of pastes were measured at specific intervals. The expansion of the cement pastes was calculated according to the formula given in Chinese Standard (JC/T 313-2009).

2.2.2. Determination of the Content of Periclase in LHC Pastes. The XRD (Smart Lab, Rigaku, Tokyo, Japan) internal standard method [27] was employed to determine the content of periclase in LHC clinkers and the hydrated paste samples. The internal standard substance and target component were ZnO (AR, $2\theta = 36.50°$) and periclase ($2\theta = 42.90°$). The pure phase periclase crystal was obtained by calcining magnesium carbonate basic at 1000°C for 1.0 hour. The content of ZnO was 1%, and the content of periclase was 1%, 2%, 4%, 6%, 8%, and 10%, respectively. Step scanning was conducted from 35° to 45°, using a step size of 0.02° and a scan speed of 1°/min. The XRD results were fitted by Jade6.0 to obtain the characteristic peak intensity. Using I_{MgO}/I_{ZnO} as abscissa axis and MgO percentage as ordinate axis to draw standard working curve, the equation is as follows:

$$Y = 0.49927 + 1.60929X\left(R^2 = 0.9968\right). \qquad (1)$$

The ovendry paste samples were ground into a powder through an 80 μm sieve and then well mixed with 1% ZnO. The hydrated paste was calcined at 950°C for 3.0 hours to remove the bound water to achieve the normalization of the quantification data.

2.2.3. Determination of Particle Size of Periclase in Clinkers and Analysis of Periclase in LHC Pastes. The median sizes of periclase in the LHC clinkers were got by the statistical methods based on the images of FE-SEM (Nava NanoSEM 450, FEI, Oregon State, USA). The FE-SEM was also used to analyse the hydration process of the periclase in cement paste. For BSEM image analysis, the dried slice samples of the cement clinker and the hydrated cement pastes were impregnated with epoxy resin in order to block in the hole. After the epoxy hardened, the samples were polished with 240, 400, 600, 2400, and 4000 grit abrasive paper and then polished using polishing liquid to the mirror surface by Automatic Grinder Polisher (EcoMet 250, Buehler, Illinois, USA).

3. Results and Discussion

3.1. Particle Size and Content of Periclase in LHC Clinkers. Figure 2 shows the BSEM images of LHC clinkers calcined at 1350°C for 2 hours, 1400°C for 1.0 hour, 1400°C for 1.5 hours, 1400°C for 2.0 hours, 1450°C for 1.0 hour, and 1450°C for 2.0 hours. According to the EDS analysis, the deep color particle is periclase in Figure 2. The shape of periclase in LHC clinkers was mostly polygonal or circular with clear boundaries, and it did not vary with calcination conditions changing. The median sizes and contents of periclase in LHC clinkers are shown in Table 3. The average particle size of

TABLE 1: Chemical compositions of raw materials (by weight, %).

Materials	LOI*	SiO_2	Al_2O_3	Fe_2O_3	CaO	MgO	K_2O	Na_2O	SO_3	Ratio
Limestone	41.76	4.03	0.72	0.50	50.29	1.65	0.10	0.02	0.04	67.04
Silica	1.66	87.16	2.41	5.02	1.59	1.52	0.12	0.09	0.07	10.23
Dolomite	43.00	5.04	1.11	0.66	29.60	19.80	0.20	0.08	0.03	15.09
Aluminum ore	14.72	30.48	28.36	21.64	1.86	1.46	0.73	0.24	0.00	5.77
Copper slag	0.00	26.58	6.57	45.80	12.37	5.78	0.24	0.08	0.06	1.87
Gypsum	6.36	2.88	0.32	0.25	33.14	0.33	—	—	42.54	—

*Loss on ignition.

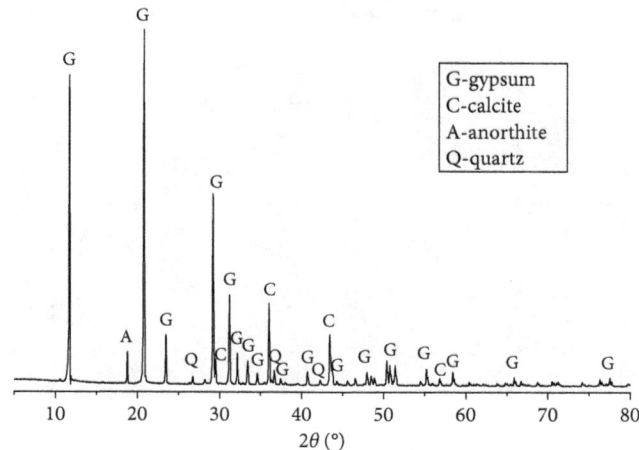

FIGURE 1: X-ray diffraction (XRD) pattern of gypsum.

TABLE 2: Theoretic mineral composition of LHC clinkers (%).

	C_3S	C_2S	C_3A	C_4AF	KH	SM	IM
Clinker	34.99	40.01	3.01	15.00	0.80	2.64	0.87

periclase in LHC clinkers calcined at 1400°C and 1450°C for 1.0 hour is 2.60 and 3.92 μm, respectively. The size of periclase in LHC clinkers calcined at 1350°C, 1400°C, and 1450°C for 2.0 hours is 2.94, 3.00, and 4.00 μm. Higher calcination temperature tends to increase the average particle size of periclase in LHC clinkers. The average particle size of periclase in LHC clinkers calcined at 1400°C for 1.0, 1.5, and 2.0 hours is 2.60, 2.80, and 3.00 μm, respectively. Longer holding time will enlarge the size of periclase in LHC clinkers. LHC clinkers calcined at 1350°C for 2.0 hours and at 1450°C for 2.0 hours contain the largest and smallest contents of periclase, being 4.15% and 3.56%. Higher calcination temperature or longer holding time tends to reduce slightly the content of periclase in LHC clinkers.

3.2. Expansion of LHC Pastes.

Figure 3 shows expansion of LHC pastes cured in water at 20°C, 30°C, 40°C, and 80°C. It can be seen from Figures 3(a), 3(b), and 3(c), at the same curing temperature, the expansion of cement pastes with 1400°C calcination temperature was the largest when the LHC clinker were calcined for 2.0 hours at 1350°C, 1400°C, and 1450°C, respectively. Both of cement pastes calcined at 1350°C and 1400°C for 2.0 hours and 1.0 hour began to

shrink when they were cured at 20°C for 200 days, and the expansion of cement pastes was −0.0102% and −0.00467%. While the cement clinkers were calcined at 1400°C for 1.0 hour, 1.5 hours, and 2.0 hours, respectively, the expansion of cement paste calcined at 1400°C for 2.0 hours cured at 20°C was bigger than others, the expansion of cement paste calcined at 1400°C for 1.5 hours cured at 30°C was larger than others, and the expansion of cement paste calcined at 1400°C for 1.0 hours cured at 40°C and 80°C was the largest. So it could be concluded that different calcination time periods had little effect on the expansion of LHC pastes. When cement paste calcined at 1400°C for 1 hour was cured in 40°C and 80°C water, the maximum expansion of the cement paste was up to 0.064% and 0.131%. At 20°C, 30°C, and 40°C curing conditions, the expansion of cement paste calcined at 1450°C for 2.0 hours was larger than that of cement paste calcined at 1450°C for 1.0 hour. The expansion of cement paste prepared at 1450°C for 1.5 hours was close to 0% when pastes were cured at 20°C for 240 days. Combining Table 3 and Figure 3, we can conclude that when the calcination temperature of the LHC clinker is lower, the cement paste has greater expansion with the smaller particle size of the periclase, and the LHC has higher content of periclase.

3.3. Hydration of Periclase in LHC Pastes

3.3.1. XRD Internal Standard Method for the Determination of Periclase.

Figure 4 listed the XRD patterns of the dried cement pastes of LHC with 2.60, 3.92, and 4.00 μm periclase

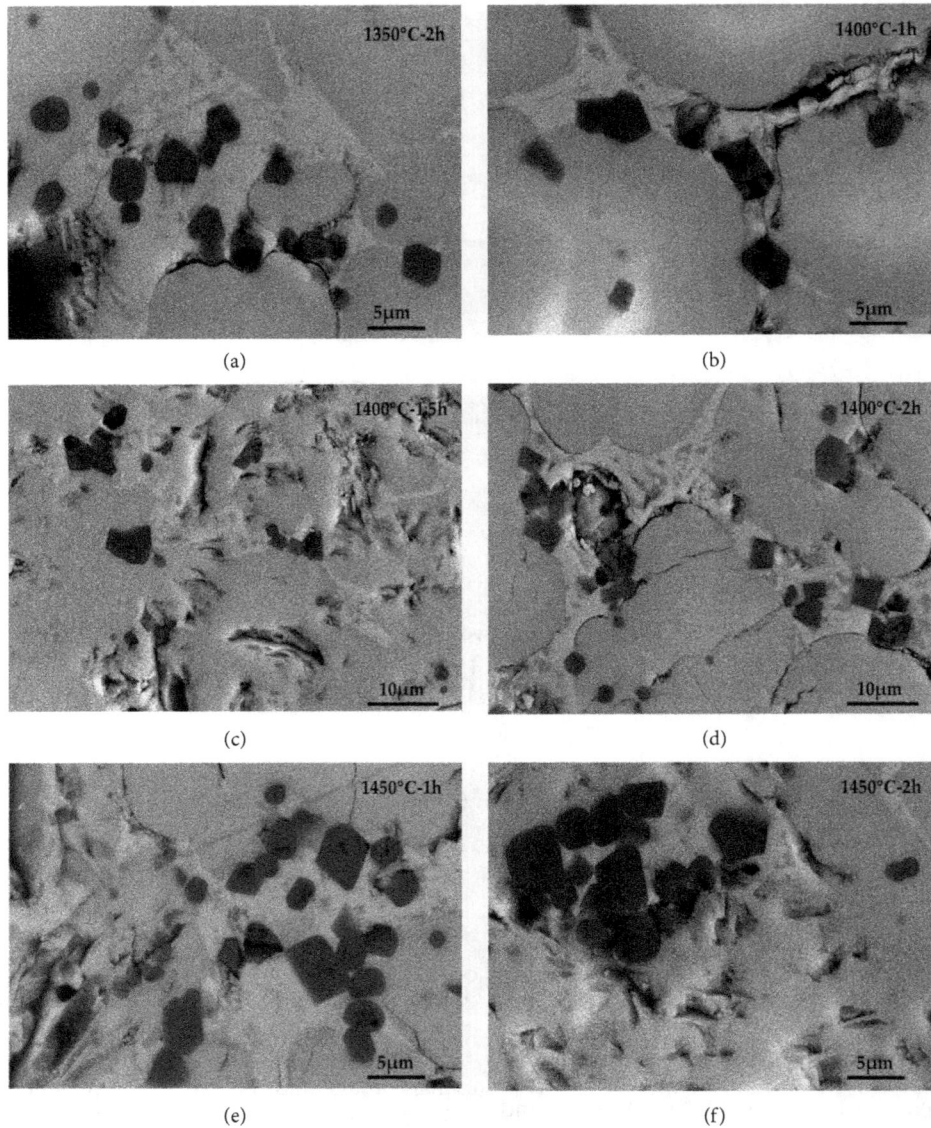

Figure 2: BSEM images of LHC clinkers calcined at 1350°C for 2.0 hours (a), 1400°C for 1.0 hour (b), 1400°C for 1.5 hours (c), 1400°C for 2.0 hours (d), 1450°C for 1.0 hour (e), and 1450°C for 2.0 hours (f) (the deep color particles represent periclase).

Table 3: The particle size and the content of periclase in LHC clinkers calcined at different conditions.

Calcination conditions	1.0 hour		2.0 hours			1400°C		
	1400°C	1450°C	1350°C	1400°C	1450°C	1.0 hour	1.5 hours	2.0 hours
Particle size of periclase (μm)	2.60	3.92	2.94	3.00	4.00	2.60	2.82	3.00
Content of periclase (wt./%)	3.91	3.72	4.13	3.65	3.56	3.91	3.81	3.65

size which were hydrated at 20°C, 40°C, and 80°C water for 240 days. According to Figure 4(a), it can be seen that there was no obvious diffraction peak of brucite generated until cement pastes were cured at 20°C for 240 days. As Figure 4(b) shows that the very obvious characteristic peak of brucite was observed when the cement paste with 4.00 μm periclase particle was cured at 40°C and 80°C for 240 days, and the characteristic peak intensity of $Mg(OH)_2$ became higher and higher with the increase of curing temperature.

The XRD internal standard method was used to determine the content of periclase in cement pastes. The hydration degrees of periclase are shown in Table 4. It can be seen from Table 4 that the periclase in cement pastes had not hydrated greatly when the cement pastes were cured at 20°C for 7 days, and the hydration degrees of periclase with particle sizes of 2.60, 3.92, and 4.00 μm were 0.27%, 6.80%, and 6.69%, respectively. Before cured for 90 days, the hydration degree of periclase in cement paste was smaller with

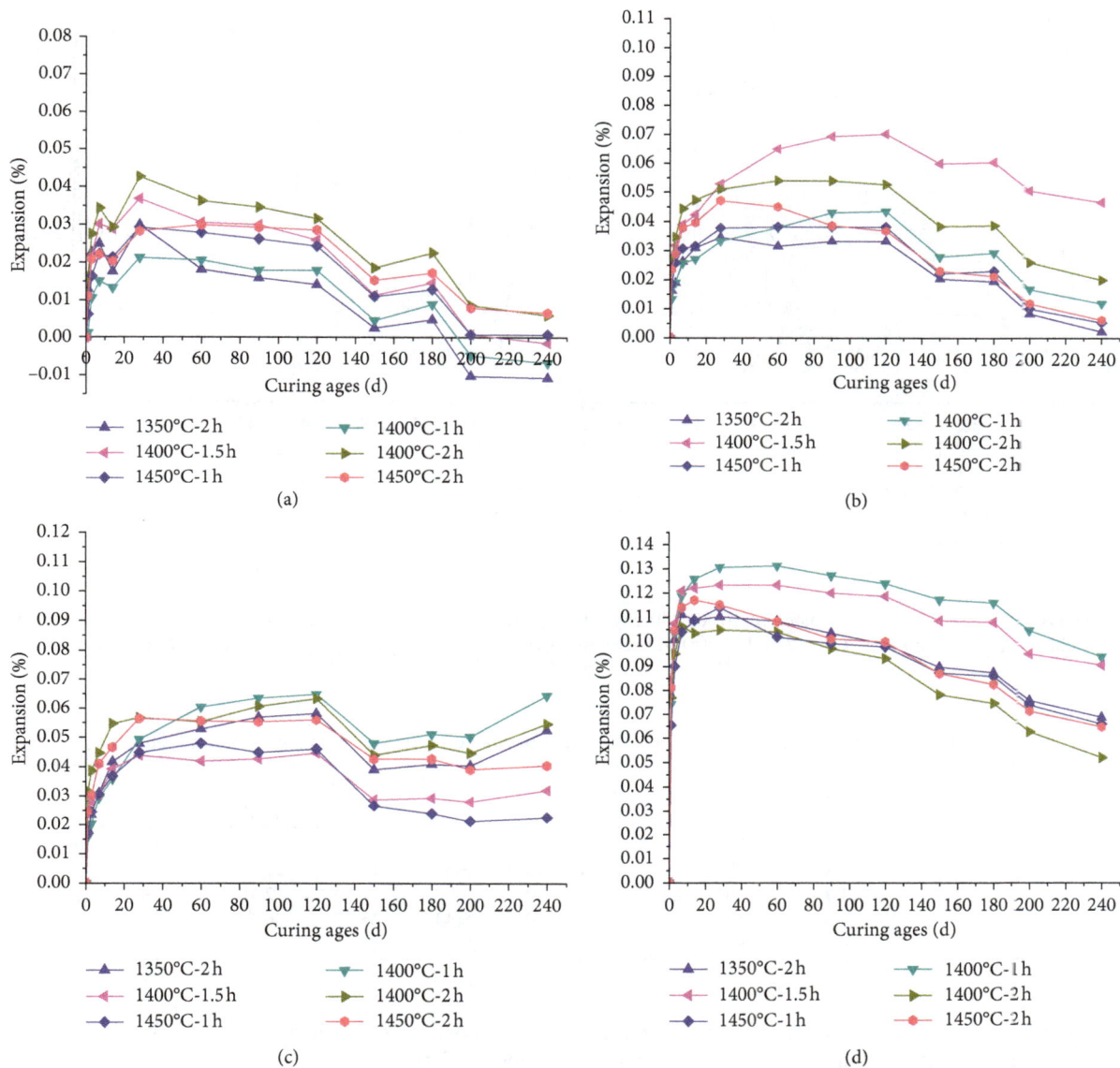

FIGURE 3: The expansion of LHC pastes prepared at different calcination conditions: (a) 20°C, (b) 30°C, (c) 40°C, and (d) 80°C.

the particle size of periclase decreasing. After 90 days, the periclase of 2.60 μm has the highest hydration degree although the periclase size is the smallest in three kinds of cement pastes. At the age of 240 days, the hydration degrees of periclase with particle sizes of 2.60, 3.92, and 4.00 μm were 30.73%, 24.36%, and 28.20%, respectively. When the LHC pastes were cured at 40°C and 80°C for 60 days, the hydration degrees of periclase with 4.00 μm particle size decreased rapidly with the delay of the curing ages, and the hydration degrees were 36.63% and 61.92%, respectively. After 60 days, the hydration degree increased a little with the prolongation of curing ages.

Combining Figure 4 and Table 4, the conclusion can be drawn that the hydration speed of periclase of cement paste cured at 20°C for 240 days was faster with the smaller particle size of the periclase in cement paste obtained at lower calcination temperature. The periclase hydration needed

much more time when cement paste was cured at 20°C because of the slow hydration speed. And the hydration degree of periclase with a particle size of 4.00 μm in cement paste cured at 20°C for 240 days was 28.20% which was much lower than that of periclase cured at 80°C with 67.15%. As mentioned in references [10, 14, 19], the hydration degree of periclase became smaller with the larger particle size of periclase and the increase of curing temperature.

3.3.2. BSEM Images of Cement Paste. BSEM images of cement paste with 4.00 μm periclase particles that cured at 20°C, 40°C, and 80°C for 60, 180, and 240 days, respectively, are shown in Figure 5. P represents the periclase particles, and B indicates the Mg(OH)$_2$ particles. Figure 5(a) shows that there was no obvious hydration marks around the periclase in the cement paste when it was cured at 20°C for 60

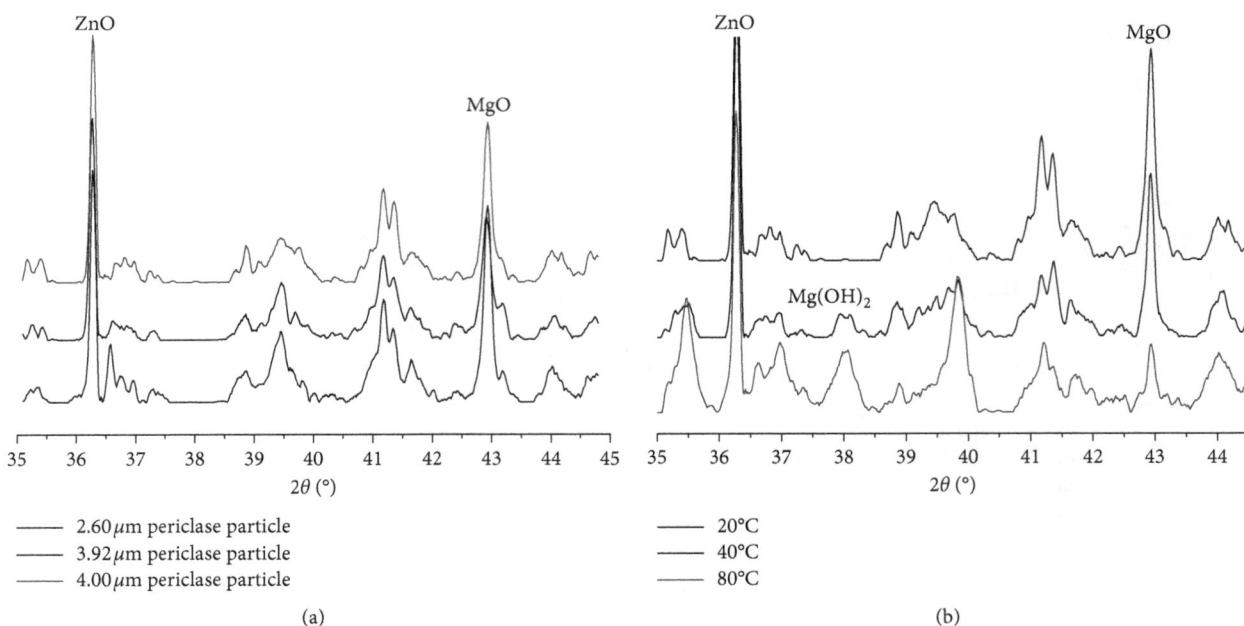

FIGURE 4: The XRD patterns of cement pastes cured for 240 days at 20°C and 80°C, respectively. (a) The cement pastes with 2.60, 3.92, and 4.00 μm periclase particle cured at 20°C for 240 days; (b) the cement paste with 4.00 μm periclase particle cured at 20°C, 40°C, and 80°C for 240 days.

TABLE 4: Hydration degree of periclase in LHC cement pastes cured for 7, 28, 60, 90, 180, and 240 d (%).

Particle size of periclase (μm)	Curing temperature, T (°C)	The hydration degree of periclase (%)						
		0 d	7 d	28 d	60 d	90 d	180 d	240 d
2.60	20	0.00	0.27	3.77	4.85	16.71	29.38	30.73
3.92	20	0.00	6.80	11.05	11.33	15.86	19.55	24.36
4.00	20	0.00	6.69	11.63	12.79	16.28	25.29	28.20
4.00	40	0.00	8.43	25.87	36.63	39.24	45.93	45.93
4.00	80	0.00	32.56	48.55	61.92	63.66	66.57	67.15

days, which indicated that the periclase did not react with water in the early curing ages. In addition, the periclase wrapped in the silicate phase had not hydrated.

From Figure 5(b), it can be seen that when the cement paste was cured at 20°C for 180 days, the color of a few regions around the periclase particles in the backscattered electron image was deepened. Thus, the periclase of cement paste reacted with water to produce a little $Mg(OH)_2$ that can be judged while the cement paste was cured at 20°C for 180 days. But the color of the surrounding area of periclase particles of the cement paste cured for 240 days was much deeper, and a small amount of brucite was generated at 20°C of water.

There was obvious production of $Mg(OH)_2$ around periclase in cement paste cured at 40°C for 240 days, which can be seen from Figure 5(d). The periclase in cement paste had hydrated a lot in the 80°C water-curing condition. The small-size periclase had hydrated completely, and a large amount of dense brucite was generated around the large particles of the periclase to form a protective layer around the periclase, which slowed down the hydration speed of periclase.

According to the aforementioned phenomenon, it can be concluded that the periclase hydrated slowly at 20°C. The periclase will hydrate quickly to produce more brucite with the increasing of curing ages and temperature.

4. Conclusions

In order to study the effect of the particle size of periclase on the expansion and hydration of low-heat cement, the expansion of cement paste, hydration degree of periclase, and particle size of periclase were evaluated by testing the length of specimens, XRD internal standard method, and BSEM, respectively. The main conclusions in this paper can be drawn as follows:

(1) The particle size of periclase was bigger with the holding time increasing at the same calcination temperature. When the residence time was kept consistent, the higher the calcination temperature of the cement linker was, the greater the size of the periclase particle was. And the particle size of periclase in the LHC clinker calcined at 1450°C for 2 h was the largest (4.00 μm).

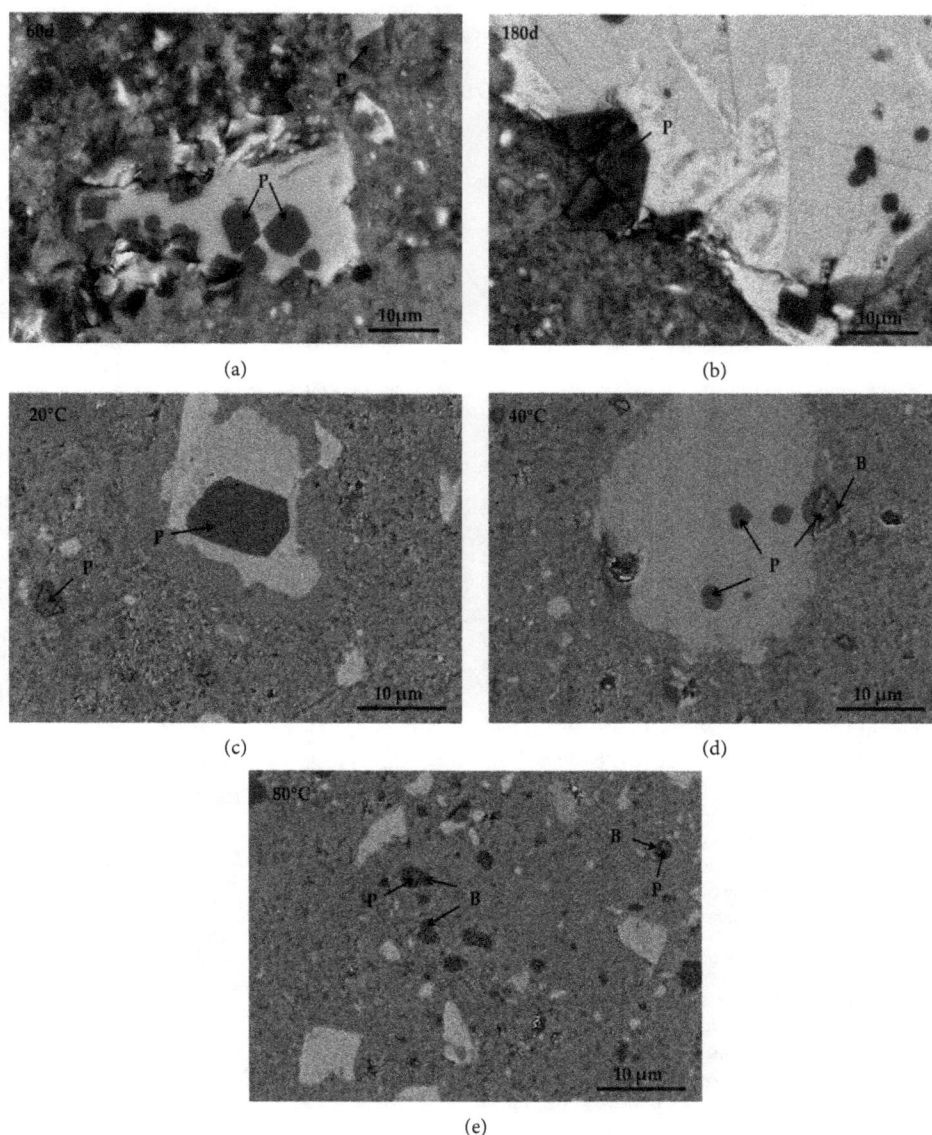

FIGURE 5: BSEM images of cement paste with 4.00 μm periclase particles cured at different temperatures for 60, 180, and 240 days. (a) Curing at 20°C for 60 days; (b) curing at 20°C for 180 days; (c) curing at 20°C for 240 days; (d) curing at 40°C for 240 days; (e) curing at 80°C for 240 days.

(2) The cement paste has greater expansion than others when the particle size of the periclase in LHC clinker became smaller. The expansion of cement paste with the 4.00 μm particle size of periclase was smaller than the others with the smaller periclase particle size when the cement paste was cured for 240 days at 20°C, 30°C, and 40°C, respectively. For the same curing age, the expansion of the cement pastes cured at 80°C was greater than that of pastes cured under other curing conditions. Shrinkage of the cement paste cured at 20°C water occurred first.

(3) The hydration speed of periclase in cement paste cured at 20°C for 240 days was higher with the smaller particle size of the periclase. And periclase needed much more time to hydrate when cement paste was cured at 20°C because of the slow hydration speed. The speed of periclase hydration was faster under 80°C than that at 20°C and 40°C, and the hydration degree of periclase with the 4.00 μm particle size was 32.56% under 80°C cured for 7 days while the periclase had not hydrated at 20°C cured for 7 days.

(4) The periclase hydrated a lot in the 80°C water-curing condition for 240 days. And the small-size periclase had hydrated completely, and a large amount of dense brucite was generated around the large particles of the periclase when the curing temperature was 80°C.

Conflicts of Interest

The authors declare no conflicts of interest.

Authors' Contributions

Man Yan and Chen Wang designed and conducted the experimental program. Min Deng provided and designed the project. Zhiyang Chen helped to do experiment and gave many writing suggestions. All authors contributed to the analysis and conclusion.

Acknowledgments

This work was supported by the State Key Laboratory of Materials-Oriented Chemical Engineering of Nanjing Tech University and granted financial resource from the National Key Research and Development Plan of China (No. 2016YFB0303601).

References

[1] T. B. Sui, S. H. Guo, and K. Z. Liu, "Research on high belite cement: Part II," in *Proceeding of 4th Beijing International Symposium on Cement and Concrete*, Beijing, China, October 1998.

[2] D. P. Chen, E. Sakai, M. Daimon, and Y. Ohba, "Carbonation of low heat portland cement paste precured in water for different time," *Journal of University of Science and Technology Beijing, Mineral, Metallurgy, Material*, vol. 14, no. 2, pp. 178–184, 2007.

[3] S. Nagaoka, M. Mizukoshi, and T. Kuroda, "Property of concrete using belite-rich cement ternary blended cement (special issue on structural materials)," *Journal of the Society of Materials Science, Japan*, vol. 43, no. 491, pp. 936–942, 1994.

[4] H. Q. Yang, Y. C. Wang, and S. H. Zhou, "Anti-crack performance of low-heat Portland cement concrete," *Journal of Wuhan University of Technology*, vol. 22, no. 3, pp. 555–559, 2007.

[5] Y. M. Yang, J. L. Cai, and Z. P. Chen, "Production quality control of moderate heat Portland cement and analysis of application effect in roller compacted concrete," *Applied Mechanics and Materials*, vol. 423–426, pp. 1090–1095, 2013.

[6] T. B. Sui, L. Fan, Z. J. Wen, and J. Wang, "Low energy and low emission cement with high performance and low hydration heat and its concrete application," *Materials China*, vol. 28, pp. 46–52, 2009.

[7] X. B. Wang and Z. J. Wen, "Low-heat Portland cement and its application in large hydropower projects," *Cement*, vol. 11, pp. 22–25, 2014.

[8] C. Du, "A review of magnesium oxide in concrete a serendipitous discovery in China leads to new concrete for dam construction," *Concrete International*, vol. 27, pp. 45–50, 2005.

[9] J. Feng, G. Zhang, X. Luo, and C. Zhang, "Modelling autogenous expansion for magnesia concrete in arch dams," *Frontiers of Architecture and Civil Engineering in China*, vol. 2, no. 3, pp. 211–218, 2008.

[10] L. W. Mo, M. Deng, M. S. Tang, and A. Al-Tabbaa, "MgO expansive cement and concrete in China: Past, present and future," *Cement and Concrete Research*, vol. 57, pp. 1–12, 2014.

[11] P. W. Gao, X. L. Lu, F. Geng et al., "Production of mgo-type expansive agent in dam concrete by use of industrial by-products," *Building and Environment*, vol. 43, no. 4, pp. 453–457, 2008.

[12] H. X. Chen, D. X. Li, Q. Ye, Y. Q. Wang, and Z. H. Lou, "Study of effects of two expansion sources of ettringite and periclase in cement," *Journal of the Chinese Ceramic Society*, vol. 28, pp. 1–5, 2000.

[13] Z. H. Lou, Q. Ye, H. X. Chen, Y. Q. Wang, and J. J. Shen, "Hydration of MgO in clinker and its expansion property," *Journal of the Chinese Ceramic Society*, vol. 26, pp. 430–436, 1998.

[14] L. W. Mo, M. Deng, and M. S. Tang, "Effects of calcination condition on expansion property of MgO-type expansive agent used in cement-based materials," *Cement and Concrete Research*, vol. 40, no. 3, pp. 437–446, 2010.

[15] P. K. Mehta, D. Pirtz, and G. J. Komandant, "Magnesium oxide additive for producing selfstress in mass concrete," in *Proceedings of 7th International Congress on the Chemistry of Cement*, Paris, France, 1980.

[16] Q. Song, Q. Zhang, and X. U. Delong, "Effect of magnesia in clinker on shrinkage and hydration of cement paste," *Journal of the Chinese Ceramic Society*, vol. 45, pp. 644–650, 2017.

[17] L. Y. Wei, S. H. Zhao, S. H. Liu, W. Lan, and X. M. Guan, "Effect of preparation conditions on the shape and distribution of MgO in clinker," *Journal of Wuhan University of Technology*, vol. 35, pp. 27–32, 2013.

[18] S. H. Zhao, L. Y. Wei, S. H. Liu, L. Wang, and X. M. Guan, "Effect of high temperature liquid phase on the formed of periclase in clinker," *Bulletin of the Chinese Ceramic Society*, vol. 33, pp. 1599–1603, 2014.

[19] P. K. Mehta, "Cement Standards-Evolution and Trends, ASTM STP 663," ASTM International, West Conshohocken, PA, USA, 1978.

[20] China Industrial Standard (CIS), *Common Portland Cements; GB/T 175-2007/XG 2-2009*, CIS, Beijing, China, 2009, in Chinese.

[21] China Industrial Standard (CIS), *Moderate-Heat Portland Cement, Low-Heat Portland Cement; GB/T 200-2017*, CIS, Beijing, China, 2017, in Chinese.

[22] Z. C. Ma, Y. Yao, X. B. Wang, L. Yun, Z. J. Wen, and X. Y. Xu, "Preparation and performance of high-magnesia moderate heat cement," *Journal of Wuhan University of Technology*, vol. 36, pp. 1–6, 2014.

[23] P. W. Gao, S. Y. Xu, X. Chen, J. Li, and X. L. Lu, "Research on autogenous volume deformation of concrete with MgO," *Construction and Building Materials*, vol. 40, pp. 998–1001, 2013.

[24] L. W. Mo, M. Deng, and A. Wang, "Effects of MgO-based expansive additive on compensating the shrinkage of cement paste under non-wet curing conditions," *Cement and Concrete Composites*, vol. 34, no. 3, pp. 377–383, 2012.

[25] L. L. Xu, M. Deng, and X. Zhao, "Compensating shrinkage of cement paste by new MgO-based expansive material," *Journal of Building Materials*, vol. 8, pp. 67–70, 2005.

[26] L. W. Mo, M. Liu, A. Al-Tabbaa, and M. Deng, "Deformation and mechanical properties of the expansive cements produced by inter-grinding cement clinker and MgOs with various reactivities," *Construction and Building Materials*, vol. 80, pp. 1–8, 2015.

[27] Y. M. Xu, L. L. Xu, and W. W. Li, "Quantitative determination of periclase and periclase hydration in high magnesium cement clinker with XRD internal standard method," *Journal of Nanjing Tech University*, vol. 36, pp. 30–35, 2014.

11

Effect of Using Micropalm Oil Fuel Ash as Partial Replacement of Cement on the Properties of Cement Mortar

Kwangwoo Wi,[1] Han-Seung Lee,[1] Seungmin Lim [ID],[1] Mohamed A. Ismail,[2] and Mohd Warid Hussin[3]

[1]School of Architecture and Architectural Engineering, Hanyang University, Ansan, Gyeonggi-do, Republic of Korea
[2]Department of Civil Engineering, Miami College of Henan Univerisity, Kaifeng, Henan, China
[3]Construction Research Centre (UTM CRC), Institute for Smart Infrastructure and Innovative Construction, Universiti Teknologi Malaysia, 81310 UTM, Johor Bahru, Johor, Malaysia

Correspondence should be addressed to Seungmin Lim; smlim09@hanyang.ac.kr

Academic Editor: Wei Zhou

This study investigates the effects of micropalm oil fuel ash (mPOFA) on compressive strength and pore structure of cement mortar. Various experimental techniques, such as compression test, isothermal calorimetry, mercury intrusion porosimetry, and X-ray diffraction, are performed to figure out the effect of using mPOFA as partial replacement of cement on the hydration of cement and determine its optimal replacement level to increase mechanical property of the mortar specimens. 10 wt.% of cement replacement with mPOFA is found to give the highest level of compressive strength, achieving a 23% increase over the control specimens after 3 days of curing. High K_2O contents in mPOFA stimulate C_3S in cement to form C-S-H at early ages, and high surface area of mPOFA acts as a nucleus to develop C-S-H. Also, small mPOFA particles and C-S-H formed by pozzolanic reaction fill the pores and lead to reduction in large capillary pores. In XRD analysis, a decrease in $Ca(OH)_2$ and SiO_2 contents with age confirmed a high pozzolanic reactivity of mPOFA.

1. Introduction

The production of cement consumes an enormous amount of raw materials and energy [1]. 3.2–6.3 GJ of energy and 1.7 tons of raw materials are consumed for every ton of clinker, resulting in the emission of approx. 850 kg of CO_2 [2, 3]. As a result, cement production currently accounts for about 7% of global CO_2 emissions [4], and this rate is expected to be increased, which demands to limit the production of cement clinker. This leads many scientific studies to be conducted for the development of alternative construction materials [5, 6] by utilizing industrial residues or agricultural residues [7]. Industrial residues, such as silica fume and fly ash, are widely used as supplementary cementitious materials (SCMs) due to their high pozzolanic reactivities [8–11]. There are also continuous efforts to develop agriculturally sourced pozzolanic materials, such

as rice husk ash (RHA) [12, 13], bamboo leaf ash (BLA) [14], corn cob ash (CCA) [15, 16] and palm oil fuel ash (POFA) [6,17–20].

As shown in Figure 1, POFA is generated by the combustion of agricultural residues, such as palm fibers and palm kernel shells in biomass thermal power plants [21]. Following the combustion, around 5% of the total solid is turned into POFA, which consists mainly of silica and alumina [22]. However, most of POFA is dumped as waste, which leads to soil contamination and other associated environmental problems [23]. Furthermore, the use of POFA is limited in cement and concrete industries because the surface area available for a reaction is too small when it is incorporated into cement-based composite materials. There are few studies [24, 25] that investigated how to increase the fineness of POFA, but the increased volume of plasticizer for the large amount of unburned carbon raises a problem [26].

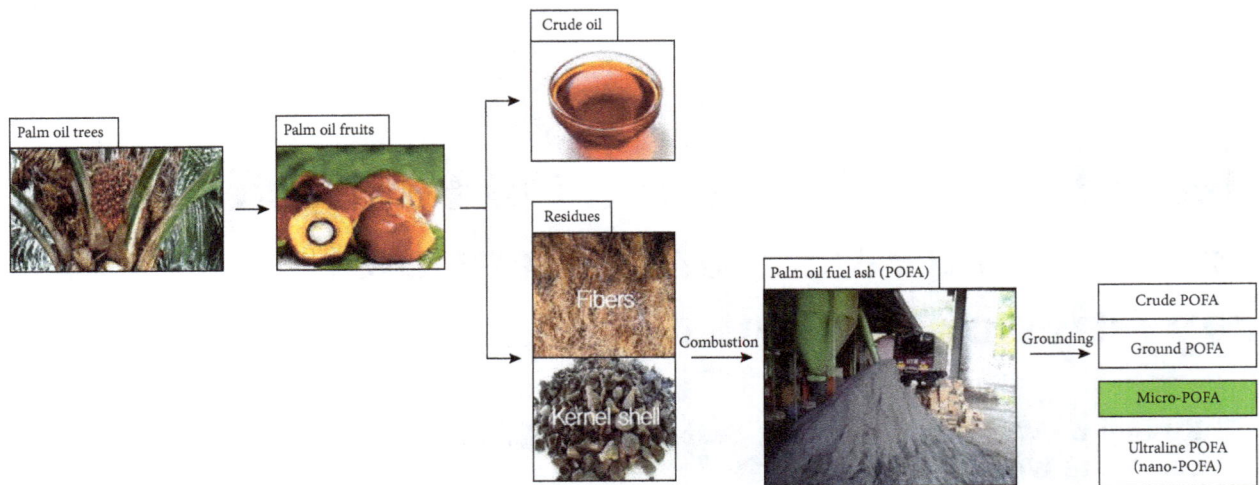

FIGURE 1: Production of palm oil fuel ash (POFA).

Further improvement of the fineness through an additional grinding process can solve the issue, after which the unburned carbon can be removed by drying the POFA in an oven at 500°C [25, 27].

Recent studies about POFA mostly focus on the effects of POFA on the durability and general properties of cement-based composite materials, such as the compressive strength of mortar and concrete [28, 29], the expansion rate [30], the heat of hydration [31–33], and the resistance to sulfate attack [34]. However, there is still lack of studies about the hydration properties and reactivity of very fine, micro-POFA (mPOFA) in a cement matrix. This study examines the hydration properties and reactivity of mPOFA in a cement matrix to determine the optimal replacement rate.

2. Experimental Program

2.1. Materials

2.1.1. Cement. Type I Ordinary Portland cement (OPC) was used in this study, which is defined by KS L 5201 [35]. Its physical properties and chemical composition are summarized in Table 1.

2.1.2. Micropalm Oil Fuel Ash (mPOFA). POFA was produced by the combustion of palm oil kernel shells and palm oil fibers at a palm oil mill, which is located in Johor, Malaysia. This POFA powder was dried in an oven at 110 ± 5°C for 24 h and then passed through a 300 μm sieve to remove coarse particles. The under-sized powder was further ground until 90% passed through a 45 μm sieve according to ASTM C618 [36]. This finely ground POFA is hereafter referred to as mPOFA. Its physical properties and chemical composition are shown in Table 1.

Figure 2 is an SEM image of mPOFA particles. It shows that mPOFA particles have sharp angles and irregular shapes as a result of being reduced in size by ball milling. There are also pores on the surface of mPOFA particles, which formed during the additional grinding [37].

TABLE 1: Physical properties and chemical composition of cement and mPOFA.

Chemical composition	Cement (%)	mPOFA (%)
SiO_2	19.47	70.9
Al_2O_3	5.24	5.63
Fe_2O_3	2.69	3.51
CaO	61.8	3.78
MgO	3.72	3.61
SO_3	2.49	—
Na_2O	0.18	0.39
K_2O	0.87	5.66
LOI*	2.6	10.1
Physical properties		
Specific gravity (kg/m^3)	3.15	2.3
Specific surface area (cm^2/g)	3578	6091

*LOI = loss on ignition.

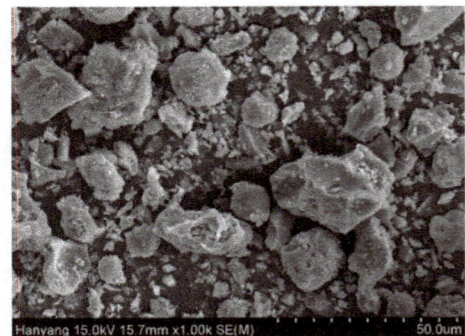

FIGURE 2: SEM image of mPOFA.

The XRD result of the mPOFA is presented in Figure 3. It confirms that mPOFA is a Si-rich material with main crystalline phases α-quartz and cristobalite. This XRD pattern corresponds to the XRD results in the literature [1, 8, 26], which reported that POFA has about 70% amorphous content through semiquantitative and quantitative XRD analysis using the Rietveld method [27, 28].

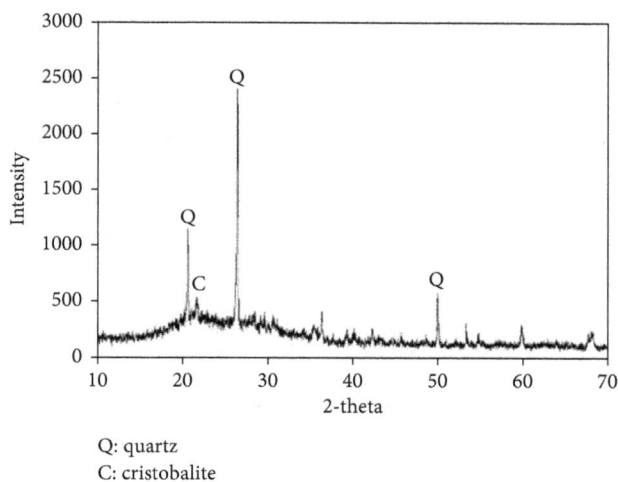

Q: quartz
C: cristobalite

FIGURE 3: XRD pattern of mPOFA.

Thus, the mPOFA used in this study is expected to contain a similar amount of amorphous content.

2.1.3. Fine Aggregates. Siliceous sand was used as fine aggregates. Its physical properties are shown in Table 2.

2.2. Mix Proportions. Table 3 shows the mix proportion of mortar samples, which were prepared according to KS L ISO 679 [38]. Each sample was prepared by replacing 0, 10, 20, 30, or 40 wt.% of cement with mPOFA at a water-to-binder (cement + mPOFA) ratio of 50%. After casting, each mortar sample was cured at 20°C for 24 h prior to demolding. The samples were then cured in water at 20 ± 1°C until reaching the age for each test. A separate set of samples without sand, cement paste, was prepared under the same conditions for X-ray diffraction (XRD) and scanning electron microscopy (SEM).

2.3. Test Methods

2.3.1. Compressive Strength. The compressive strength was measured on prismatic mortar specimens with 40 mm × 40 mm × 160 mm in accordance with KS L ISO 679 using the universal testing machine (UTM) after 3, 7, 14, and 28 days of curing. Testing was carried out on six specimens at each age, and the average value was reported.

2.3.2. Heat of Hydration. The heat of hydration was monitored for 72 h according to ASTM C1702 [39] (external mixing), using an isothermal conduction calorimeter (MMC-511SV, Tokyo Rico., LTD.). The samples were externally mixed before inserting them in the respective channels. It is noted that the data for the first 45 min were not considered for analysis to avoid the heat associated with mixing and lacing the externally prepared samples and to allow them to stabilize with the set temperature of 22°C.

2.3.3. Pore Structure. Small pieces (approx. 2 g in total) of mortar samples were soaked in acetone for 24 hours in a vacuum desiccator to stop hydration after 3 and 28 days of curing and then dried in an oven at 60°C for 24 h to remove any water. The pore structure of each sample was then examined using MIP (Autopore IV 9520) with maximum applied pressure of mercury, 400 MPa.

2.3.4. Analysis of Hydration Products. The crystal phases present in cement paste samples cured for 3 and 28 days were analyzed using XRD with a 2θ range between 10° and 70° at a scan rate of 5°/min.

3. Results and Discussion

3.1. Compressive Strength of Cement Mortars Containing mPOFA. Figure 4 shows the results of the compressive strength of mortar specimens at varying curing times and mPOFA contents. The specimens with an incorporation of 10% mPOFA by weight (POFA10) consistently produce the highest compressive strength, with a 33% increase over the control at 28 days of curing. The compressive strength of POFA20 was also consistently higher than the control specimen at all ages; however, there was a decrease in compressive strength if the replacement rate of mPOFA was increased any further.

The results show that the optimal replacement rate of mPOFA is 10% by weight, and the maximum replacement rate of mPOFA is 20% by weight. When mPOFA is replaced with cement at certain levels, the small mPOFA particles fill the pores between cement particles, thereby affecting the development of compressive strength [28, 34]. It is thought that the small particle size, silica content, and amorphous nature of mPOFA enabled it to act as nuclei in the cement matrix, which in turn stimulates the formation of C-S-H gel [19]. mPOFA is also a pozzolanic material, so $Ca(OH)_2$ that is formed by the cement hydration undergoes a pozzolanic reaction with the silica in mPOFA to produce secondary C-S-H [1, 27, 28]. However, the strength development worsened steadily at higher levels of mPOFA. By the replacement of cement with 40% of mPOFA, compressive strength at 28 days decreases by 77% in comparison with the control. Less cement content in specimens with higher levels of mPOFA could result in the reduction in compressive strength.

3.2. Heat of Hydration. Figure 5 shows the heat evolution rate of samples measured for 72 h after the initial mixing. When cement particles contact with water, it begins to generate heat (Stage 1); furthermore, it passes through a dormant period (Stage 2). After a few hours, C_3S begins to react actively and the heat evolution rate reaches a peak, referred to as second peak, at the end of the acceleration period (Stage 3) [40].

The heat evolution of POFA10 was the highest in Stage 3, with the peak starting earlier than with the control. This is thought to be caused by the alkali ions of mPOFA (K^+ and Na^+) promoting hydrolysis by stimulating C3S and forming

TABLE 2: Physical properties of fine aggregates.

Maximum size (mm)	Surface dry density (g/cm³)	Absolute volume (%)	Fineness modulus (F.M)	Absorption rate (%)	Specific weight (kg/m³)
5.0	2.6	61.2	2.87	1.02	1,590

TABLE 3: Mix proportion of control mortar and mortar specimens containing mPOFA.

Sample	W/B (%)	Replacement ratio (%)	Cement (g)	POFA (g)	Sand (g)	Water (g)
Control	50	0	450	—	1350	225
POFA10*	50	10	405	45	1350	225
POFA20	50	20	360	90	1350	225
POFA30	50	30	315	135	1350	225
POFA40	50	40	270	180	1350	225

*POFA10 = replacement ratio of cement with mPOFA is 10%.

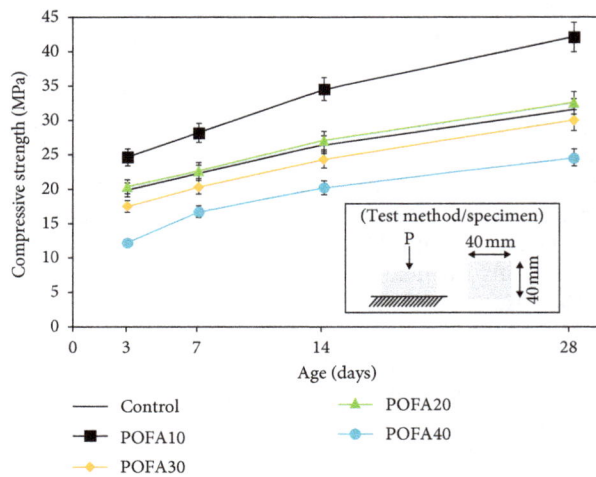

FIGURE 4: Compressive strength of mortars containing mPOFA.

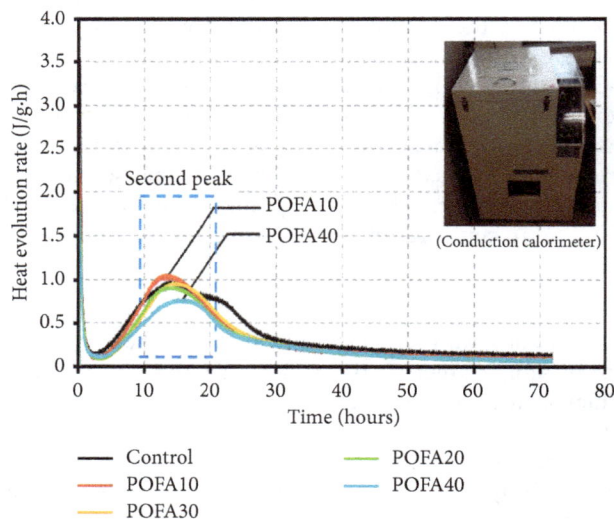

FIGURE 5: Heat evolution rate of mortars containing mPOFA.

C-S-H during the initial period [41, 42]. In addition, an increase in the mPOFA replacement rate decreased the size of the peak in Stage 3 and delayed the time of its occurrence [31].

Figure 6(a) shows the cumulative heat evolution over time, and Figure 6(b) shows the relative cumulative heat evolution according to the mPOFA contents, compared with the control. This reveals that the control had a higher cumulative heat (267 J/g) than POFA10 (233 J/g) and POFA40 (184 J/g), which confirmed that the cumulative heat decreases when the mPOFA content is increased. This result agrees well with that in previous research [33], which reported that 40% mPOFA replacement reduced the cumulative heat by 31% compared with the control. Again, high K_2O content of mPOFA affects the hydration of cement through the stimulation of C_3S. However, in view of the total amount of hydration heat, samples containing mPOFA showed lower cumulative heat evolution. This is because the total amount of cement for releasing heat lacked with the increasing replacement rate of mPOFA.

3.3. Pore Structure of Cement Mortar Containing mPOFA.
Figure 7 shows the pore size distribution of mortar samples, in which one can see that large capillary pores (0.05–10 μm in size) in POFA10 and POFA20 are either less or equal in quantity to those in the control after 3 days of curing. In contrast, POFA30 and POFA40 had either more or an equal number of large capillary pores as the control. Consequently, it is thought that the mPOFA particles effectively fill the pores in mortar specimens with 10–20% mPOFA replacement [28], which is consistent with their greater compressive strength.

After curing for 28 days, the amounts of pores larger than 0.1 μm in diameter decreased in all samples compared with 3 days of curing, but the pores smaller than 0.1 μm increased in all samples except the control. This is attributed to the formation of hydration products in the large capillary pores, which increases the number of midsized (0.01–0.5 μm diameter) capillary pores. The cumulative pore volume of the mortar samples containing mPOFA is shown in Figure 8. The volume of pores 0.1 μm in diameter or larger increased with mPOFA replacement at 3 days of curing. In addition, the gel pores of 0.01 μm in diameter or less were only observed in mortars with 10 or 20% mPOFA replacement. With 28 days of curing, the cumulative pore volume of most of the samples decreased, and the volume of pores greater

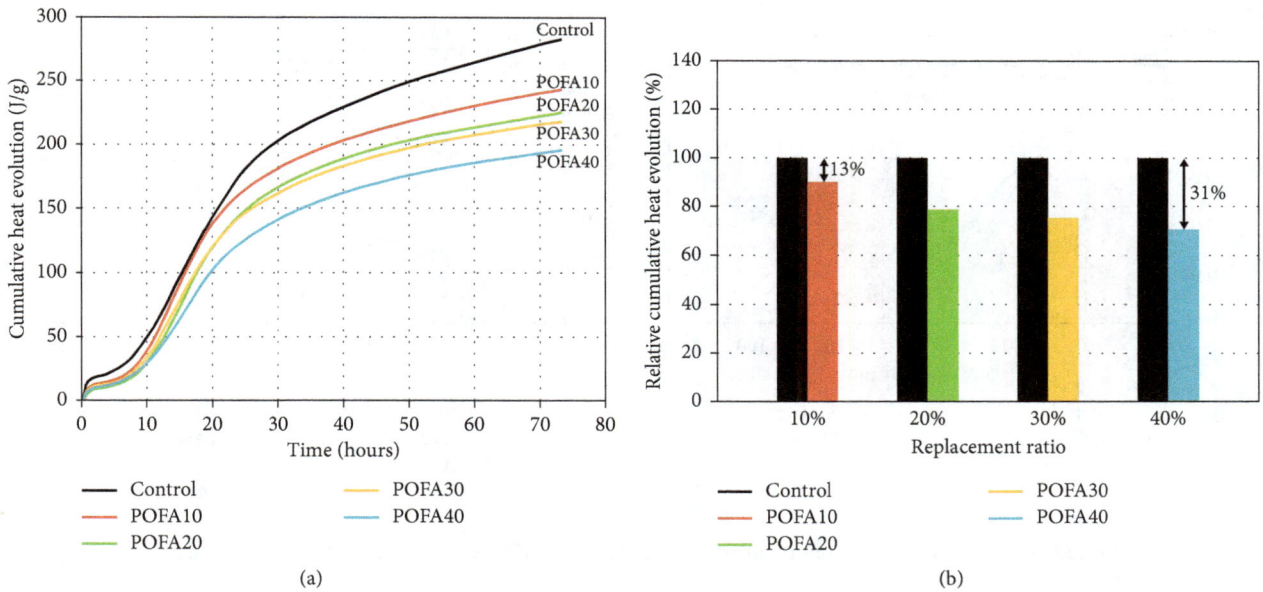

FIGURE 6: (a) Cumulative heat evolution of mortars containing mPOFA. (b) Relative cumulative heat evolution of mortars containing mPOFA.

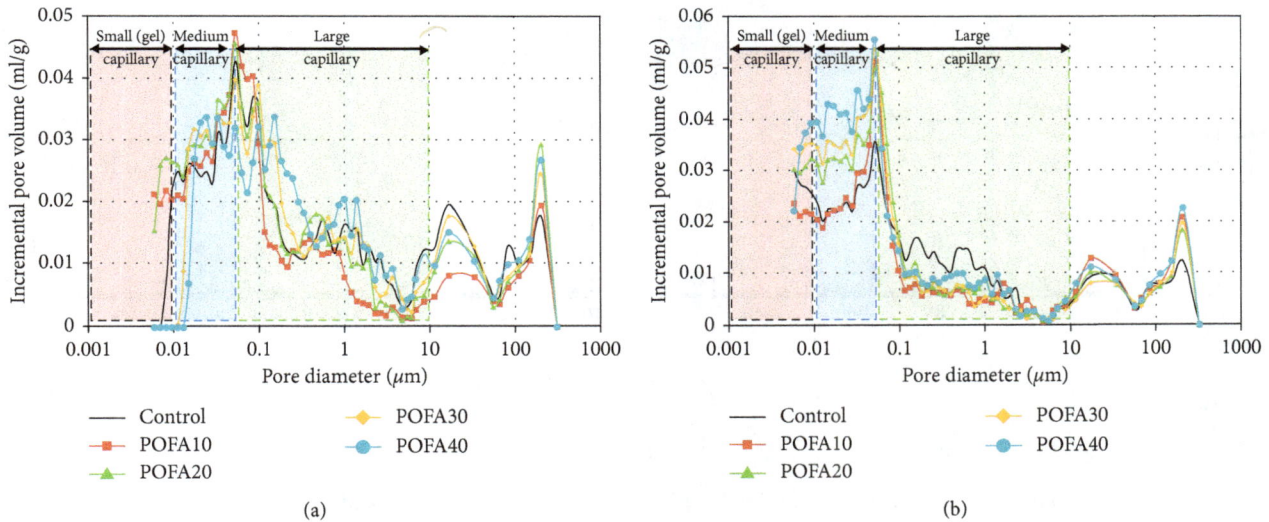

FIGURE 7: Pore distribution of mortars containing mPOFA aged for (a) 3 and (b) 28 days.

than 0.1 μm in diameter was greater in the control than that in the samples with mPOFA.

The pore size distribution of concrete is closely related to its permeability which governs durability. The pores can be categorized as gel, medium, large capillary pores, or air voids depending on their size [40]. The gel pores, which constitute the internal porosity of C-S-H, were observed in POFA10 and POFA20 after 3 days of curing. It is thought that mPOFA affects the development of C-S-H at early ages.

3.4. Phase Analysis of Cement Paste Containing mPOFA.
The XRD results are shown in Figure 9. Ca(OH)$_2$ and SiO$_2$ peaks were observed in all samples at 3 days of curing. As the

replacement level of mPOFA increased, the Ca(OH)$_2$ peak decreased and the SiO$_2$ peak increased owing to the replacement of cement with mPOFA; that is, the volume of Ca^{2+} needed to produce Ca(OH)$_2$ decreased because mPOFA is predominantly silica. After 28 days of curing, Ca(OH)$_2$ and SiO$_2$ peaks were again observed in all samples; however, the peak of Ca(OH)$_2$ decreased compared with 3 days. This is because Ca(OH)$_2$ and SiO$_2$ were consumed to produce C-S-H by pozzolanic reaction.

3.5. Role of mPOFA in Cement Hydration.
A schematic diagram of the role of mPOFA in a cement hydration is provided in Figure 10, which is drawn based on the

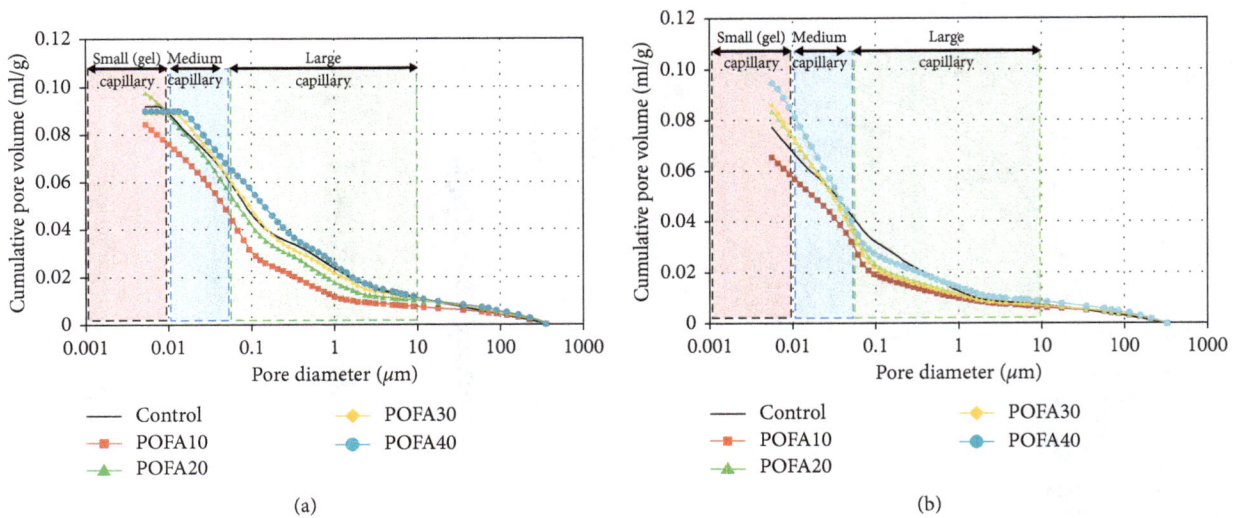

FIGURE 8: Cumulative pore volume of mortars containing mPOFA aged for (a) 3 and (b) 28 days.

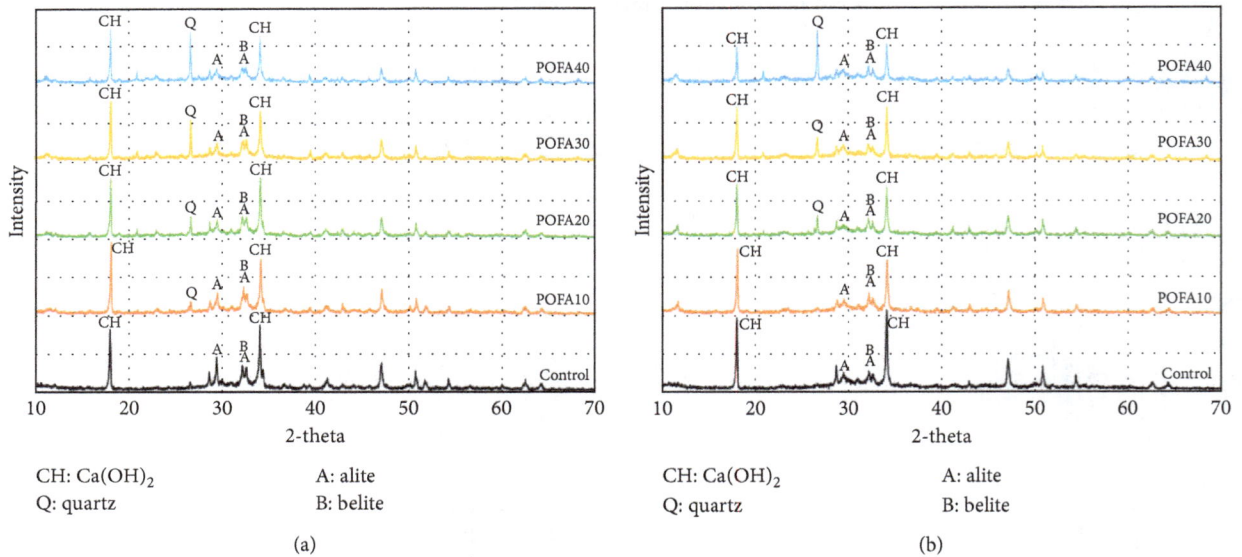

FIGURE 9: XRD patterns of pastes containing mPOFA aged for (a) 3 and (b) 28 days.

experimental results. When mPOFA particles contact with water, alkali ions (K^+, Na^+), which weakly bound to mPOFA, are easily eluted, thereby making the mPOFA surface negatively charged with electrons. The eluted alkali ions subsequently stimulate the hydrolysis of C_3S in cement clinker, which increases the concentration of Ca^{2+} ions in solution [43] as confirmed by isothermal calorimetry.

As the elution of alkali ions is the ongoing process, the concentration of alkali ions in solution steadily increases and a Si-rich zone is formed around the surface of the mPOFA particle. Over time, SiO_4^{4-} ions in the Si-rich zone are eluted by osmotic pressure and then react with Ca^{2+} to form C-S-H. It is thought that mPOFA particles play a role in this process by providing a nucleus for C-S-H growth due to their high specific surface area [44]. After further curing, $Ca(OH)_2$ and

SiO_2 were consumed to produce C-S-H through pozzolanic reaction, which probably blocks the pores and thus makes a dense microstructure.

4. Conclusions

This study examines the effects of an agricultural waste, mPOFA, on changes in compressive strength and pore structure of cement mortar. An optimal replacement rate, 10 wt.% of cement with mPOFA is determined for maximizing its utility as a supplementary cementitious material. Based on various experimental results, the following conclusions are drawn.

A 10 wt.% replacement of cement with mPOFA yields the highest compressive strength at all ages, with a 33% increase in strength over the control at 28 days. This could be

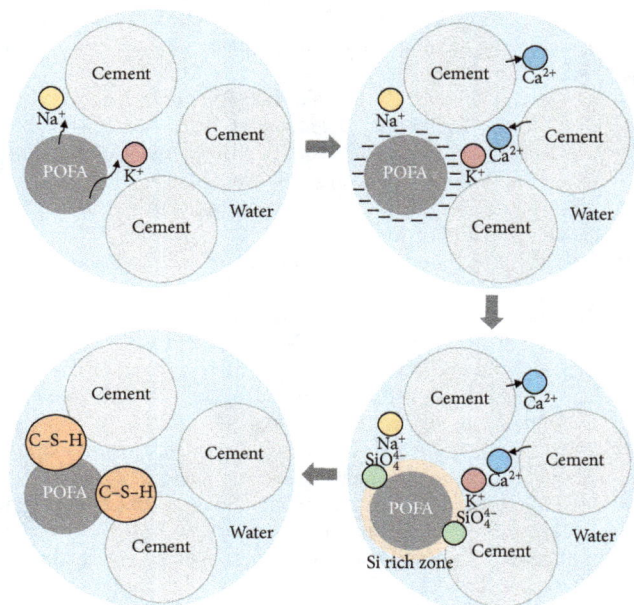

FIGURE 10: Role of mPOFA in cement hydration.

due to mPOFA filling the space between cement particles and promoting the formation of C-S-H.

Increasing the amount of mPOFA reduces the compressive strength, but the difference in strength between POFA40 and the control decreased by 27% with a prolonged curing of 28 days. This is attributed to the Si-rich mPOFA particles reacting with $Ca(OH)_2$ formed during cement hydration to produce secondary C-S-H.

The rate of heat evolution generally decreased with increasing mPOFA content, but POFA10 exhibited a higher rate and earlier peak than the control in Stage 2. This could be the result of alkali ions eluted from mPOFA stimulating the hydrolysis of C_3S.

The incorporation of 10% and 20% mPOFA produced fewer large capillary pores, which is most likely the result of filling the spaces between cement particles by mPOFA particles and newly formed C-S-H.

High levels of K_2O in mPOFA stimulate the hydrolysis of C_3S in cement, and mPOFA particles with high surface area acts as nuclei for hydration. Consequently, pores are probably filled with C-S-H formed by pozzolanic reaction, which makes a dense microstructure and refines the pore structure.

Conflicts of Interest

The authors declare no conflict of interest.

Acknowledgments

This research was supported by basic science research program through the National Research Foundation (NRF) of Korea funded by the Ministry of Science, ICT and Future Planning (No. 2015R1A5A1037548).

References

[1] T. Sinsiri, W. Kroehong, C. Jaturapitakkul, and P. Chindaprasirt, "Assessing the effect of biomass ashes with different finenesses on the compressive strength of blended cement paste," *Materials and Design*, vol. 42, pp. 424–433, 2012.

[2] H. G. Oss and A. C. Padovani, "Cement manufacture and the environment: part I: chemistry and technology," *Journal of Industrial Ecology*, vol. 6, no. 1, pp. 89–105, 2002.

[3] H. G. Oss and A. C. Padovani, "Cement manufacture and the environment part II: environmental challenges and opportunities," *Journal of Industrial Ecology*, vol. 7, no. 1, pp. 93–126, 2003.

[4] V. M. Malhotra, "Global warming, and role of supplementary cementing materials and superplasticisers in reducing greenhouse gas emissions from the manufacturing of portland cement," *International Journal of Structural Engineering*, vol. 1, no. 2, pp. 116–130, 2010.

[5] V. Sata, C. Jaturapitakkul, and K. Kiattikomol, "Influence of pozzolan from various by-product materials on mechanical properties of high-strength concrete," *Construction and Building Materials*, vol. 21, no. 7, pp. 1589–1598, 2007.

[6] E. Aprianti, P. Shafigh, S. Bahri, and J. N. Farahani, "Supplementary cementitious materials origin from agricultural wastes–A review," *Construction and Building Materials*, vol. 74, pp. 176–187, 2015.

[7] E. Aprianti, "A huge number of artificial waste material can be supplementary cementitious material (SCM) for concrete production–a review part II," *Journal of Cleaner Production*, vol. 142, pp. 4178–4194, 2017.

[8] S. Chatterji, N. Thaulow, and P. Christensen, "Puzzolanic activity of byproduct silica-fume from ferro-silicom production," *Cement and Concrete Research*, vol. 12, no. 6, pp. 781–784, 1982.

[9] P. K. Mehta and O. E. Gjørv, "Properties of portland cement concrete containing fly ash and condensed silica-fume," *Cement and Concrete Research*, vol. 12, no. 5, pp. 587–595, 1982.

[10] S. Diamond, "The utilization of flyash," *Cement and Concrete Research*, vol. 14, no. 4, pp. 455–462, 1984.

[11] P. Chindaprasirt, C. Jaturapitakkul, and T. Sinsiri, "Effect of fly ash fineness on microstructure of blended cement paste," *Construction and Building Materials*, vol. 21, no. 7, pp. 1534–1541, 2007.

[12] G. R. De Sensale, "Effect of rice-husk ash on durability of cementitious materials," *Cement and Concrete Composites*, vol. 32, no. 9, pp. 718–725, 2010.

[13] M. F. M. Zain, M. N. Islam, F. Mahmud, and M. Jamil, "Production of rice husk ash for use in concrete as a supplementary cementitious material," *Construction and Building Materials*, vol. 25, no. 2, pp. 798–805, 2011.

[14] E. Villar-Cociña, E. V. Morales, S. F. Santos, H. Savastano, and M. Frías, "Pozzolanic behavior of bamboo leaf ash: characterization and determination of the kinetic parameters," *Cement and Concrete Composites*, vol. 33, no. 1, pp. 68–73, 2011.

[15] D. A. Adesanya and A. A. Raheem, "Development of corn cob ash blended cement," *Construction and Building Materials*, vol. 23, no. 1, pp. 347–352, 2009.

[16] D. A. Adesanya and A. A. Raheem, "A study of the permeability and acid attack of corn cob ash blended cements," *Construction and Building Materials*, vol. 24, no. 3, pp. 403–409, 2010.

[17] W. Kroehong, T. Sinsiri, and C. Jaturapitakkul, "Effect of palm oil fuel ash fineness on packing effect and pozzolanic reaction of blended cement paste," *Procedia Engineering*, vol. 14, pp. 361–369, 2011.

[18] N. H. A. S. Lim, M. A. Ismail, H. S. Lee, M. W. Hussin, A. R. M. Sam, and M. Samadi, "The effects of high volume nano palm oil fuel ash on microstructure properties and hydration temperature of mortar," *Construction and Building Materials*, vol. 93, pp. 29–34, 2015.

[19] M. A. A. Rajak, Z. A. Majid, and M. Ismail, "Morphological characteristics of hardened cement pastes incorporating nano-palm oil fuel ash," *Procedia Manufacturing*, vol. 2, pp. 512–518, 2015.

[20] M. M. U. Islam, K. H. Mo, U. J. Alengaram, and M. Z. Jumaat, "Mechanical and fresh properties of sustainable oil palm shell lightweight concrete incorporating palm oil fuel ash," *Journal of Cleaner Production*, vol. 115, pp. 307–314, 2016.

[21] W. Tangchirapat, C. Jaturapitakkul, and P. Chindaprasirt, "Use of palm oil fuel ash as a supplementary cementitious material for producing high-strength concrete," *Construction and Building Materials*, vol. 23, no. 7, pp. 2641–2646, 2009.

[22] J. H. Tay and K. Y. Show, "Use of ash derived from oil-palm waste incineration as a cement replacement material," *Resources, Conservation and Recycling*, vol. 13, no. 1, pp. 27–36, 1995.

[23] A. S. M. A. Awal and S. K. Nguong, "A short-term investigation on high volume palm oil fuel ash (POFA) concrete," in *Proceedings of 35th Conferenece on Our World in Concrete and Structure*, pp. 85–92, Singapore, August 2010.

[24] J. H. Tay, "Ash from oil-palm waste as a concrete material," *Journal of Materials in Civil Engineering*, vol. 2, no. 2, pp. 94–105, 1990.

[25] W. Tangchirapat, T. Saeting, C. Jaturapitakkul, K. Kiattikomol, and A. Siripanichgorn, "Use of waste ash from palm oil industry in concrete," *Waste Management*, vol. 27, no. 1, pp. 81–88, 2007.

[26] C. Chandara, E. Sakai, K. A. M. Azizli, Z. A. Ahmad, and S. F. S. Hashim, "The effect of unburned carbon in palm oil fuel ash on fluidity of cement pastes containing superplasticizer," *Construction and Building Materials*, vol. 24, no. 9, pp. 1590–1593, 2010.

[27] W. Kroehong, T. Sinsiri, C. Jaturapitakkul, and P. Chindaprasirt, "Effect of palm oil fuel ash fineness on the microstructure of blended cement paste," *Construction and Building Materials*, vol. 25, no. 11, pp. 4095–4104, 2011.

[28] C. Jaturapitakkul, J. Tangpagasit, S. Songmue, and K. Kiattikomol, "Filler effect and pozzolanic reaction of ground palm oil fuel ash," *Construction and Building Materials*, vol. 25, no. 11, pp. 4287–4293, 2011.

[29] W. Tangchirapat and C. Jaturapitakkul, "Strength, drying shrinkage, and water permeability of concrete incorporating ground palm oil fuel ash," *Cement and Concrete Composites*, vol. 32, no. 10, pp. 767–774, 2010.

[30] A. A. Awal and M. W. Hussin, "The effectiveness of palm oil fuel ash in preventing expansion due to alkali-silica reaction," *Cement and Concrete Composites*, vol. 19, no. 4, pp. 367–372, 1997.

[31] A. A. Awal and M. W. Hussin, "Effect of palm oil fuel ash in controlling heat of hydration of concrete," *Procedia Eng*, vol. 14, pp. 2650–2657, 2011.

[32] C. Chandara, K. A. M. Azizli, Z. A. Ahmad, S. F. S. Hashim, and E. Sakai, "Heat of hydration of blended cement containing treated ground palm oil fuel ash," *Construction and Building Materials*, vol. 27, no. 1, pp. 78–81, 2012.

[33] A. A. Awal and I. A. Shehu, "Evaluation of heat of hydration of concrete containing high volume palm oil fuel ash," *Fuel*, vol. 105, pp. 728–731, 2013.

[34] C. Jaturapitakkul, K. Kiattikomol, W. Tangchirapat, and T. Saeting, "Evaluation of the sulfate resistance of concrete containing palm oil fuel ash," *Construction and Building Materials*, vol. 21, no. 7, pp. 1399–1405, 2007.

[35] KS L 5201, *Portland Cement*, Korean Standards Association, Seoul, Rebublic of Korea, 2013.

[36] ASTM C 618, *Standard Specification for Coal Fly Ash and Raw or Calcined Natural Pozzolan for Use in Concrete*, ASTM International, West Conshohocken, PA, USA, 2012.

[37] M. R. Karim, M. F. M. Zain, M. Jamil, and F. C. Lai, "Fabrication of a non-cement binder using slag, palm oil fuel ash and rice husk ash with sodium hydroxide," *Construction and Building Materials*, vol. 49, pp. 894–902, 2013.

[38] KS L ISO 679, *Methods of Testing Cements-Determination of Strength*, Korean Standards Association, Seoul, 2010.

[39] ASTM C1702, *Standard Test Method for Measurement of Heat of Hydration of Hydraulic Cementitious Materials Using Isothermal Conduction Calorimetry*, ASTM International, West Conshohocken, PA, USA, 2009.

[40] S. Mindess, J. F. Young, and D. Darwin, *Concrete*, Prentice-Hall, Upper Saddle River, NJ, USA, 2nd edition, 2003.

[41] I. Jawed and J. Skalny, "Alkalies in cement: a review: II. Effects of alkalies on hydration and performance of Portland cement," *Cement and Concrete Research*, vol. 8, no. 1, pp. 37–51, 1978.

[42] O. Mendoza, C. Giraldo, S. S. Camargo, and J. I. Tobón, "Structural and nano-mechanical properties of Calcium Silicate Hydrate (CSH) formed from alite hydration in the presence of sodium and potassium hydroxide," *Cement and Concrete Research*, vol. 74, pp. 88–94, 2015.

[43] S. H. Lee, "Pozzolan reaction," *Cement*, vol. 158, p. 40, 2003.

[44] L. P. Singh, S. R. Karade, S. K. Bhattacharyya, M. M. Yousuf, and S. Ahalawat, "Beneficial role of nanosilica in cement based materials–A review," *Construction and Building Materials*, vol. 47, pp. 1069–1077, 2013.

12

Experimental Study on the Shear Strength of Cement-Sand-Gravel Material

Jie Yang[ID],[1] Xin Cai[ID],[1,2] Qiong Pang[ID],[3] Xing-wen Guo[ID],[2] Ying-li Wu[ID],[3] and Jin-lei Zhao[ID][4]

[1]College of Water Conservancy and Hydropower Engineering, Hohai University, Nanjing 210098, China
[2]College of Mechanics and Materials, Hohai University, Nanjing 210098, China
[3]Nanjing Hydraulic Research Institute, Nanjing 210024, China
[4]Jiangsu Surveying and Design Institute of Water Resources Co., Ltd., Yangzhou 225127, China

Correspondence should be addressed to Xin Cai; xcai@hhu.edu.cn

Academic Editor: Georgios I. Giannopoulos

An experimental study on the shear strength development of cement-sand-gravel (CSG) material was carried out using triaxial compression tests. The effects of the cementing agent content, aggregate content, and gradation on the shear strength of CSG material were analyzed. The shear strength remarkably increased with increasing cementing agent content and aggregate content for a given confining pressure. The increase in shear strength with increasing cementing agent content far exceeded that with increasing aggregate content. However, the stress-strain curves and shear strength changed only slightly when the aggregate gradation for CSG material was adjusted. Based on the test data, a strength criterion for CSG material is proposed as a function of the cementing agent content, aggregate content, and shear strength of the aggregate gradation.

1. Introduction

Like roller-compacted concrete (RCC), cement-sand-gravel (CSG) material consists of water, aggregate (rockfill material, sandy gravel material, etc.), and cementing agents such as Portland cement and fly ash. As compared to RCC, the advantages of CSG material include a lower requirement of cementing agent content, its compatibility with local aggregate, and less stringent temperature control requirements. CSG materials with varying cementing agent contents, aggregate contents, and gradations have been utilized in various infrastructure applications, such as embankments, soil treatments, reinforcement for small rural hydropower structures, and, most commonly, in dam construction [1].

A strength requirement is a basic premise in engineering applications of geotechnical materials; thus, examination of the strength characteristics of geotechnical materials is extremely important. Since the 1990s, scholars have been researching cemented sand. Some researchers [2, 3] have obtained results on the strength characteristics of CSG materials from a series of compressive strength tests. The results of previous research indicate that the compressive strength of CSG material increases with increased cementing agent content, the optimal water-cement ratio is 1.2, and the strength is maximized when the fines content lies within the range of 25–30%. Kongsukprasert et al. [4] studied the effects of several factors, including the water content, cementing agent content, dry density, and curing period, on the shear strength of CSG material using triaxial compression tests conducted at a confining pressure of 19.8 kPa. Although extensive, the effects of the confining pressure on the shear strength of CSG material were not considered in these previous studies. Wu et al. [5] analyzed the effects of curing age on the shear strength of CSG material via triaxial testing and subsequently used the test data to establish a shear strength criterion as a function of the curing age and confining pressure. Sun et al. [6] conducted triaxial tests on CSG materials with cementing agent content of less than 60 kg/m³; additionally, static and dynamic triaxial tests on CSG materials with cementing agent content exceeding 60 kg/m³ were conducted by Fu et al. [7]. These studies mainly focused on the effects of the cementing agent content

TABLE 1: Aggregate gradation for Material II.

Name	Smaller than 1 mm	1–5 mm	5–10 mm	10–20 mm	20–40 mm
No. 2 gradation	17%	12%	22%	15.5%	33.5%
No. 3 gradation	16.25%	8.75%	21.25%	28.75%	25%

on the shear strength of CSG material under different confining pressures. Amini and Hamidi [8] analyzed the effects of the cementing agent content on the cohesion c and internal friction angle φ in the Mohr–Coulomb criterion under drained and undrained conditions using triaxial compression testing. However, a shear strength criterion based on the cementing agent content and confining pressure was not proposed in these studies. Li et al. [9] conducted triaxial compression tests on artificial cemented sand, which is a type of material similar to CSG material, and proposed several novel strength criteria based on the experimental results for varying cementing agent content. Because the size of the aggregate in CSG material is significantly different from that in cemented sand, it is unclear whether a strength criterion developed for artificial cemented sand can be directly applied to CSG material. Clough et al. [10] and Wang [11] analyzed the effects of the aggregate content on the shear strength of CSG material for different confining pressures, but did not propose an aggregate content-based strength criterion. A review of the literature shows that research regarding the effects of aggregate gradation on the shear strength of CSG material is insufficient. For CSG dams, because the geological conditions and requirements of each dam project differ, the cementing agent content, aggregate content, and gradation of CSG material also vary for different dams.

In this study, triaxial compression tests were conducted to assess the effects of the cementing agent content, aggregate content, and gradation on the shear strength of CSG material. Additionally, a new strength criterion for CSG material is proposed based on the results. The purpose of the proposed strength criterion is to provide a basis for the construction of a reasonable constitutive model suitable for various types of CSG materials and to meet the engineering requirements for various infrastructure applications, including CSG dam construction.

2. Materials and Methods

2.1. Raw Materials.
Two types of CSG materials, hereinafter referred to as Material I and Material II, were examined by means of drained triaxial shear tests.

2.1.1. Material I
Cement: 32.5 grade ordinary silicate cement from the Anhui Digang Hailuo Cement Co., Ltd.

Crushed stone: particle sizes less than 5 mm (3%), 5–10 mm (20%), 10–20 mm (35%), and 20–40 mm (42%), sourced from a Nanjing suburb [5].

Sand: particle size of approximately 0–4.75 mm, medium-coarse sand crushed from limestone.

Water: tap water.

TABLE 2: Test programs on Material I to investigate effects of cementing agent content and aggregate content.

Sequence number	Cementing agent content (kg/m^3)	Aggregate content (kg/m^3)	Aggregate gradation
1	20	2130	No. 1
2	40	2130	No. 1
3	60	2110	No. 1
4	60	2130	No. 1
5	80	2090	No. 1
6	80	2110	No. 1
7	80	2130	No. 1
8	100	2130	No. 1

The ratio of sand to crushed stone is 1 : 4, which was the same as that used in the experimental study by Sun et al. [5]. In this paper, the aggregate gradation for Material I is termed No. 1.

2.1.2. Material II
Cement: 42.5 grade ordinary silicate cement from the Anhui Digang Hailuo Cement Co., Ltd.

Fly ash: Type I fly ash from the Nanjing market.

Sand and gravel: One aggregate gradation (No. 2) and a second gradation (No. 3), which are listed in Table 1.

Water: tap water.

2.2. Mix Proportions of CSG Material and Test programs.
For Material I, the water-cement ratio was 1.0 [12], and the cementing agent contents were 20 kg/m^3, 40 kg/m^3, 60 kg/m^3, 80 kg/m^3, and 100 kg/m^3; the aggregate contents, including the crushed stone and sand contents, were 2090 kg/m^3, 2110 kg/m^3, and 2130 kg/m^3. Samples of Material I, which vary in cementing agent content and aggregate content, were subjected to drained triaxial shear tests under various confining pressures (300 kPa, 600 kPa, 900 kPa, and 1200 kPa). To confirm the adequacy of the strength of CSG material in this paper, a mixture of Material I, with a cementing agent content of 60 kg/m^3 and an aggregate content of 2110 kg/m^3, was used in samples subjected to additional drained triaxial shear tests carried out under varying confining pressures. Test programs conducted on Material I, to investigate the effects of cementing agent content and aggregate content, are presented in Table 2.

For Material II, the ratio of cement to coal ash was 1 : 1, and the water-cement ratio was 1.0 [12]. One aggregate gradation (No. 2) samples of Material II were subjected to drained triaxial shear tests under various confining pressures (300 kPa, 600 kPa, 900 kPa, and 1200 kPa) and varying cementing agent content (20 kg/m^3, 80 kg/m^3, and 100 kg/m^3). In addition, to assess the effects of aggregate gradation on the strength characteristics of CSG, a second gradation (No. 3) was tested in Material II samples. They

TABLE 3: Test programs of Material II to investigate effects of cementing agent content and aggregate gradation.

Sequence number	Cementing agent content (kg/m^3)	Aggregate content (kg/m^3)	Aggregate gradation
1	20	2130	No. 2
2	80	2130	No. 2
3	80	2130	No. 2
4	100	2130	No. 2

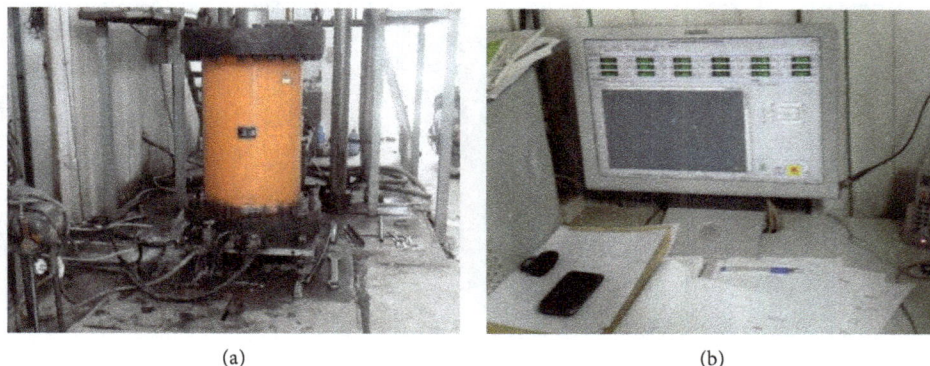

(a)   (b)

FIGURE 1: TYD-1500 triaxial tester, which should be listed as (a) large-scale dynamic triaxial tester; (b) data acquisition system.

(a)   (b)

FIGURE 2: (a) Mixed materials; (b) samples of CSG material.

were subjected to drained triaxial shear tests under different confining pressures (300 kPa, 600 kPa, 900 kPa, and 1200 kPa). Test programs conducted on Material II, to investigate the effects of cementing agent content and aggregate gradation, are presented in Table 3.

2.3. Equipment and Test Methods Used in Drained Triaxial Shear Tests.

Drained triaxial shear tests of CSG material samples were conducted using a TYD-1500 dynamic triaxial tester, as is shown in Figure 1.

The mixing materials used to produce CSG material samples for a series of large-scale triaxial tests are shown in Figure 2(a). The materials were compacted in steel molds of 30 cm in diameter and 70 cm in height (Figure 2(b)). The samples were cured in a laboratory at a temperature of $20 \pm 2°C$ for 28 days.

The triaxial tests for determination of the shear strength of the CSG materials were conducted in accordance with China Standard SL237-1999 [13]. The samples were first

saturated and then subjected to one of the four levels of confining pressures (300, 600, 900, or 1200 kPa) for 10 min prior to axial loading. Axial loading at a strain rate of 2 mm/min was then applied and stopped when the axial strain reached 15%.

To improve the accuracy of the results, two samples were prepared and tested for each test group. To prevent damage to the tester due to particles falling from damaged samples, the samples were covered with rubber sleeves that were securely fastened.

3. Results and Discussion

3.1. Shear Strength versus Cementing Agent Content.

The results of the drained triaxial shear tests performed on the samples of Material I and Material II are presented in Figures 3 and 4. As these figures show, when the cementing agent content is low, the $q - \varepsilon_a$ curves of CSG material comprise of three stages: an initial stress increase, a slowing

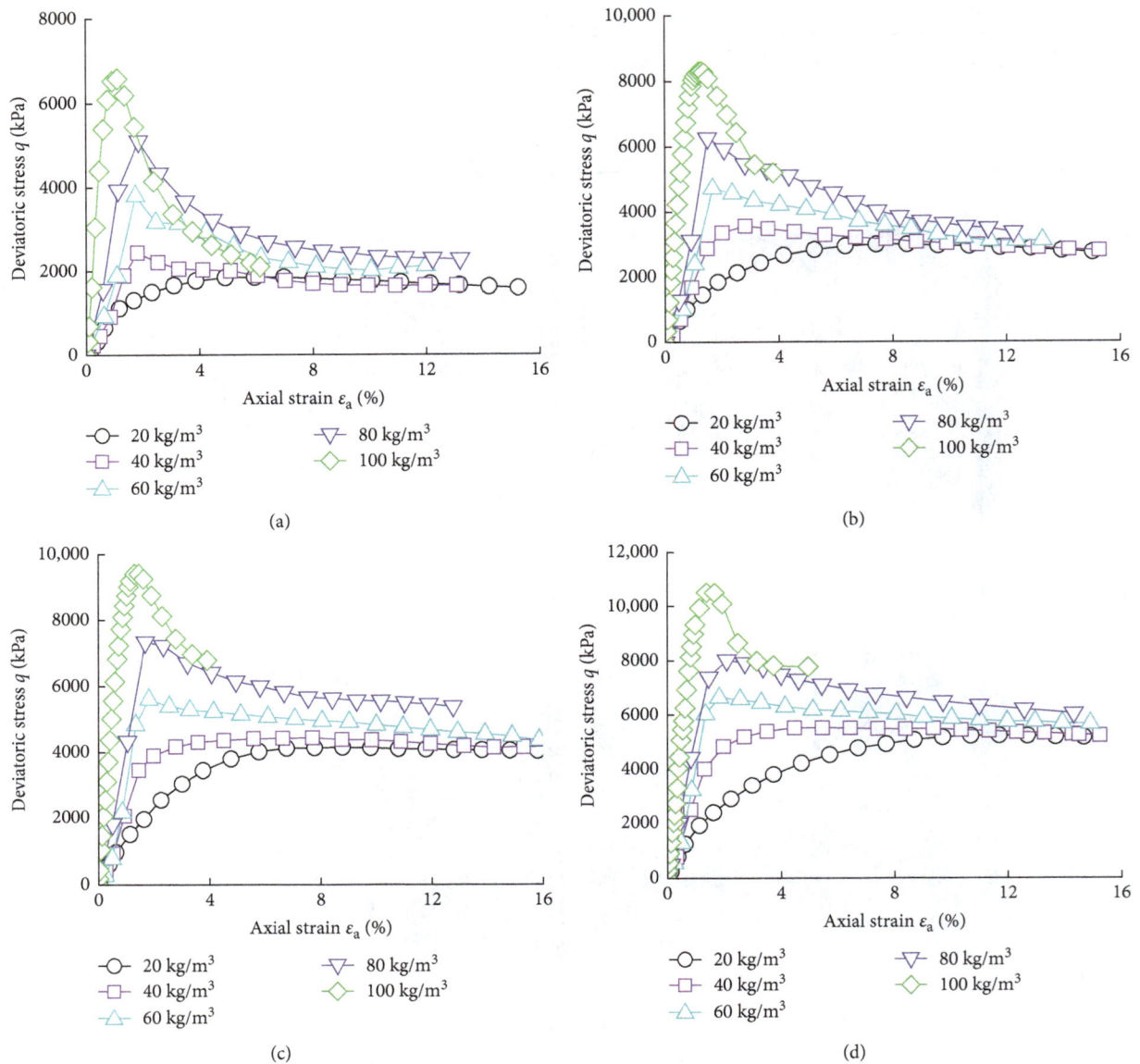

FIGURE 3: Stress-strain curves of Material I with regard to the effect of cementing agent content under various confining pressures: (a) 300 kPa; (b) 600 kPa; (c) 900 kPa; (d) 1200 kPa.

stress increase, and a peak stress similar to that of the rockfill material in the CSG material. The influence of the cementing agent content on the strain-softening behavior of the material is apparent. The stress-strain curves consist of five stages: an initial stress increase, a slowing stress increase, a peak stress, plastic softening, and a residual strength that approaches that of RCC material when the cementing agent content is increased to $100 \, kg/m^3$. The maximum stress and stress at a given axial strain both significantly increase with increasing cementing agent content at each confining pressure considered in this study (300 kPa, 600 kPa, 900 kPa, and 1200 kPa). This is because cementation between the particles in the CSG material increases with the cementing agent content, thus causing the internal bearing capacity mechanism to change from friction between particles, as in rockfill material, to gradually increasing internal cohesive

strength. This is consistent with the results of Li et al. [9], who reported that the cementing agent content is the main factor influencing the strength of artificial cemented sand, which is similar to CSG material.

Figure 5 illustrates the shear strength, which is the maximum stress shown in the curves in Figures 3 and 4 under varying confining pressure and cementing agent content. As Figure 5 shows, the shear strength of CSG material ranges from 1200 to 12,000 kPa and increases with increasing cementing agent content and confining pressure. The relationship between the peak strength and confining pressure is approximately linear for a given cementing agent content. This is consistent with the observations of Fu et al. [7] and other researchers [5] who have reported that increasing the cementing agent content is highly effective in enhancing the shear strength of CSG and various other

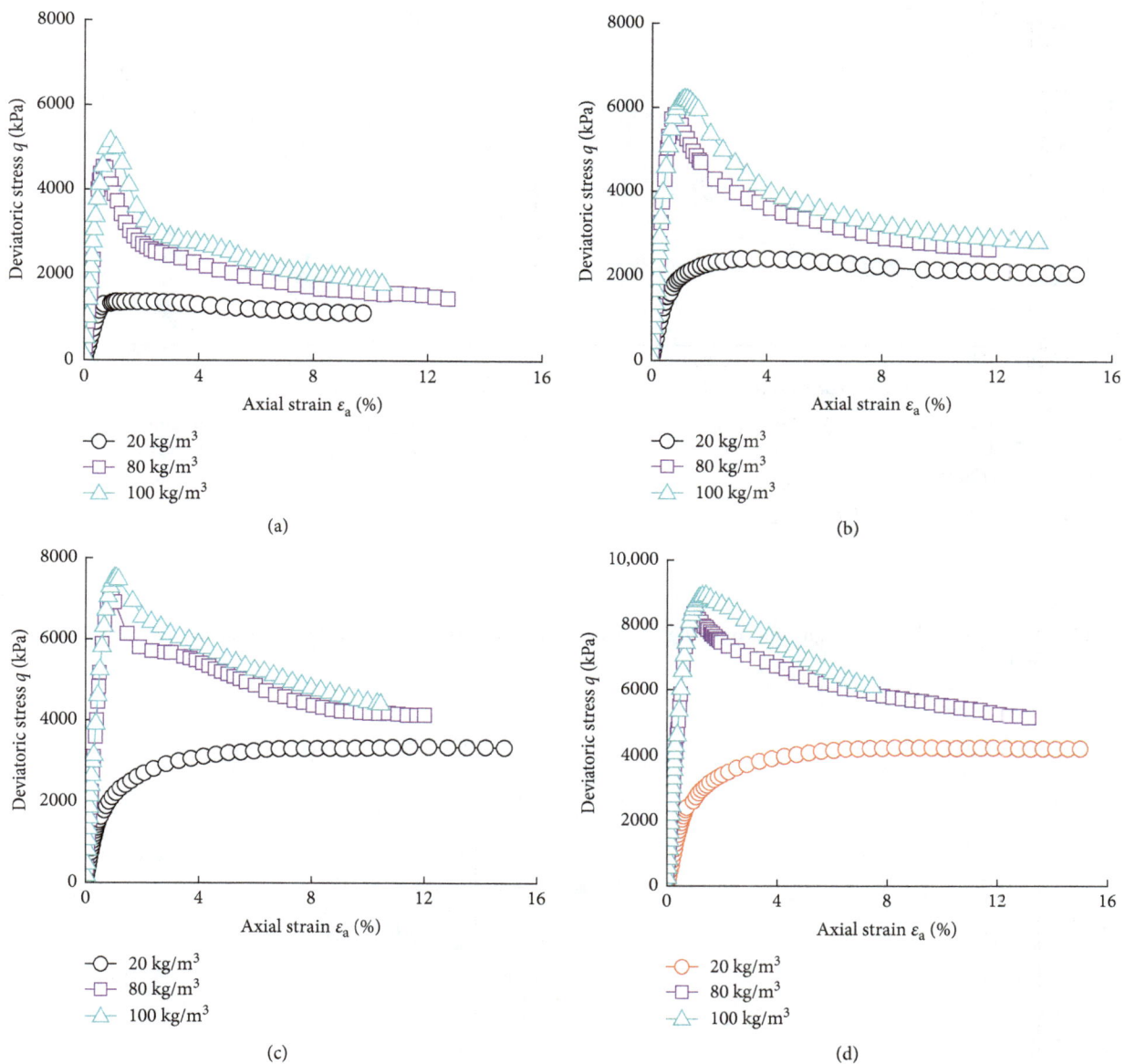

FIGURE 4: Stress-strain curve of Material II with regard to the effect of cementing agent content under various confining pressures: (a) 300 kPa; (b) 600 kPa; (c) 900 kPa; (d) 1200 kPa.

materials, such as cemented sand and polyurethane foam adhesive rockfill materials.

3.2. Shear Strength versus Aggregate Content. Figure 6 shows the stress-strain curves for CSG materials with respect to the aggregate content, obtained via drained triaxial shear testing. As Figure 6 shows, the aggregate content has little effect on the shape of the stress-strain curve, but the peak stress increases as the aggregate content increases. This is attributed to an increase in the aggregate content reinforcing the internal bearing capacity of the CSG material, which is a result of increased particle contact area.

Figure 7 shows the shear strength, which is the maximum stress in the curves shown in Figure 6, for varying confining pressure and aggregate content. As these figures show, the shear strength increases with increased confining pressure and aggregate content. This is consistent with the results of Wang [11] regarding the changes in the strength characteristics of CSG material with relative density and a confining pressure below 300 kPa. In comparison with the cementing agent content, the aggregate content yields less effect on the shear strength of CSG material.

3.3. Shear Strength versus Aggregate Gradation. Figure 8 shows the stress-strain curves for CSG materials with different aggregate gradations, obtained via drained triaxial shear testing. The effects of aggregate gradation on the stress-strain behavior of CSG material are not notable for any of the confining pressures considered (300 kPa, 600 kPa, 900 kPa, and 1200 kPa). This is similar to the minimal effect of

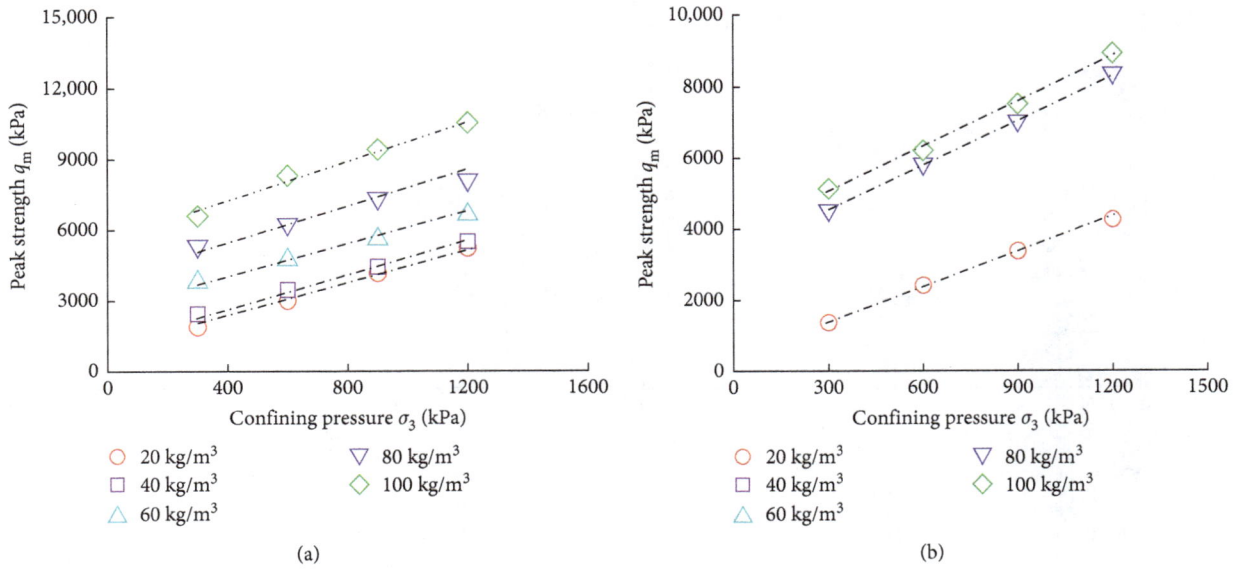

FIGURE 5: Relation curves of peak strength and confining pressure of CSG material with different cementing agent contents: (a) Material I and (b) Material II.

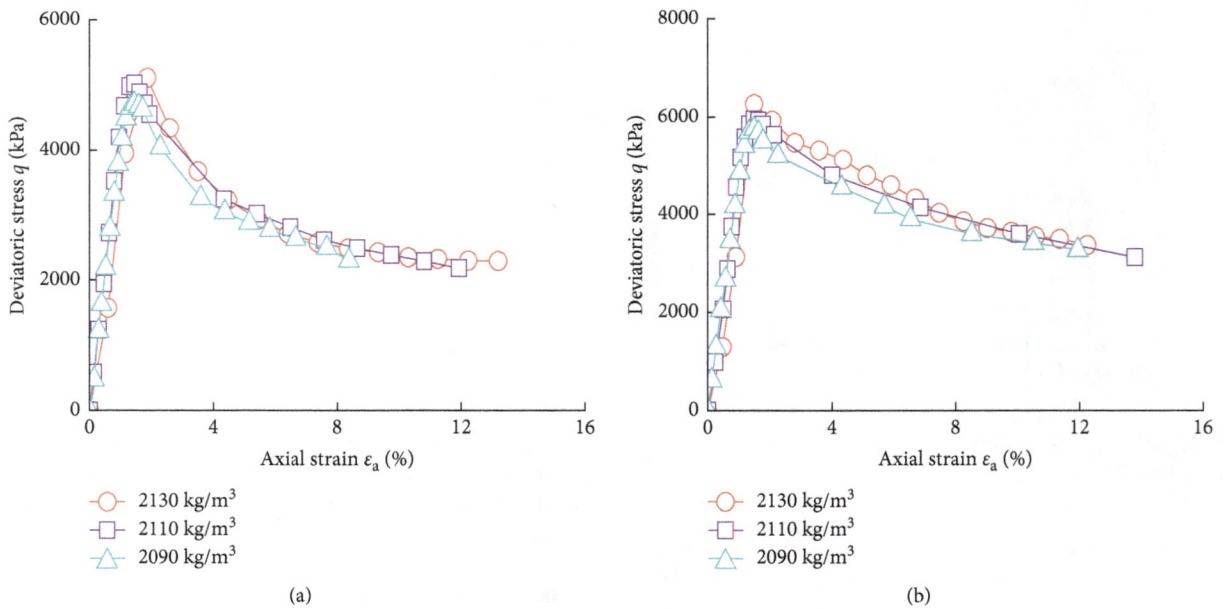

FIGURE 6: Continued.

(c)

(d)

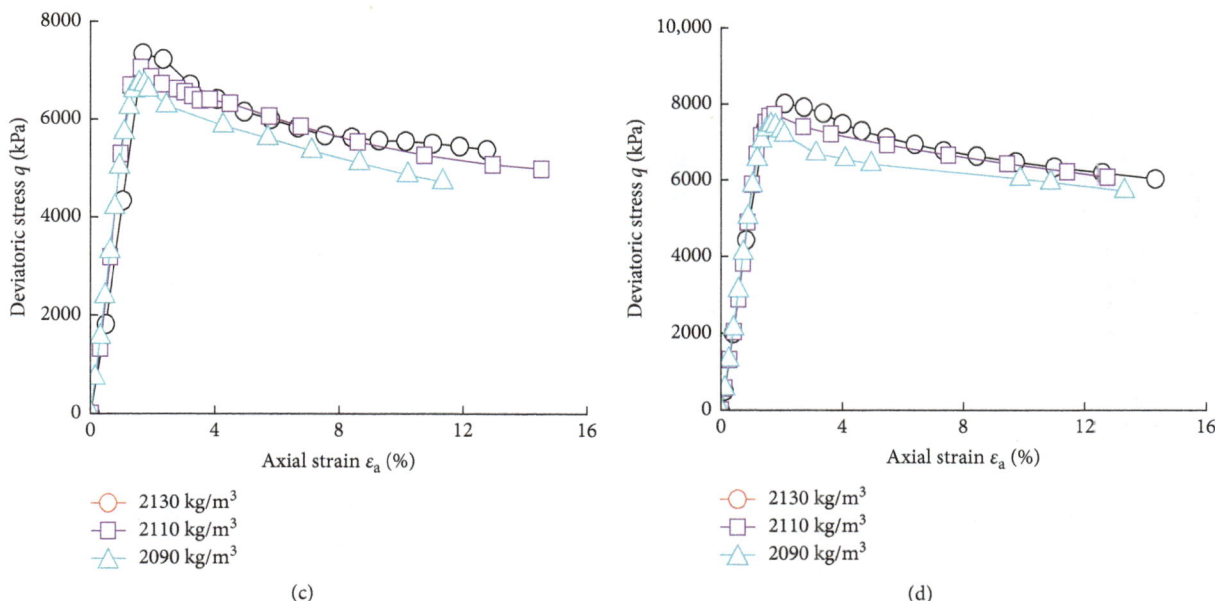

FIGURE 6: Stress-strain curve of CSG material considering the effect of aggregate content under various confining pressures: (a) 300 kPa; (b) 600 kPa; (c) 900 kPa; (d) 1200 kPa.

gradation on the strength of rockfill material when the aggregate content ranges from 60 to 70%.

4. Strength Criterion

The Mohr–Coulomb theory, which serves as the basis for the strength criterion for CSG material in this study, is commonly used to describe the stress-strain response of materials [5–8]. It can be expressed as follows:

$$\tau_f = c + \sigma \tan \varphi, \qquad (1)$$

where c is the cohesion of the material, φ is the angle of internal friction, τ_f is the shearing stress, and σ is the normal stress.

To describe the relationships between the peak strength and confining pressure for the varying cementing agent content and aggregate content shown in Figures 6 and 7, the Mohr–Coulomb criterion represented by the principal stress in (1) can be expressed as follows:

$$q_m = \frac{2c \cos \varphi}{1 - \sin \varphi} + \frac{2 \sin \varphi}{1 - \sin \varphi}\sigma_3, \qquad (2)$$

where q_m is the peak strength, σ_3 is the confining pressure for drained triaxial shear testing, c is the cohesion, and φ is the angle of the internal friction (shearing resistance).

Based on (2) and the shear strength test results obtained for differing cementing agent content and aggregate content, values for the cohesion c and angle of internal friction (shearing resistance) φ were extracted for this analysis, as is shown in Tables 4 and 5.

The Mohr–Coulomb theory is based on the assumption that the cohesion and angle of shearing resistance in (1) are constant. However, for CSG materials used in practical engineering applications, the cohesion and angle of shearing

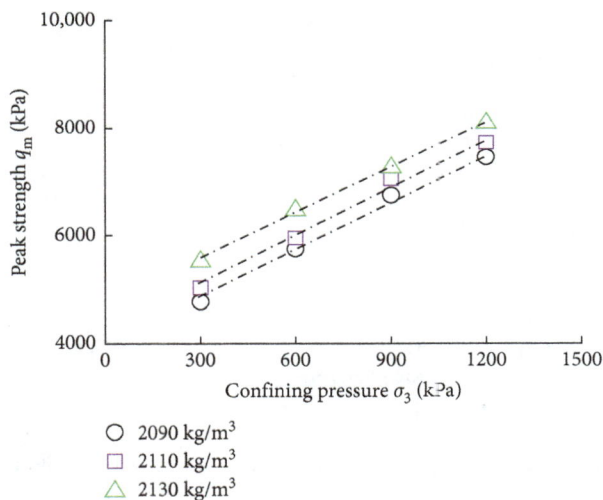

FIGURE 7: Relation between peak strength and confining pressure under different aggregate contents.

resistance vary according to the cementing agent content and aggregate content. This means that the original Mohr–Coulomb strength theory expression is not suitable for CSG materials with varying cementing agent content and aggregate content. Thus, a new strength criterion for the shear strength of CSG material is proposed. This criterion is a function of the cementing agent content and aggregate content.

4.1. Cohesion c. Based on the cohesion values obtained for different cementing agent contents (listed in Table 4), the relationship between cohesion and the cementing agent content can be expressed as follows:

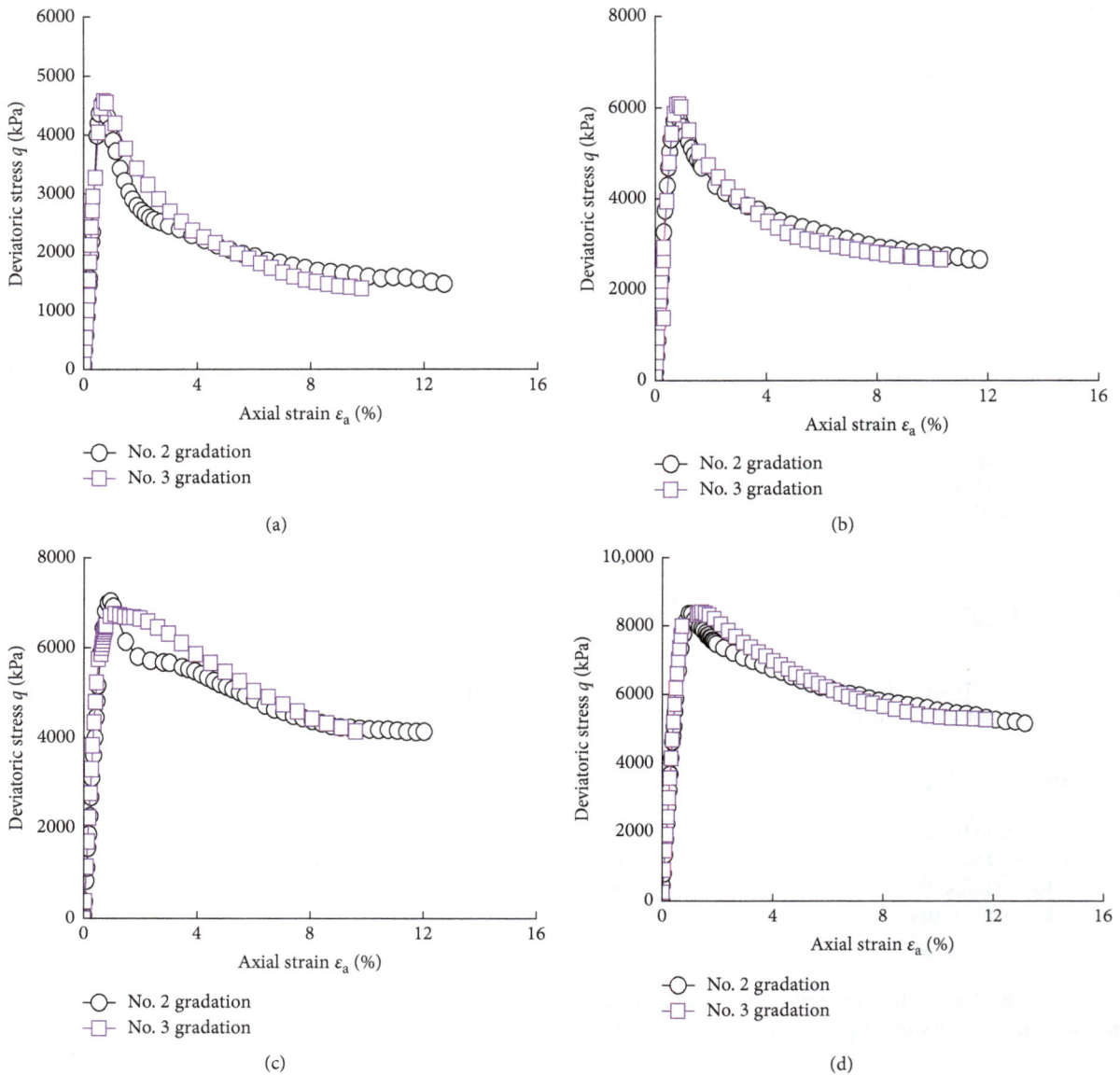

FIGURE 8: Stress-strain curve of CSG material with regard to the effect of aggregate gradation under various confining pressures: (a) 300 kPa; (b) 600 kPa; (c) 900 kPa; (d) 1200 kPa.

$$c = H_0 C_c, \tag{3}$$

where H_0 is the parameter related to the composition of the CSG material and C_c is the cementing agent content.

When the cementing agent content in (3) is low, the cohesion of the CSG material is near zero, which is close to the cohesion of rockfill material, as calculated by Sun et al. [5]. Figure 9 shows a comparison of the test data and results calculated using (3) for CSG material [5, 6, 8], PFA-reinforced rockfill material [14], cemented sand [9], and cemented soil [15] for different cementing agent contents. As Figure 9 shows, the calculated results for CSG material, PFA-reinforced rockfill material, and cemented sand fit the experimental results well; this confirms that (3) yields a reasonable description of the cohesion of those cemented and bonded materials as a function of the cementing agent content. However, because the soil

in the cemented soil studied by Baxter et al. [15] had some viscosity and a cohesion value greater than zero, (3) is not suitable for this type of cemented soil.

Based on the cohesion values obtained for different aggregate contents (Table 5), curves of cohesion as a function of aggregate content were developed, as shown in Figure 10. These curves show that the cohesion of CSG material increases with increasing aggregate content. However, compared to the influence of the cementing agent content, the influence of the aggregate content on the cohesion of CSG is lower.

The relationship between cohesion and aggregate content can be formulated as follows:

$$c = H_g \rho_g, \tag{4}$$

where H_g is the parameter related to the type of aggregate in the CSG material and ρ_g is the aggregate content. By

TABLE 4: Cohesion and internal friction angle under different cementing agent contents.

Cementing agent content, C_c (kg/m^3)	Cohesion, c (kPa)	Internal friction angle (°)
20	256.1	40.3
40	405.2	38.9
60	691.9	38.0
80	1031.2	39.3
100	1280.1	41.4

TABLE 5: Cohesion and internal friction angle under different aggregate contents.

Aggregate contents, ρ_g (kg/m^3)	Cohesion, c (kPa)	Internal friction angle (°)
2090	934.4	38.8
2110	989.2	39.1
2130	1031	39.3

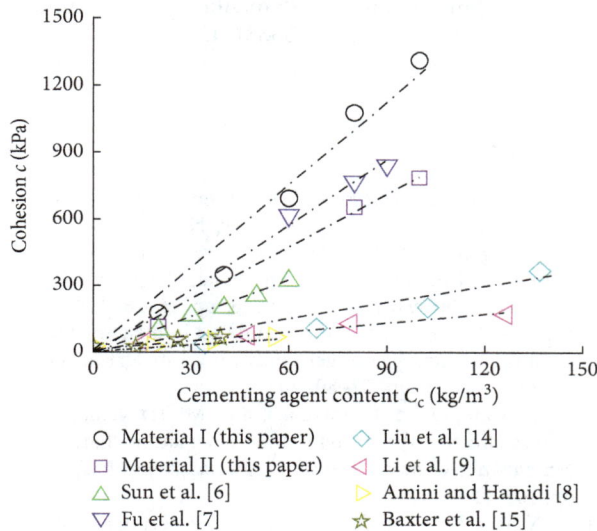

FIGURE 10: Relation between cohesion and aggregate content.

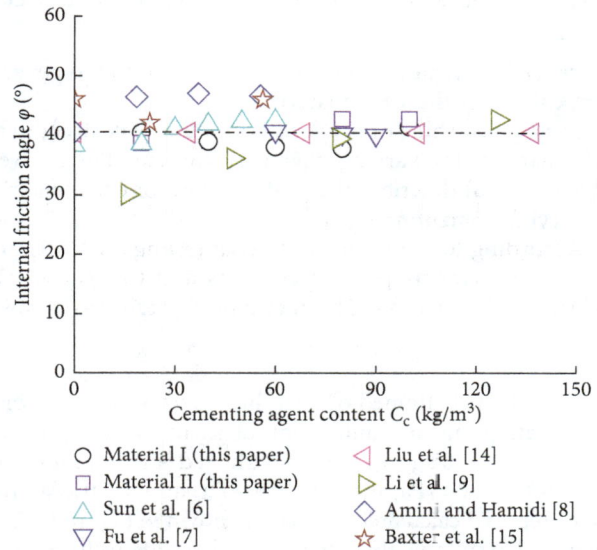

FIGURE 9: Relation between cohesion and cementing agent content.

combining (3 and 4), the following expression for cohesion c as a function of the cementing agent content and aggregate content can be obtained:

$$c = H_z \rho_g C_c, \qquad (5)$$

where H_z is the parameter related to the composition and type of aggregate for CSG material. According to the results of the drained triaxial shear tests described above, $H_z = 0.005$.

4.2. Internal Friction Angle φ. Figure 11 illustrates the internal friction angle values obtained for CSG material [5, 6, 8], PFA-reinforced rockfill material [14], cemented sand [9], and cemented soil [15] for different cementing agent contents. As Figure 11 shows, the internal friction angle of CSG material, PFA-reinforced rockfill material, cemented sand, and cemented soil ranges from 30° to 50°, which moderately differs from the range of 25° to 65° for

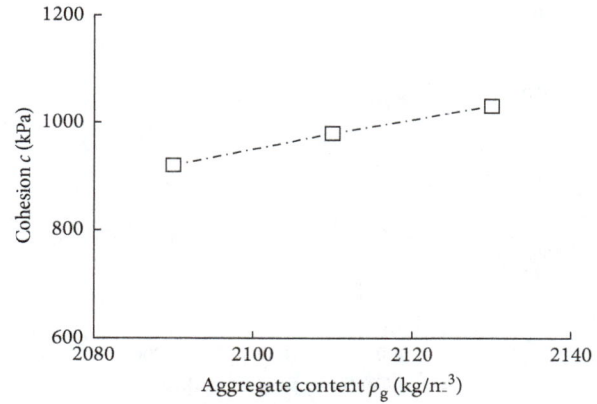

FIGURE 11: Relation between internal friction angle and cementing agent content.

gravel [16]. The reason for this difference is that the cementing agents in the cemented materials limit the slippage angle of the aggregate. The internal friction angle of CSG material and PFA-reinforced rockfill material for various cementing agent contents is approximately 39.5°. Similarly, the internal friction angle of CSG material is approximately 39° for various aggregate contents, as presented in Table 5. Based on these results, the internal friction angle value for CSG material is taken as 39.3° for a range of cementing agent contents and aggregate contents.

4.3. Strength Criterion according to Cementing Agent Content and Aggregate Content. Based on the results summarized above, the following expression for the shear strength of CSG material as a function of the cementing agent content and aggregate content is proposed:

$$q_m = \frac{2H_z \rho_g C_c \cos \varphi}{1 - \sin \varphi} + \frac{2 \sin \varphi}{1 - \sin \varphi} \sigma_3, \qquad (6)$$

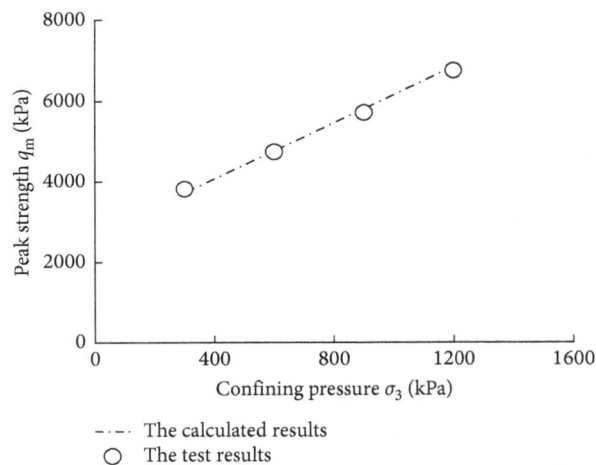

FIGURE 12: Comparison of the test results and calculated results.

where H_z is a parameter associated with the type and composition of the CSG material.

For a given value of c, (6) describes the peak strength of CSG material for varying aggregate content. For a given value of ρ_g, (6) describes the peak strength of CSG material for varying cementing agent content.

According to the results of triaxial testing on Material I for various cementing agent contents and (6), $H_z = 0.005$ and $\varphi = 39.3°$. Equation (6) can also be expressed as follows:

$$q_m = 0.021\rho_g C_c + 3.47\sigma_3. \quad (7)$$

To verify (6), drained triaxial shear tests on samples with a cementing agent content of 60 kg/m^3 and aggregate content of 2110 kg/m^3 were conducted under confining pressures of 300 kPa, 600 kPa, 900 kPa, and 1200 kPa. The test results and calculated results are shown in Figure 12. The calculated results fit the experimental results well, thereby demonstrating that (6) can be used to describe the shear strength of CSG material as a function of the cementing agent content and aggregate content.

5. Conclusions

The effects of the cementing agent content, aggregate content, and gradation on the shear strength of CSG material were investigated by means of drained triaxial shear testing. The conclusions drawn from the test results can be summarized as follows.

(a) The influence of the cementing agent content on the shear strength of CSG material is much more significant than the influence of the aggregate content and gradation.

(b) The cohesion of CSG increases with increasing cementing agent content and aggregate content, whereas the internal friction angle changes only slightly. The effects of the aggregate gradation on cohesion and the internal friction angle are negligible.

(c) A strength criterion for CSG material is proposed based on an analysis of the strength characteristics of

the material as a function of the cementing agent content, aggregate content, and aggregate gradation. Overall, the strength model fits the test data well of CSG material and can provide evidence for numerical calculation of CSG dam.

Conflicts of Interest

The authors declare that there are no actual or potential conflicts of interests regarding the publication of this paper.

Acknowledgments

This study was supported by the projects including the National Science and Technology Pillar Program during the 12th Five-Year Plan Period (2012BAD10B02), the General Programs of the National Natural Science Foundation of China (51179061), and the Fundamental Research Funds for the Central Universities (2014B36814).

References

[1] J. M. Raphael, *The Optimum Gravity Dam, Rapid Construction of Concrete Dams*, ASCE, New York, NY, USA, 1970.

[2] X. J. Tang and S. Y. Lu, "Preliminary research on mechanical behaviors of cemented rockfill material," *Engineering Journal of Wuhan University*, vol. 30, no. 6, pp. 15–18, 1997.

[3] L. L. Liu, J. X. He, and L. Liu, "Study on influencing factors of compressive strength of gel sandy gravel materials and law," *Concrete*, vol. 3, pp. 77–80, 2013.

[4] L. Kongsukprasert, F. Tatsuoka, and M. Tateyama, "Several factors affecting the strength and deformation characteristics cement-mixed gravel," *Soils and Foundations*, vol. 45, no. 3, pp. 107–124, 2005.

[5] M. X. Wu, B. Du, and Y. C. Yao, "An experimental study on stress-strain behavior and constitutive model of hardfill material," *Science China Physics, Mechanics and Astronomy*, vol. 54, no. 11, pp. 2015–2024, 2011.

[6] M. Q. Sun, C. S. Peng, and Y. L. Li, "Triaxial test of over lean cemented material," *Advances in Science and Technology of Water Resource*, vol. 27, no. 4, pp. 46–49, 2007.

[7] H. Fu, S. S. Chen, and H. Q. Han, "Experimental study on static and dynamic properties of cemented sand and gravel," *Chinese Journal of Geotechnical Engineering*, vol. 37, no. 2, pp. 357–362, 2015.

[8] Y. Amini and A. Hamidi, "Triaxial shear behavior of a cement-treated sand-gravel mixture," *Journal of Rock Mechanics and Geotechnical Engineering*, vol. 6, no. 5, pp. 455–465, 2014.

[9] D. L. Li, X. R. Liu, and X. S. Liu, "Experimental study on artificial cemented sand prepared with ordinary Portland cement with different contents," *Materials*, vol. 8, no. 7, pp. 3960–3974, 2015.

[10] G. Clough, N. Sitar, R. Bachus, and N. Rad, "Cemented sands under static loading," *Journal of the Geotechnical Engineering Division*, vol. 107, pp. 799–817, 1981.

[11] Q. Wang, "Experimental study on behavior of strength and deformation of cemented coarse-grained soil," Dalian University of Technology, Dalian, China, Master thesis, 2010.

[12] N. Liu, J.-S. Jia, F. Jin, Z.-M. Liu, and SL678–2014, *Technical Guideline for Cemented Granular Material Dams*, Beijing, China, 2014.

[13] S.-X. Sheng, Y. Dou, X.-Z. Tao, SL237-1999 et al., *Specification of Soil Test*, Beijing, China, 1999.

[14] P. Liu, H. L. Liu, and Y. Xiao, "Experimental research on mechanical properties of PFA-reinforced rockfill materials," *Rock and Soil Mechanics*, vol. 36, no. 3, pp. 749–754, 2015.

[15] C. Baxter, M. Sharma, and K. Moran, "Use of (A=0) as a failure criterion for weakly cemented soils," *Journal of Geotechnical and Geoenvironmental Engineering*, vol. 137, no. 2, pp. 161–170, 2011.

[16] Y. Xiao, *Three-dimensional strength and particle breakge state dependent constitutive model for rockfill materials*, Ph.D. thesis, Hohai University, Nanjing, China, Ph.D. thesis, 2014.

13

Effect of Metallic Inclusions on the Compressive Strength of Cement-Based Materials

Tomáš Ficker (ID)

Faculty of Civil Engineering, Brno University of Technology, CZ-602 00 Brno, Czech Republic

Correspondence should be addressed to Tomáš Ficker; tftf@seznam.cz

Academic Editor: Antonio Caggiano

In the concrete foundations, materials come into contact with bedrocks. The surfaces of bedrocks are often covered by sharp protrusions called asperities. Although geotechnical engineers have developed a reliable theory for assessing the mechanical stability of rocky terrains, the stability of transition zones between concrete and sharp asperities remains unsolved. Due to the large pressures that exist in these transition zones, the invasive influence of sharp asperities on the integrity of the concrete raises a question about possible changes of the mechanical properties of concrete materials used in foundations. These circumstances have inspired experiments in which metallic needles of various lengths have been embedded into cement-based materials to assess the influence of the needles on the compressive strength. This influence has been quantified, and the critical limits identifying the changes of material integrity have been determined. It has been conjectured that sharp rock asperities or needle-like rods of steel reinforcement in concrete may cause similar changes of material integrity as the metallic needles used in the experiments performed.

1. Introduction

In concrete research, there is a continuous interest in mechanical processes acting in the transition zones between hydrated cement matrices and aggregates [1–6]. The aggregates (sand grains and gravels) in the cement matrix (sometimes called the cement stone) play the role of fillers that do not usually increase the compressive strength of the cement-based materials, although high-quality aggregates (e.g., granite, basalt, and gneiss) very often show higher compressive strength than the cement stone itself. The reason is the imperfect transition zones between aggregates and cement stone. These imperfect zones represent critical regions (interfaces) in which microcracks often initiate. Such structural perturbations may finally lead to fatal failure, i.e., complete breakdown of the structure, if the material is faced with a higher stress. As a consequence, a lower compressive strength of these composite materials may be observed when compared with cement stone, which is a substance similar to hydrated cement paste.

Transition zones in concrete exist not only near to aggregate particles but also near to the steel reinforcement bars. Another example of the transition zone is the interface between the cement stone and rock protrusions (large asperities) in foundations. Figure 1 shows a scheme of a pillar foundation along with the photograph of the real situation. The pillar construction consists of several parts. Part B represents a leveling layer usually made of plain concrete that fills the rock irregularities (sharp asperities) that were formed during excavation work. The leveling layer B is not usually too thick, and thus the asperity layer may often occupy a considerable portion of the volume of the leveling layer. The heights of asperities may sometimes compete with the height of the leveling concrete layer. So it is not surprising that the ratio between the height of some highest asperities and the height of the leveling layer (A/B) may reach even more than 30%. Part C represents the base of the pillar, and D is the main body of the pillar. Parts C and D are made of reinforced concrete, whereas B is of plain concrete. The height E of the reinforcement approaches the height D of the pillar itself, so that the height ratio E/D may often exceed ~80%. Each part of the pillar is made subsequently after hardening of the preceding part, i.e., each part may be investigated separately as an element that is loaded individually by forces exerted by higher parts

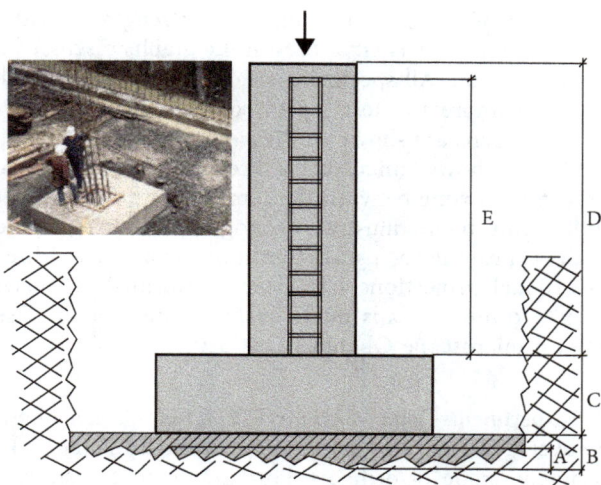

FIGURE 1: A scheme of pillar construction combined with the photo of the real situation.

of the construction. This especially concerns the lowest part B, called the leveling layer, since the asperities and their transition zones could introduce undesirable perturbations inside the volume of the leveling layer.

There is no reason to suppose that the transition zones accompanying the rock asperities and the metallic reinforcement behave very differently from the transition zones of gravel aggregates. All these zones may introduce certain structural imperfections that may influence the compressive strength in some cases. The sharp peaks of asperities in foundations and the steel reinforcement bars in fact represent needle-like perturbations in cement-based materials. Yet, there is a certain specificity of the transition zones bridging rock asperities and cement stone. The cement stone does not fully integrate rock asperities into its volume in contrast to the steel reinforcement bars that are fully integrated. The cement stone only touches the asperity surfaces, and the stability of such interface is prevalently determined by the mechanical wedging of both these materials.

When mechanical stability of concrete foundations in rocky terrains is explored, two aspects should be taken into account, namely, the stability of the asperity-concrete interface and the stability of rocky terrain itself.

The stability of asperity-concrete interface is determined by the mechanical wedging of asperities with concrete. The strength of this wedging depends on the asperity roughness, their mechanical strength, and the strength of concrete itself. For this reason, not only the studies of various aspects of compressive and tensile strength of concrete materials have been accomplished in our laboratory [7–9] but also the asperity roughness has been the subject of our interest in recent times [10–12].

In the geotechnical literature, there is also a continuous interest in strength properties of rock-concrete interfaces. Recently, the rock-concrete interface has been investigated by using bimaterial disks exposed to Brazilian tests [13]. These authors have identified three typical fracture patterns:

interface fracture, tensile fracture, and their combination. Numerical simulations of shearing of unbounded rock-concrete planar and saw-tooth triangular joints have been realized using the shear box genesis approach. The results of these simulations have indicated that the used approach reproduces suitably the shear experimental data [14]. Rock-concrete joints with various roughness profiles have been subjected to three-point bending and four-point shear tests to investigate the propagation of interfacial cracks. The results have documented that, on the basis of the known initial fracture toughnesses of the rock, the concrete, and the rock-concrete interface, it is possible to predict the propagation of interfacial cracks in these materials and their interface [15–18]. New failure criteria for the rock, concrete, and their joints have been proposed [19, 20]. A model describing the rock deformation occurring during shearing within the rock-concrete assembly has been presented [21]. Various failure mechanisms of concrete retaining structures situated on rock foundations have been studied and shown that sliding is possible only when weakened subhorizontal surfaces with low shear parameters are present near the interface between the structure and foundation [22, 23]. The experimental study of the shear behavior of rock-concrete joints with roughness in the idealized forms of regular and irregular triangles has been performed and shown that the irregular joints have evidenced a much greater relative ductility than the regular joints [24]. A greater increase in the fracture toughness with shear loading has been observed with rock-concrete joints [25]. Shear tests have been performed with planar rock-concrete interfaces to investigate the influences of interface roughness on shear strength [26, 27].

From this overview of published results, it is clear that the research of the strength properties of rock-concrete interfaces is solely focused on shearing phenomena occurring between these two materials. Shear strength of these interfaces is predominantly investigated but practically nothing is known about the possible acting of asperities on the compressive strength of concrete materials inside foundations. The present paper is devoted just to this topic.

As the stability of the bedrock is concerned, it should be mentioned that rocky massive often consists of layers that are also mutually jointed by asperity wedging. The rock joints usually form volume networks in rock masses. An example of such a network of rock joints is shown in Figure 2. The shear strength τ of slopes consisting of rock joints may be estimated, e.g., by the empirical formula introduced by Barton [28] and Barton and Choubey [29]:

$$\tau = \sigma_n \tan\left[\text{JRC} \cdot \log_{10}\left(\frac{\text{JCS}}{\sigma_n}\right) + \alpha_{\max}\right], \quad (1)$$

where σ_n is the normal effective stress, JRC represents the joint roughness coefficient, JCS is the compressive strength of the joint walls, and $\alpha_{\max}$ represents a critical angle for plane joints with smooth surfaces (without asperities).

When the mechanical stability of transition zones between concrete and bedrock is considered, a question arises, namely, whether the invasive acting of sharp asperities on

FIGURE 2: The network of rock joints uncovered as a result of excavation work at lime quarry Hády near to the city of Brno in the Czech Republic.

concrete structure may modify the compressive strength of this material. There are no theoretical and experimental studies that would answer this question. Practitioners in geotechnics and in civil engineering usually assume a priori that asperities cannot significantly influence the mechanical properties of concrete materials. However, this question deserves attention and thorough examination at least within a model situation in which asperities could be simulated by sharp metallic needles embedded in the base of concrete specimens that would experience compressive stress.

The present contribution examines such situations. The samples made of cement-based materials with embedded metallic needles are subjected to compressive tests to verify the mechanical integrity of these materials. This contribution represents a first attempt to answer the question about the possible invasive acting of asperity-like needles in concrete materials.

2. Specimens Made of Ordinary Portland Cement Paste

It is necessary to answer the basic question whether the invasive action of small needles can modify the compressive strength of cement-based materials. For this purpose, 50 specimens (3 cm × 3 cm × 3 cm) of hydrated ordinary Portland cement paste (water-to-cement ratio = 0.3 by weight) were prepared in normal laboratory conditions. They were divided into 5 groups. Four groups contained specimens with embedded steel needles of different lengths (2 mm, 4 mm, 7 mm, and 10 mm) while the fifth group consisted of needleless specimens. After 28 days of hydration (20°C, 100% RH), the specimens were subjected to compressive tests (loading rate 0.2 MPa/s), and the resulting values were averaged within each group of specimens. Although the European standard (EN 12390-3) suggests using the loading rate of 0.6 MPa/s for compressive tests, we used a lower loading rate of 0.2 MPa/s to better satisfy the condition of static loading. In spite of the fact that the stress-controlled rate of compressive tests does not usually enable observation of the postpeak behavior of specimens, these tests make it possible to obtain the peak stress that represents the sought compressive strength. Each group of specimens consisted of 10 specimens which enabled calculation of the statistical

uncertainty of the resulting compressive strength. Statistical errors were plotted as error bars in the graphs discussed in the next section. All specimens successfully passed the destructive compressive tests up to complete failure. Several such specimens are shown in Figure 3.

Figure 4 shows a microscopic three-dimensional detail of the transition zone between the cement paste and the steel needle. This figure illustrates topographical irregularities which appear in a usual transition zone. The three-dimensional projections have been performed by means of 3,937 confocal snaps made by the camera of the laser confocal microscope Olympus Lext 3100.

2.1. Experiments with Ordinary Portland Cement Paste. The averaged values of compressive strength of the tested specimens made of ordinary Portland cement paste have been plotted against the relative needle lengths ρ (Figure 5):

$$\rho = \frac{h}{H} \times 100\% = \frac{\text{needle length}}{\text{sample height}} \times 100\%. \qquad (2)$$

The experimental data in Figure 5 indicate a clear dependence of the compressive strength on the lengths of the needles. An exponential function f_c,

$$f_c = f_0 \exp\left(-\frac{\rho}{\rho_0}\right) + f_1, \qquad (3)$$

has been fitted to the data in Figure 5, and the following optimized parameters have been obtained:

$$f_0 = 645.972 \text{ MPa},$$
$$\rho_0 = 2.356\%, \qquad (4)$$
$$f_1 = 63.493 \text{ MPa}.$$

The high absolute value of the computed correlation coefficient ~0.999 confirms that the functional type of the fitting function f_c has been chosen suitably.

The decreasing behavior of the compressive strength of the tested specimens in Figure 5 looks reasonable since longer needles in the specimens may introduce larger perturbations in the material and thus the compressive strength diminishes. However, when the values of the compressive strength are compared to the strength of the needleless specimens (the dashed horizontal line in Figure 5), it becomes evident that some needles reduce whereas others increase the compressive strength of this material. There is a certain crossing point determined by the strength curves of the needle specimens and the specimens without needles. The crossing point determines a limiting relative needle length. Below this point, the short needles increase the strength, whereas above it the longer needles decrease the strength. The critical value of the relative length belonging to the crossing point amounts to 14.28%. This means that the needles shorter than 14.28% of the specimen height may contribute to a better compressive strength of the hydrated cement paste. Such short needles fastened on the steel bases (Figure 3) introduce only negligible perturbations in the material and act as a steel reinforcement of the lower parts of the specimens.

FIGURE 3: Hydrated cement paste specimens containing steel needles after the final destructive tests.

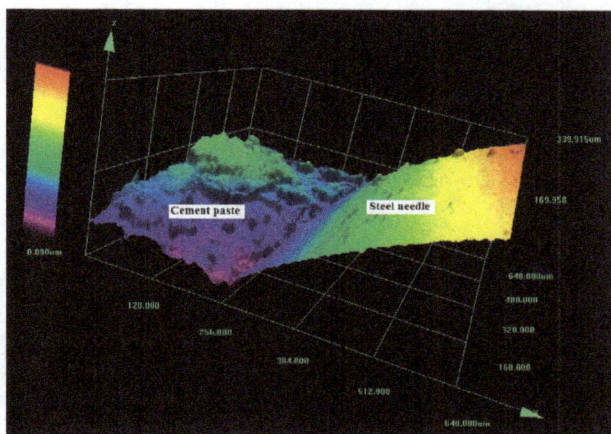

FIGURE 4: Microscopic three-dimensional topography of the transition zone between cement paste and the steel needle. The laser confocal microscope Olympus Lext 3100 has been used. The format of the image: $640\,\mu m \times 640\,\mu m$, 1024 pixels × 1024 pixels, vertical step $\Delta z = 0.09\,\mu m$, total height $334\,\mu m$, and magnification 20x.

The longer needles that encompass more than 15% of the specimen height may introduce such large perturbations in the hydrated cement paste that this material inevitably lowers its strength. The mentioned lowering of compressive strength does not fall to zero value but it approaches a figure close to $f_1 \approx 63.493\,MPa$. This represents a decrease of 1.5 MPa (2.3%) relative to the compressive strength of the unperturbed needleless material. Although the drop of 2.3% might seem to be close to material variability, the error bars shown in Figure 5 do not support such an interpretation. In addition, the 16 MPa increase of the compressive strength caused by short needles with relative length $\rho \approx 8.5\%$ represents 24.6% increase owing to the compressive strength of the unperturbed needleless specimens, and such a large increase cannot with certainty be interpreted as the consequence of material variability.

The experiments with hydrated ordinary Portland cement paste showed that the presence of steel needles of various lengths in the volume of this material causes changes of compressive strength. When short thin steel needles ($8.5\% < \rho < 15\%$) were embedded into the volume of the specimens, the compressive strength was higher

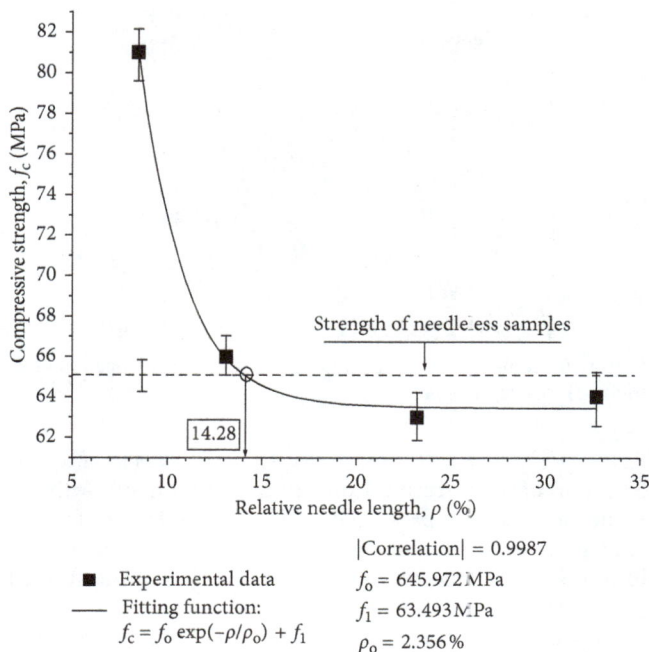

FIGURE 5: Compressive strength of hydrated ordinary Portland cement paste with embedded steel needles of various lengths.

than that of needleless specimens, whereas longer needles ($\rho > 15\%$) decreased the strength under the value of needleless specimens.

This surprising behavior of cement paste raised a question whether similar behavior can also be expected with other cement-based materials, e.g., with cement mortars. For this reason, a new series of larger specimens made of cement mortars (plain "concretes" with fine aggregates) were prepared to study strength changes caused by the needles.

3. Specimens Made of Cement Mortar

To simulate the action of sharp needles on the compressive strength of cement mortar, a series of specimens containing steel needles of various lengths were prepared. The cement mortar was mixed from sand with grains 0.1–0.4 mm and cement Cem I 42.5R produced in the Czech Republic. The water-to-cement-to-sand ratio was $1:1.943:7.773$ by weight, i.e., the water-to-cement ratio was 0.515 by weight. Forty-eight specimens ($10\,cm \times 10\,cm \times 10\,cm$) were molded under normal laboratory conditions. They were divided into 8 groups, each of which contained 6 specimens. Seven groups contained specimens with embedded steel needles of different lengths (5 mm, 7 mm, 11 mm, 19 mm, 29 mm, 40 mm, and 58 mm), while the eighth group consisted of needleless specimens. After 28 days of hydration in water (20°C), the specimens were subjected to compressive tests (loading rate 0.6 MPa/s), and the resulting values were averaged within each group of specimens. All the specimens successfully passed the destructive compressive tests up to complete failure. For illustration, three such specimens are shown in Figure 6 along with the steel needles.

FIGURE 6: Cement mortar specimens containing steel needles after the final destructive tests.

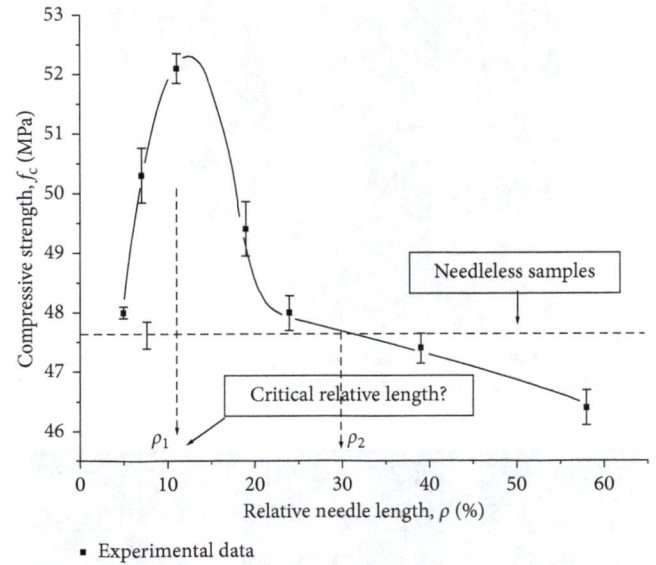

FIGURE 7: Compressive strength of cement mortar with embedded steel needles of various lengths.

3.1. *Experiments with Specimens Made of Cement Mortar.* Some preliminary results concerning experiments with cement mortars were presented in the form of the conference contribution [12] but the present study of cement mortars brings a more extensive and deeper research of the studied phenomenon.

The averaged values of compressive strength of the eight investigated groups of specimens are shown in Figure 7. The same coordinate system has been used as with the cement paste, i.e., the graph of the function $f_c(\rho)$ has been plotted.

By comparing Figures 5 and 7, it is observed that the behavior of the compressive strength of cement mortar and cement paste specimens is similar. However, the larger specimens of cement mortar have enabled to investigate a new region $\rho \in (0, 10)\%$, where the curve of compressive strength has showed an increasing tendency. The small samples of cement paste (3 cm × 3 cm × 3 cm) do not show such a region since the corresponding data belonging to the interval $\rho \in (0, 10)\%$ have not been measured in this domain due to the extremely short length of needles ($l < 1$ mm). Nevertheless, both the curves in Figures 5 and 7 show a decreasing sequence in the domain of larger needles $\rho > 10\%$. The compressive strength of the cement mortar is shifted towards lower values as compared with the hydrated cement paste because of the different water-to-cement ratios. The water-to-cement ratio of cement paste is 0.3 but the ratio of cement mortar is 0.515, i.e., almost two times higher, and this inevitably results in lower compressive strength of the mortar.

A detailed analysis of the graph in Figure 7 makes it possible to distinguish further characteristic features of the investigated phenomenon. The decrease of strength of cement mortar specimens is very rapid at first ($\rho \in (10\%, 20\%)$) but for quite large needles $\rho > 20\%$, the decrease of compressive strength is considerably reduced. In addition, the overall behavior of compressive strength may be characterized by two critical points. The first of them represents a critical needle length at which the specimens show the highest value of strength, i.e., $\rho_1 \approx 10\%$. The second critical point $\rho_2 \approx 30\%$ is specified by the crossing of the graph of compressive strength with the horizontal dashed line which represents the compressive strength of the needleless mortar. To the left of the second critical point $\rho_2 \approx 30\%$, there are specimens showing higher strength, whereas to the right the specimens show lower strength as compared with the normal needleless cement mortar.

When looking for an adequate explanation of the investigated phenomenon, it is necessary to consider the role of the needle length. It is observed that the short needles strengthen the cement mortar, whereas the longer needles reduce its compressive strength. The shorter needles may act as a certain type of reinforcement in the lower parts of the specimens. The longer needles may represent larger perturbations in the mortar specimens and as such they weaken the structure rather than reinforcing it.

Since the decreasing sequences $\rho > 10\%$ of the strength curves visible in Figures 5 and 7 seem to be of the same functional type, the former fitting function specified by equation (3) has also been used for cement mortar (naturally, the coefficients f_o, ρ_o, and f_1 resulted from the fitting procedure applied to mortar data assume different values in comparison with the fitting coefficients of cement paste):

$$f_c = f_o \exp\left(-\frac{\rho}{\rho_o}\right) + f_1, \quad \rho > \rho_1,$$

$$\rho = \frac{h}{H} \times 100\%, \tag{5}$$

$$H = 100 \text{ mm},$$

$$f_o = 12.843 \text{ MPa},$$
$$\rho_o = 12.985\%, \tag{6}$$
$$f_1 = 46.403 \text{ MPa}.$$

The symbol ρ_1 is the first critical length that has been mentioned in the foregoing paragraph and is marked in Figure 7. The resulting optimized graph of function (5) is shown in Figure 8. Again, the correlation coefficient is very high (0.990), which supports the soundness of the chosen fitting function (equation (5)). It is possible to conclude that the solid needles embedded in the volume of the cement

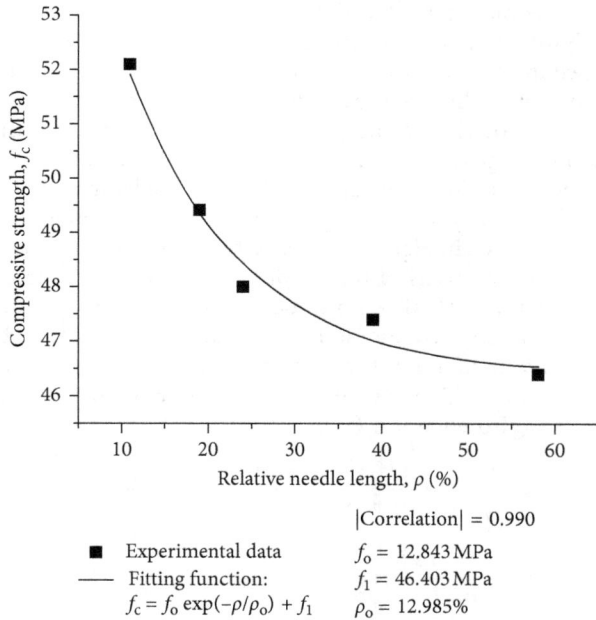

FIGURE 8: The decreasing sequence of compressive strength of cement mortar in dependence on relative needle length. The experimental data are identical to those presented in Figure 7.

mortar and cement paste can change the compressive strength of these materials. The strength changes are very similar in both these materials.

4. Discussion

As has been mentioned, the weakening of the structure of cement-based materials (cement paste and cement mortar) caused by the action of longer needles may be explained on the basis of the larger structural perturbations that the needles introduce in the structure. This idea is supported by the fact that the weakening of the structural strength can be described by exponential functions (3) and (5) analogously to the structural weakening caused by porosity perturbations P that are also expressed by the exponential function [30]:

$$f_c = f_o \exp\left(-\frac{P}{P_o}\right). \tag{7}$$

This functional (mathematical) analogy leads to the assumption that porosity perturbations and metallic perturbations are generically related. If this idea is correct, then there should be a chance to rewrite the exponential decrease specified by equations (3) or (5) into a functional form resembling the model of porosity perturbations (equation (7)). For this purpose, a ratio M between the perturbed volume v and the total volume V_o of the material is introduced:

$$M = \frac{v}{V_o} \times 100\,(\%) = \frac{n \cdot h \cdot \pi \cdot r^2}{H^3} \times 100\%,$$

$$v = n \cdot h \cdot \pi \cdot r^2, \tag{8}$$

$$V_o = H^3 = 1000\,\mathrm{cm}^3,$$

where v is the volume of metallic needles in one mortar cube, $n = 30$ is the number of needles in one mortar cube, and $r = 2$ mm is the radius of the needles used. The ratio M might be called the inclusion volume ratio of cement mortar. The inclusion volume ratio M introduced in the present paper corresponds to the well-known definition of porosity P.

The fraction ρ/ρ_o in the exponential function (5) can be replaced by the fraction M/M_o and a modified exponential function may be derived as follows:

$$\frac{\rho}{\rho_o} = \frac{h}{\rho_o \cdot H} = \frac{n \cdot h \cdot \pi \cdot r^2}{H^3} \times \frac{H^2}{\rho_o \cdot n \cdot \pi \cdot r^2}$$

$$= \frac{v}{V_o} \times \frac{H^2}{\rho_o \cdot n \cdot \pi \cdot r^2} = \frac{M}{M_o}, \tag{9}$$

$$M_o = \frac{\rho_o \cdot n \cdot \pi \cdot r^2}{H^2} = 0.489 \approx 0.49\%, \tag{10}$$

$$f_c = f_o \exp\left(-\frac{M}{M_o}\right) + f_1, \quad \text{for } M \geq 0.38\%,$$

$$\rho > \rho_1 = \frac{h_1}{H} \implies M > M_1 = \frac{h_1}{H} \times \frac{n \cdot \pi \cdot r^2}{H^2} \times 100\% \approx 0.38\%. \tag{11}$$

If all these considerations and derivations are correct, then the fit of the exponential pattern (11) to the cement mortar data should result in the same values of f_o and f_1 as in equation (6), and the parameter M_o should be equal to 0.49 as in equation (10). The results of the fitting procedure are presented in Figure 9. The procedure of least squares has provided the following parameters:

$$f_o = 12.843\,\mathrm{MPa},$$

$$M_o = 0.495\% \approx 0.49\%, \tag{12}$$

$$f_1 = 46.403\,\mathrm{MPa},$$

which are in good agreement with equations (6) and (10). The high correlation coefficient 0.990 has again confirmed the soundness of the fitting procedure.

Finally, it should be mentioned that there are some differences between the functions (7) and (11). First of all, the concept of inclusion volume ratio M in contrast to porosity P is restricted by a numerical extent. For example, f_c in equation (11) cannot approach a real value f_o of the needleless cubes since function (11) describes only the region to the right of the first critical point $\rho > \rho_1$, and for the left interval $\rho < \rho_1$, this pattern is inapplicable. In reality, M cannot go to "infinity" since the length of the needles is restricted by the size of the cubes H. Thus, the maximum value of M is restricted by the maximum length of the needles ($h_{\max} = H$) and their numbers n (in our case $n = 30$), i.e., $M_{\max} = (n \cdot \pi \cdot r^2/H^2) \times 100\% = 3.77\%$. Therefore, the following relation has to be fulfilled for maximum $M_{\max}$:

$$f_c(M_{\max}) \approx \lim_{M \to \infty} f_c(M) = f_1. \tag{13}$$

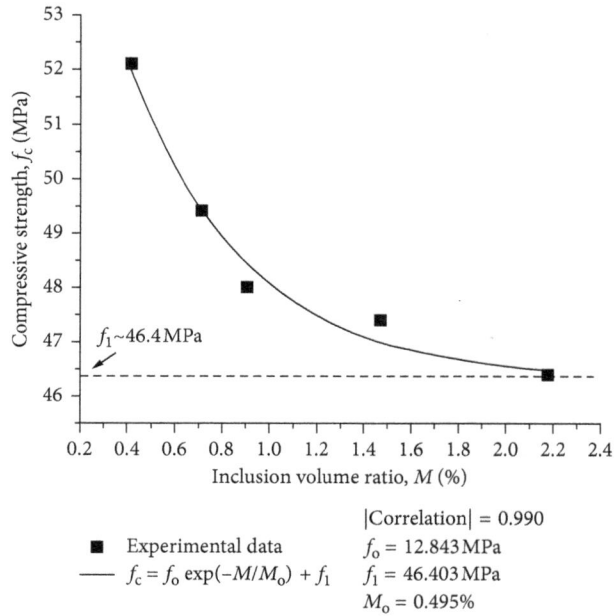

FIGURE 9: The decreasing sequence of compressive strength of cement mortar in dependence on inclusion volume ratio. The experimental data are identical to those presented in Figures 7 and 8.

This can be easily verified by inserting $M_{max} = 3.77\%$ into equation (11). The resulting value of compressive strength $f_c(M_{max}) = 46.409$ MPa corresponds almost exactly to the optimized value $f_1 = 46.403$ MPa.

All these results support our hypothesis about the analogy between the relative content of inclusions M and the porosity P. Yet, the aforementioned experimental procedures showing the responsibility of transition zones for lowering the compressive strength would rather be incomplete if the influence of the areal extent of the zones on the compressive strength were not explored. The next paragraph will shed light on the problem and will bring a desirable rationalization of the studied phenomenon.

First of all, it is evident that the thicknesses of all the transition zones between the metallic needles and the cement stone are nearly identical because all the metallic needles have the same shapes and are made of the same material, and the cement stones have been prepared with the same cement material and have hydrated under the same conditions. But the total area of the jointed surfaces (metal needles versus stone) extends when the lengths of the needles are enlarging. Thus, the phenomenon of lowering compressive strength should be dependent on the extending area of the interfaces between these two materials and, as a consequence, the exponential decrease of compressive strength should also be dependent on this area. Since the area of the bases of the needles is constant (i.e., $n \times 2\pi r^2$) and only the area of the side walls of the needles are extending, the relevant increasing area is given by the side envelopes of the needles, i.e., $s = n \times 2\pi rh$. A transition zone represents structural weakness in a material, and extending areas of zones are accompanied by a higher risk (higher probability) of structural collapse

when the material is faced with shear stresses acting along the borders of the transition zones. Since the needles are loaded in vertical positions, their side envelopes experience shear stress. Thus, the shearing of materials along the side envelopes of the needles may be the crucial mechanism leading to perturbation of the cement stone and, as a consequence, the propagation of microcracks may be directed vertically, continuing the orientations of the needle sides as shown in Figure 10. Due to the destructive shearing along the side walls of the needles, the total critical area s may be relevant also for model considerations. If this concept is correct, the active area of zones s should be implicitly included in the formerly used inclusion volume ratio M/M_o. This is documented in the following derivation (see also equation (9)):

$$\frac{M}{M_o} = \frac{v}{V_o} \times \frac{H^2}{\rho_o \cdot n \cdot \pi \cdot r^2} = \frac{n \cdot \pi \cdot r^2 \cdot h}{H^3} \times \frac{H^2}{\rho_o \cdot n \cdot \pi \cdot r^2}$$

$$= \frac{n \cdot 2\pi \cdot r \cdot h}{6H^2} \times \frac{3H \cdot r}{\rho_o \cdot n \cdot \pi \cdot r^2}$$

$$= \frac{s}{S_o} \times \frac{1}{((\rho_o \cdot n \cdot \pi \cdot r)/3H)} = \mu \times \frac{1}{\mu_o} = \frac{\mu}{\mu_o},$$

$$(14)$$

where $S_o = 6H^2 = 0.06$ m^2 is the constant surface area of one cube specimen (S_o serves as a reference quantity), $\mu_o = (\rho_o \cdot n \cdot \pi \cdot r)/(3H) = 8.1587\%$ is a constant quantity since H and r are also constants in our experiment, and $\mu = s/S_o$ is a relative dimensionless area of transition zones and will be called the "zone area ratio." To verify this theoretical scheme experimentally, a plot of the couples of data (μ, f_c) has been realized and the results are shown in Figure 11. Again, a decrease of compressive strength has been manifested, and the data have been fitted by the exponential function:

$$f_c = f_o \exp\left(-\frac{\mu}{\mu_o}\right) + f_1.$$

$$(15)$$

The fitting procedure performed by the least-squares method has yielded the following optimized numbers:

$$f_o = 12.843 \text{ MPa},$$
$$\mu_o = 8.1586\%,$$
$$(16)$$
$$f_1 = 46.403 \text{ MPa}.$$

These numbers fully confirm the derivations presented in equation (14) and, in addition, the numerical agreement between the calculated value $\mu_o = 8.1587\%$ and the fitted value $\mu_o = 8.1586\%$ documents a full harmony between the theoretical model and the experiment. The model of shear stress acting along the side walls of the vertical needles has enabled us to correlate the compressive strength with the area of the side walls, which are actually areal borders of the transition zones. The side walls form envelopes of the "cavities" that are filled with metallic needles. Because the total area s of all the side envelopes and the total volume v of all the "cavities" in one cube specimen are tightly correlated ($v = s \times r/2$, $r =$ const.), it should not be surprising that the

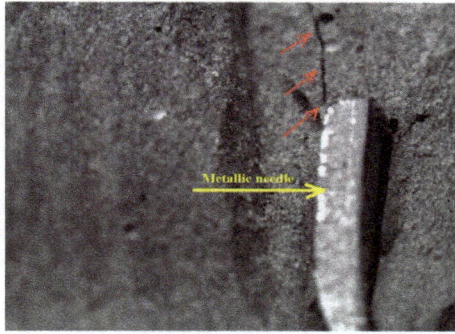

FIGURE 10: A microcrack propagation along the metallic needle embedded in hydrated cement paste. The crack continues the vertical direction of the needle side wall.

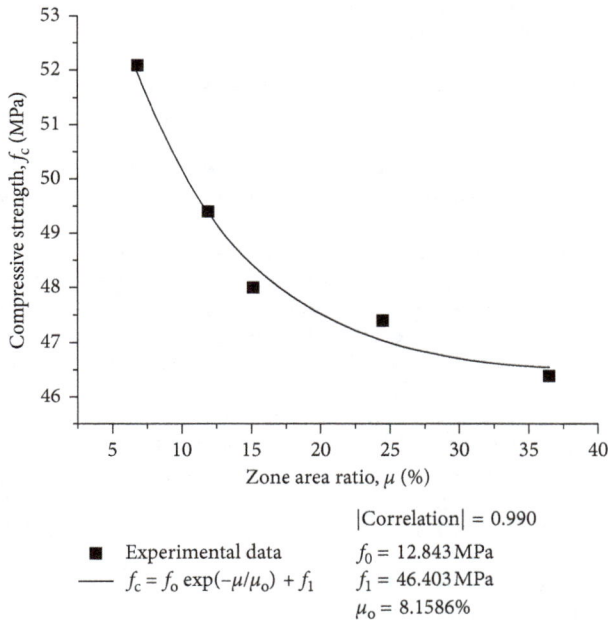

|Correlation| = 0.990

■ Experimental data
— $f_c = f_o \exp(-\mu/\mu_o) + f_1$

$f_0 = 12.843\,\text{MPa}$
$f_1 = 46.403\,\text{MPa}$
$\mu_o = 8.1586\%$

FIGURE 11: The decreasing sequence of compressive strength of cement mortar in dependence on zone area ratio. The experimental data are identical to those shown in Figures 7–9.

exponential decrease of compressive strength can be described by both the volume and area ratios, i.e., by the M and μ parameters, respectively:

$$\frac{\rho}{\rho_o} = \frac{M}{M_o} = \frac{\mu}{\mu_o}. \tag{17}$$

Since no chemical bonds are established between the metallic needles and cement stone, the studied phenomenon is of purely mechanical origin, and it may be hypothesized that after replacing the metallic needles by high-strength peaked asperities oriented vertically as the metallic needles, an analogous behavior of compressive strength might be observed. At least two conditions have to be fulfilled in order that a decrease of compressive strength may occur: (1) the compressive strength of the needle-like asperities should be larger than that of cement stone; (2) the relative asperity

length ρ should exceed the critical limit ρ_1 that is specifically optimized for transition zones between asperities and cement stone. Condition (2) may be easily fulfilled, e.g., in the leveling concrete layers mentioned in Introduction. It is not excluded that the vertically oriented steel reinforcement bars in concrete may also be considered analogously to the vertical needles in our experiments.

One important point has not been discussed so far, namely, the effect of the "population" density of needles on the compressive strength of cement-based materials. The "population" density (σ) of needles, i.e., the number of needles per unit area, is a very important factor since it determines, together with the dimensions of the needles, the volume and area ratios, i.e., $M = v/V_o$ and $\mu = s/S_o$, respectively. By varying the population density σ of needles of constant relative height ρ, both the volume M and area μ ratios are varied accordingly. This means that volume and area parameters M and μ are of the same importance as the parameters ρ. The two parameters M and μ include the influence of both the population density σ and the relative length ρ. To obtain critical limits for parameters M and μ, it is possible to use equation (17) and previously found limits $\rho_1 \approx 10\%$ and $\rho_2 \approx 30\%$ measured at a constant population density ($\sigma = 30\,\text{dm}^{-2}$):

$$M_1 = \frac{\rho_1}{\rho_o} \times M_o = \frac{10}{12.985} \times 0.495 = 0.381\%,$$

$$M_2 = \frac{\rho_2}{\rho_o} \times M_o = \frac{30}{12.985} \times 0.495 = 1.144\%,$$

$$\mu_1 = \frac{\rho_1}{\rho_o} \times \mu_o = \frac{10}{12.985} \times 8.159 = 6.283\%,$$

$$\mu_2 = \frac{\rho_2}{\rho_o} \times \mu_o = \frac{30}{12.985} \times 8.159 = 18.850\%. \tag{18}$$

The most practical parameters seem to be the inclusion volume ratios M since they can be easily determined in practice. The first critical parameter M_1 specifies the situation at which the compressive strength reaches its maximum value high above the normal compressive strength of needleless materials, and the second parameter M_2 determines the critical point at which the compressive strength drops under the normal compressive strength of needleless materials.

The parameters $M_1 \approx 0.38\%$ and $M_2 \approx 1.14\%$ can hardly be universal for all cement-based materials since different compositions of these materials inevitably change the mechanical properties of transition zones, and thus the critical limits M_1 and M_2 may change more or less their values. Nevertheless, each of these materials will have its own characteristic values for M_1 and M_2.

5. Conclusions

On the basis of the experiments presented in this paper, several conclusions may be drawn:

(i) Short needle inclusions ($\rho < \rho_2$) whose relative volume content M in cement-based materials is sufficiently small $(M < M_2)$ do not worsen the compressive strength and may even contribute to a better material integrity.

(ii) Longer needle inclusions $(\rho > \rho_2)$ with higher relative content $M > M_2$ may worsen the compressive strength of cement-based materials.

(iii) The critical parameters $\rho_{1,2}$, $M_{1,2}$, and $\mu_{1,2}$ are not universal quantities but are dependent on the type of cement-based materials and the mechanical and shape properties of the needle-like inclusions.

(iv) A hypothesis has been put forward that high-strength sharp asperities of bedrocks might act in thin concrete leveling layers similarly to needle-like inclusions. This may especially concern rock asperities composed of high-quality materials such as granite, basalt, or gneiss.

(v) There is no doubt that the steel reinforcements usually used in concrete effectively improve the values of tensile and flexural strength. However, in light of the performed experiments, it is not excluded that the steel reinforcement rods in concrete may act like the invasive steel needles modifying the compressive strength of concrete in some special cases.

Conflicts of Interest

The author declares that there are no conflicts of interest regarding the publication of this paper.

Acknowledgments

This paper was supported by the Grant Agency of the Czech Republic under grant no. 13-03403S. The grant support had started in 2013 and finished in 2017.

References

[1] H. Jin, Y. Zhou, B. Wang, and S. Zhou, "Mesoscopic finite element modeling of concrete considering geometric boundaries of actual aggregates," *Advances in Materials Science and Engineering*, vol. 2018, Article ID 7816502, 10 pages, 2018.

[2] B. Mielniczuk, M. Jebli, F. Jamin, M. S. El Youssoufi, C. Pelissou, and Y. Monerie, "Characterization of behavior and cracking of a cement paste confined between spherical aggregate particles," *Cement and Concrete Research*, vol. 79, pp. 235–242, 2016.

[3] Y. J. Peng, H. Chu, and J. W. Pu, "Numerical simulation of recycled concrete using convex aggregate model and base force element method," *Advances in Materials Science and Engineering*, vol. 2016, Article ID 5075109, 10 pages, 2016.

[4] X. H. Wang, S. Jacobsen, J. Y. He, Z. L. Zhang, S. F. Lee, and H. L. Lein, "Application of nanoindentation testing to study of the interfacial transition zone in steel fiber reinforced mortar," *Cement and Concrete Research*, vol. 39, no. 8, pp. 701–715, 2009.

[5] K. Wu, H. Shi, L. Xu, G. Ye, and G. De Schutter, "Microstructural characterization of ITZ in blended cement concretes and its relation to transport properties," *Cement and Concrete Research*, vol. 79, pp. 243–256, 2016.

[6] G. Yue, P. Zhang, Q. Li, and Q. Li, "Performance analysis of a recycled concrete interfacial transition zone in a rapid carbonization environment," *Advances in Materials Science and Engineering*, vol. 2018, Article ID 1962457, 8 pages, 2018.

[7] T. Ficker, "Fitting function for flexural strength of cement paste," in *Proceedings of World Multidisciplinary Civil Engineering-Architecture-Urban Planning Symposium (WMCAUS)*, vol. 245, Published in Book Series: IOP Conference Series-Materials Science and Engineering, Prague, Czech Republic, June 2017.

[8] T. Ficker, "Rupture strength and irregularity of fracture surfaces," in *Proceedings of World Multidisciplinary Civil Engineering-Architecture-Urban Planning Symposium (WMCAUS)*, vol. 245, Published in Book Series: IOP Conference Series-Materials Science and Engineering, Prague, Czech Republic, June 2017.

[9] T. Ficker, "Fractal properties of joint roughness coefficients," *International Journal of Rock Mechanics and Mining Sciences*, vol. 94, pp. 27–31, 2017.

[10] O. Audy and T. Ficker, "Fractal analysis of rock joint profiles," in *Proceedings of World Multidisciplinary Civil Engineering-Architecture-Urban Planning Symposium (WMCAUS)*, vol. 245, Published in Book Series: IOP Conference Series-Materials Science and Engineering, Prague, Czech Republic, June 2017.

[11] O. Audy and T. Ficker, "Evaluation of rock joint coefficients," in *Proceedings of World Multidisciplinary Civil Engineering-Architecture-Urban Planning Symposium (WMCAUS)*, vol. 245, Published in Book Series: IOP Conference Series-Materials Science and Engineering, Prague, Czech Republic, June 2017.

[12] T. Ficker and T. Komárková, "Rock joints asperities and mechanical strength of concrete," in *Proceedings of World Multidisciplinary Civil Engineering-Architecture-Urban Planning Symposium (WMCAUS)*, vol. 245, Published in Book Series: IOP Conference Series-Materials Science and Engineering, Prague, Czech Republic, June 2017.

[13] X. Chang, J. Lu, S. Wang, and S. Wang, "Mechanical performances of rock-concrete bi-material disks under diametrical compression," *International Journal of Rock Mechanics and Mining Sciences*, vol. 104, pp. 71–77, 2018.

[14] J. G. Gutiérrez-Ch, S. Senent, S. Melentijevic, and R. Jimenez, "Distinct element method simulations of rock-concrete interfaces under different boundary conditions," *Engineering Geology*, vol. 240, pp. 123–139, 2018.

[15] W. Dong, D. Yang, B. Zhang, and Z. Wu, "Rock-concrete interfacial crack propagation under mixed mode I-II fracture," *Journal of Engineering Mechanics*, vol. 144, no. 6, article 04018039, 2018.

[16] W. Dong, D. Yang, X. Zhou, G. Kastiukas, and B. Zhang, "Experimental and numerical investigations on fracture process zone of rock-concrete interface," *Fatigue and Fracture of Engineering Materials and Structures*, vol. 40, no. 5, pp. 820–835, 2016.

[17] W. Dong, Z. Wu, X. Zhou, N. Wang, and G. Kastiukas, "An experimental study on crack propagation at rock-concrete

interface using digital image correlation technique," *Engineering Fracture Mechanics*, vol. 171, pp. 50–63, 2017.

[18] W. Dong, Z. Wu, and X. Zhou, "Fracture mechanisms of rock-concrete interface: experimental and numerical," *Journal of Engineering Mechanics*, vol. 142, no. 7, article 04016040, 2016.

[19] H. Jiang, "A failure criterion for rocks and concrete based on the Hoek-Brown criterion," *International Journal of Rock Mechanics and Mining Sciences*, vol. 95, pp. 62–72, 2017.

[20] Z. J. Yang and A. J. Deeks, "Fully-automatic modelling of cohesive crack growth using a finite element-scaled boundary finite element coupled method," *Engineering Fracture Mechanics*, vol. 74, no. 16, pp. 2547–2573, 2007.

[21] V. Andjelkovic, N. Pavlovic, Z. Lazarevic, and V. Nedovic, "Modelling of shear characteristics at the concrete-rock mass interface," *International Journal of Rock Mechanics and Mining Sciences*, vol. 76, pp. 222–236, 2015.

[22] Y. A. Fishman, "Stability of concrete retaining structures and their interface with rock foundations," *International Journal of Rock Mechanics and Mining Sciences*, vol. 46, no. 6, pp. 957–966, 2009.

[23] Y. A. Fishman, "Features of shear failure of brittle materials and concrete structures on rock foundations," *International Journal of Rock Mechanics and Mining Sciences*, vol. 45, no. 6, pp. 976–992, 2008.

[24] J. K. Kodikara and I. W. Johnston, "Shear behaviour of irregular triangular rock-concrete joints," *International Journal of Rock Mechanics and Mining Sciences and Geomechanics Abstracts*, vol. 31, no. 4, pp. 313–322, 1994.

[25] K. M. Lee and O. Buyukozturk, "Fracture analysis of mortar-aggregate interfaces in concrete," *Journal of Engineering Mechanics*, vol. 118, no. 10, pp. 2031–2047, 1992.

[26] I. W. Johnston and T. S. K. Lam, "Shear behavior of regular triangular concrete/rock joints-analysis," *Journal of Geotechnical Engineering*, vol. 115, no. 5, pp. 711–727, 1989.

[27] T. S. K. Lam and I. W. Johnston, "Shear behavior of regular triangular concrete/rock joints-evaluation," *Journal of Geotechnical Engineering*, vol. 115, no. 5, pp. 728–740, 1989.

[28] N. Barton, "Review of a new shear-strength criterion for rock joints," *Engineering Geology*, vol. 7, no. 4, pp. 287–332, 1973.

[29] N. Barton and V. Choubey, "The shear strength of rock joints in theory and practice," *Rock Mechanics Felsmechanik Mécanique des Roches*, vol. 10, no. 1, pp. 1–54, 1977.

[30] P. K. Mehta and P. J. M. Monteiro, *Concrete, Microstructure, Properties and Materials*, McGraw-Hill, New York, NY, USA, 3rd edition, 2006.

14

Effect of TIPA on Chloride Immobilization in Cement-Fly Ash Paste

Baoguo Ma, Ting Zhang, Hongbo Tan ⓘ, Xiaohai Liu, Junpeng Mei, Wenbin Jiang, Huahui Qi, and Benqing Gu

State Key Laboratory of Silicate Materials for Architectures, Wuhan University of Technology, Wuhan 430070, China

Correspondence should be addressed to Hongbo Tan; thbwhut@whut.edu.cn

Academic Editor: Cristina Leonelli

Utilization of sea sands and coral aggregate for concrete in ocean construction is increasingly attracting the attention all over the world. However, the potential risk of steel corrosion resulting from chloride in these raw materials was one of the most concerned problems. To take this risk into account, chloride transporting to the surface of steel should be hindered. The formation of Friedel's salt in hydration process is widely accepted as an effective manner for this hindrance. In this study, an attempt to hasten the formation of Friedel's salt by adding triisopropanolamine (TIPA) was done in the cement-fly ash system, with intention to chemical bind chloride, and the chloride-binding capacity at 60 d age was examined. The results show that TIPA can enhance the chloride-binding capacity of cement-fly ash paste at 60 d age, and the reason is that the formation of Friedel's salt can be accelerated with addition of TIPA. The mechanism behind is revealed as follows: on the one hand, the accelerated cement hydration provides more amount of calcium hydroxide to induce the pozzolanic reaction of fly ash, which can hasten the dissolution of aluminum into liquid phase; on the other hand, TIPA can directly hasten the dissolution of aluminum in fly ash, offering more amounts of aluminum in liquid phase. In this case, the aluminum/sulfate (Al/S) ratio was obviously increased, benefiting the formation of Friedel's salt in hydration products. Such results would expect to provide useful experience to promote the chloride-binding capacity of cement-fly ash system.

1. Introduction

In recent years, marine exploitation is increasingly attracting the attention all over the world. In ocean construction, the durability of concrete is of great importance [1, 2]. However, attempt to utilize the sea sands and coral aggregate as raw materials of concrete has been done, and the potential risk of steel corrosion is one of the most concerned problems in terms of durability. In fact, this risk depends on whether the chloride could transport to the surface of the steel or not. Therefore, binding the chloride to hinder its transport would be of great importance to the durability of reinforced concrete [3–5].

There are two forms of chloride ions in ocean aggregates. One is free chloride, which means that this kind of chloride can transport around in concrete, and the other is bound chloride. The former would lead to the corrosion of

reinforcing steel if the chloride is transported to the surface of steel, while the latter has almost no risk of steel corrosion. Therefore, binding free chloride would significantly reduce the risk of steel corrosion, with great benefit to the durability of concrete structures. According to the binding mechanism, chemical reactions and physical absorption can be found in the literatures [6–8]. The former means that chloride ions can participate in the hydration reaction to form the hydration products, such as Friedel's salt (FS, $3CaO \cdot Al_2O_3 \cdot CaCl_2 \cdot 10H_2O$) and Kuzel's salt (KS, $3CaO \cdot Al_2O_3 \cdot 0.5CaCl_2 \cdot 0.5CaSO_4 \cdot 10H_2O$) [9, 10]. The latter is that chloride ions are mainly absorbed by calcium silicate hydrate (C-S-H) gel [11–13]. By contrast, the former is more effective and plays the dominant role. Furthermore, binding chloride by formation of FS and KS depends on the reaction of aluminum ions, sulfate ions, chloride ions, and calcium ions [14]. Generally, more amounts of the aluminum phase in the cement system can lead to more

TABLE 1: Chemical compositions of cement and FA.

	Loss	SiO_2	Al_2O_3	Fe_2O_3	SO_3	CaO	MgO	K_2O	Na_2O
Cement (wt.%)	3.82	24.08	4.72	2.46	2.31	58.24	1.95	1.02	0.27
FA (wt.%)	5.97	48.33	31.69	4.14	1.37	4.12	0.50	1.34	0.37

amounts of formation of FS. Taking C_3A, for example, cement with higher content of C_3A can obviously increase the content of FS in hydration products [15–17]. Addition of supplementary cementitious materials, such as fly ash (FA), can also hasten the formation of FS because aluminum can be dissolved in the process of pozzolanic reaction [18–20].

In the cement-FA system, chemicals can also hasten the pozzolanic reaction of FA and dissolution of aluminum. Paya et al. showed that grinding fly ash to finer particles can obviously hasten the hydration of fly ash, but this method will lead to energy consumption [21]. Dakhane et al. reported that pH-neutral alkali sulfates could activate fly ash, resulting in 70% reduction of clinker factor [22]. Sodium sulfate can also activate the pozzolanic reaction of fly ash, but this kind of chemical has negative effect on long-term performance of cement-based materials [23]. By contrast, TIPA can exert high efficiency to hasten the dissolution of FA and cement minerals [24–27]. On the one hand, the accelerated hydration of cement can form more calcium hydroxide (CH) to hasten the pozzolanic reaction of FA; on the other hand, TIPA can also induce the dissolution of FA to release more amounts of silicate and aluminum into solution to participate in the hydration. In this case, with addition of TIPA in the cement-FA system, pozzolanic reaction of FA and dissolution of aluminum would be accelerated and the amount of FS would be expected to increase, with great contribution to chloride-binding capacity.

In this study, the chloride-binding capacity of the cement-FA system with addition of TIPA was systemically studied. The free chloride was induced with addition of sodium chloride (NaCl), and the chloride-binding capacity of the paste cured for 60 d was examined. Hydration process of the system was investigated with analysis of hydration heat, and the hydration products were characterized with scanning electron microscope (SEM), thermogravimetric analysis (TGA), and X-ray diffraction (XRD). The reaction degree of cement and FA was evaluated with solid-state nuclear magnetic resonance (NMR). The dissolution of FA was analyzed with SEM and inductively coupled plasma (ICP) emission spectrometer. The mechanism behind was investigated in terms of hydration heat, hydration products, and dissolution of FA. Such results were expected to provide useful experience to promote the chloride-binding capacity of the cement-FA system.

2. Materials and Test Methods

2.1. Materials

2.1.1. Cement and Fly Ash. Portland cement (P.I 42.5, Wuhan Yadong Cement Co., Ltd.) in accordance with the requirements of GB175-2007 (Chinese Standard) and class F-II FA in accordance with the requirements of GB/T 1596-2005 (Chinese Standard) were used in this study.

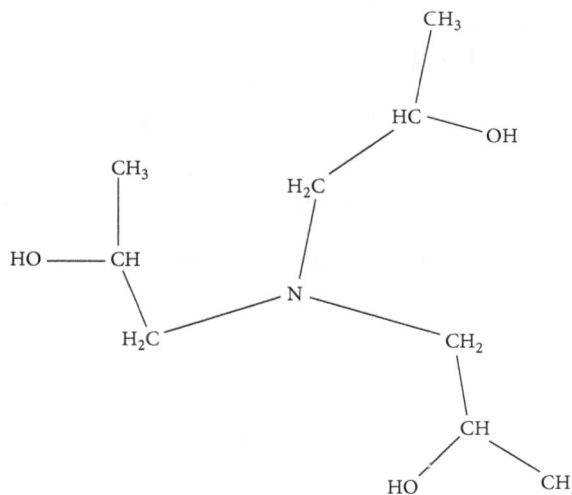

FIGURE 1: Schematic diagram of molecular structure of TIPA.

The chemical compositions of cement and FA were analyzed by XRF, and the results are given in Table 1.

2.1.2. TIPA. A reagent-grade triisopropanolamine (TIPA, anhydrous white solid, ≥95% purity, made by Aladdin Biochemical Technology Co., Ltd., Shanghai, China) was used. Additionally, the added dosage of TIPA was recorded as the solid amount. The chemical structure of TIPA is show in Figure 1.

2.1.3. Preparation of Specimens. Cement-FA paste (30% FA and 70% cement) with addition of NaCl (1.11% of cement-FA binder) and TIPA (0%, 0.03%, 0.06%, and 0.10% of cement-FA binder) was prepared with a water/binder ratio of 0.38. TIPA and NaCl were dissolved in water in advance. The fresh pastes were cast in 40 mm × 40 mm × 40 mm cubic metallic moulds, cured in a >90% RH and 20 ± 1°C chamber for 24 h, and then demoulded and further cured with the same condition. At the age of 60 days, compressive strength was measured. The samples were also broken into small pieces and immediately immersed into ethanol in order to stop hydration. The pieces were dried in a vacuum drier at 60°C. The samples were prepared for the measurement of chloride-binding capacity. Additionally, these specimens were also powdered by hand, and the powder, which could pass through a 63 μm sieve, was prepared for the phase analysis (i.e., hydration products).

2.2. Test Methods

2.2.1. Chloride-Binding Capacity. The sample (about 20 g) was dried in a vacuum drier at a temperature of 60 ± 5°C for

2 hours and then powdered by hand. The powder, which could pass through a $15\,\mu m$ sieve, was prepared for the measurement. The sample ($10\,g$) was put into a triangular flask and distilled water ($100\,g$) was added, and then it was oscillated violently for 1-2 min. The sample was soaked for 24 hours and then filtered. The filtrate ($20\,g$) was put into triangular flask. Two drops of phenolphthalein were added as a pH indicator, and then it was neutralized with dilute sulfuric acid to be colorless. After that, 10 drops of potassium chromate indicator with a concentration of 5% were added, and silver nitrate ($0.02\,mol/L$) was added continuously until brick red precipitate appears. At this time, the added volume of silver nitrate solution was marked [28].

The amount of free chloride ion can be calculated as follows:

$$P = \frac{CV_3 \times 0.03545}{(G \times V_2/V_1)} \times 100\%, \qquad (1)$$

where P is the content of free chloride ions in paste, %; C is the concentration of $AgNO_3$, mol/L; G is the weight of the paste sample, g; V_1 is the volume of water used to soak the sample, mL; V_2 is the volume of filtrate used for measurement, mL; and V_3 is the volume of consumption of silver nitrate solution, mL.

The initial content of chloride ions (P_0) in the sample could be calculated by the added amount of NaCl, and the chloride-binding rate (CBR) was calculated as follows:

$$CBR = \frac{(P_0 - P)}{P_0} \times 100\%. \qquad (2)$$

2.2.2. Compressive Strength. Three specimens of each mixture were tested, in accordance with GB/T50081-2002, and the average was the result of compressive strength.

2.2.3. Ions Dissolution of FA. Firstly, pore solution with different concentrations of TIPA (0–20 g/L) was prepared with KOH and NaOH ($K^+/Na^+ = 1:1$; pH = 13). One gram of FA was added into these solutions ($20\,g$), respectively, and then mixed. The suspension was sealed in a plastic container and cured at a temperature of $20 \pm 1°C$. For each 12 h, the containers were shocked in order to make the suspension to be even.

At the age of 60 d, the suspension was centrifuged at 3600 r/min for 10 minutes in a centrifuge, and the content of Al, Fe, and Si in supernatant solution was tested with inductively coupled plasma (ICP, Optima 4300 DV, made by Perkin Elmer Ltd., USA) emission spectrometer. Based on these results, the effect of TIPA on the dissolution of FA was investigated.

In addition, the solid was dried in a vacuum drier with the temperature of $60 \pm 5°C$, and then the surficial morphology of the FA was characterized with field emission scanning electron microscope (SEM).

2.2.4. Hydration Heat. TIPA (0–0.10 wt.% of cement-FA binder) and NaCl (1.11 wt.% of cement-FA binder) were added into water in advance, and then the solution and cement-FA (30% FA and 70% cement) were mixed together with a water/binder ratio of 0.5. Hydration heat was obtained with an isothermal calorimetry (TAM AIR, C80, SETARAM, France).

2.2.5. Phase Analysis. The effect of TIPA on hydration products was investigated with scanning electron microscope (SEM), thermogravimetric analysis (TGA), solid-state nuclear magnetic resonance (NMR), and X-ray diffraction (XRD).

(1) XRD. XRD data were collected with an X-ray diffractometer (XRD, D/Max-RB) (Cu Kα radiation) at room temperature ($2\theta = 5–70°$, step 0.03° with 3 s/point).

(2) SEM. Field emission scanning electron microscope (FE-SEM, QUANTA FEG 450, FEI Co, USA) was used for SEM microstructural characterization.

(3) TG-DTG. TGA was conducted with the comprehensive thermal analyzer (German-resistant STA449F3). The heating rate was 10°C/min, using nitrogen as purging gas, and the temperature ranged from the room temperature to 1000°C. CH was decomposed at the temperature ranging from 400 to 500°C, and calcium carbonate resulting from the carbonation of CH in the process of preparing the samples was decomposed at 500–750°C, as shown in the following equation [23]:

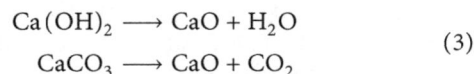

$$\begin{aligned} Ca(OH)_2 &\longrightarrow CaO + H_2O \\ CaCO_3 &\longrightarrow CaO + CO_2 \end{aligned} \qquad (3)$$

The total content of CH in hydration products can be calculated as follow:

$$M_{Ca(OH)_2} = \frac{74 \times M_{H_2O}}{18} + \frac{74 \times M_{CO_2}}{44}. \qquad (4)$$

where $M_{Ca(OH)_2}$: the mass of calcium hydroxide; M_{H_2O}: at 400–500°C, the weight loss resulting from water; and M_{CO_2}: at 500–750°C, the weight loss resulting from carbon dioxide.

Additionally, the weight loss at the temperature range from 50 to 200°C, due to evaporation of free water, dehydration of C-S-H gel, decomposition of ettringite (AFt), and dehydration of FS, is of great interest. To be more precise, the first peak at the low temperature is related to the evaporation of free water and dehydration of C-S-H gel, and the second peak with higher temperature is involved in evaporation of free water, dehydration of C-S-H gel, and decomposition of AFt; the third one is about dehydration of FS [29]. Such results can provide supplementary evidence to illustrate the effect of TIPA on the formation of FS.

(4) NMR. To further verify the effect of TIPA on hydration of cement-FA paste, the hydration products were characterized with ^{29}Si MAS NMR. As reported in the literatures [30], six peaks can be found in NMR spectrum of hydrated cement-FA paste; Q^1 (chain-end groups), Q^2 (middle-chain groups), and Q^2 (1Al) (middle-chain groups where one of the adjacent tetrahedral sites is occupied by Al^{3+}) represent the Si-O tetrahedron in hydration products; and Q^0 represents

the Si-O tetrahedron in unhydrated cement minerals, and Q^3 and Q^4 represent the Si-O tetrahedron in FA. Due to different chemical surroundings of Si sites in cementitious materials, the polymerization degree of Si-O tetrahedron and Al/Si ratios in C-S-H can be evaluated.

^{29}Si-NMR (solid-state nuclear magnetic resonance) was conducted with a Bruker Advance III400 spectrometer operating at 79.5 MHz. The rotation frequency was 5 kHz and the delay time was 10 s. Tetramethylsilane was used as a standard for ^{29}Si. The data were processed with commercial solid-state NMR software package. It was firstly fitted, and then the phasing and baseline were corrected, followed by subsequently iterative fitting. During the deconvolution of ^{29}Si-NMR spectra, the peak shapes were constrained with the Gaussian function. The main chain length (MCL) of C-S-H gel and the ratio of Si in C-S-H substituted by Al were calculated as follows [31]:

$$\text{MCL} = \frac{\left\{2I\left(Q^1\right) + 2I\left(Q^2\right) + 3I\left[Q^2\left(\text{Al}\right)\right]\right\}}{Q^1},$$

$$\frac{\text{Al}}{\text{Si}} = \frac{0.5I\left[Q^2\left(\text{Al}\right)\right]}{\left\{I\left(Q^1\right) + I\left(Q^2\right) + I\left[Q^2\left(\text{Al}\right)\right]\right\}}. \tag{5}$$

Reaction degree of FA and cement was also calculated as follows:

$$A_{\text{FA}}\left(\%\right) = \frac{1 - I\left(Q^3 + Q^4\right)}{I_0\left(Q^3 + Q^4\right)},$$

$$A_{\text{C}}\left(\%\right) = \frac{1 - I\left(Q^0\right)}{I_0\left(Q^0\right)}, \tag{6}$$

where $I\left(Q^0\right)$, $I\left(Q^1\right)$, $I\left(Q^2\right)$, and $I\left[Q^2\left(\text{Al}\right)\right]$ represent the integrated intensities of signals Q^0, Q^1, Q^2, and $Q^2\left(\text{Al}\right)$ in hydrated cement-FA paste, respectively; $I_0\left(Q^0\right)$, $I_0\left(Q^3\right)$, and $I_0\left(Q^4\right)$ represent the integrated intensities of signals Q^0, Q^3, and Q^4 in unhydrated cement-FA mixture.

3. Results and Discussion

3.1. Chloride Solidification. The effect of TIPA on the CBR of cement-FA paste cured for 60 days was examined, and the results are shown in Figure 2. As can be seen from the figure, in comparison with the reference (i.e., without TIPA), CBR at the age of 60 days is increased with addition of TIPA; with the dosage of 0.06%, CBR reaches 48.40%, with an increase of 11.1%; when the dosage is further increased to 0.10%, CBR increases to 53.47%, with an increase of 22.8%.

The binding of chlorine ions in the cement-FA system can be divided into chemical binding and physical adsorption. It can be inferred that addition of TIPA probably hastens the hydration of cement-FA system. On the one hand, the addition of TIPA can accelerate the dissolution of the ferric phase in cement minerals, which in most cases exists on the surface of the mineral particles due to the lower melting point. Accordingly, the dissolution of other phase can also be hastened, resulting in formation of more amount

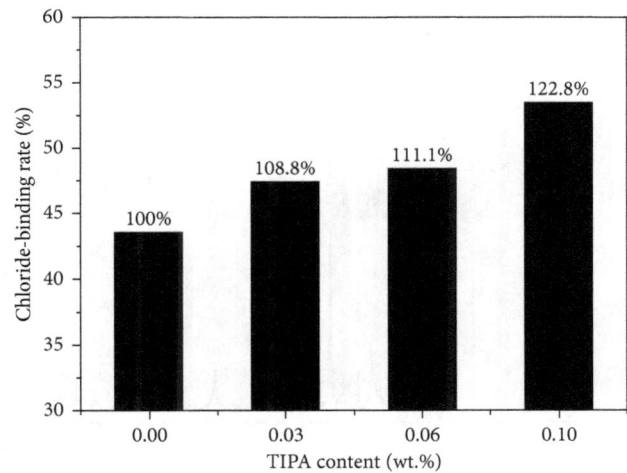

FIGURE 2: Effect of TIPA on the chloride-binding rate of 60 d.

of CH, which can induce the pozzolanic reaction of FA [32]. On the other hand, the addition of TIPA can also hasten the dissolution of FA, and in this case, the more amounts of aluminum would exist in the liquid phase and participate in hydration. Most likely more amounts of FS or KS could be formed [33].

Additionally, physical adsorption of chloride resulting from C-S-H gel can also contribute to chloride immobilization. With addition of TIPA, formation of C-S-H gel can be promoted, and more amount of C-S-H gel can exert stronger ability to adsorb and wrap chloride ions.

Based on discussion above, the improvement in CBR of the cement-FA system with addition of TIPA at 60 d age is closely related to the hydration of the system and the formation of FS and KS, which would be further illustrated in the following text.

3.2. Analysis of Hydration Products. To verify the contribution of FS or KS to CBR, the hydration products were characterized with XRD, TG, and SEM.

3.2.1. XRD. Cement-FA paste hydrated for 60 days with addition of TIPA (0.06%) was discussed with XRD, and the results are shown in Figure 3. As can be seen from the figure, the peaks of CH, AFm, AFt, and FS can be observed obviously. By contrast, the addition of TIPA increases the peak intensity of AFt and FS and reduces the intensity of AFm. This result indicates that TIPA can hasten the formation of AFt and FS at 60 d age.

3.2.2. TG-DTG. The mass loss curve of the sample is shown in Figure 4(a). As it can be seen from the TGA curve, the mass loss at the temperature range from 0 to 400°C is related to the loss of free water, decomposition of C-S-H gel, hydrated calcium sulphoaluminate and FS; that for 400–500°C is involved in the decomposition of CH; and that for 500–750°C is due to the decomposition of carbon

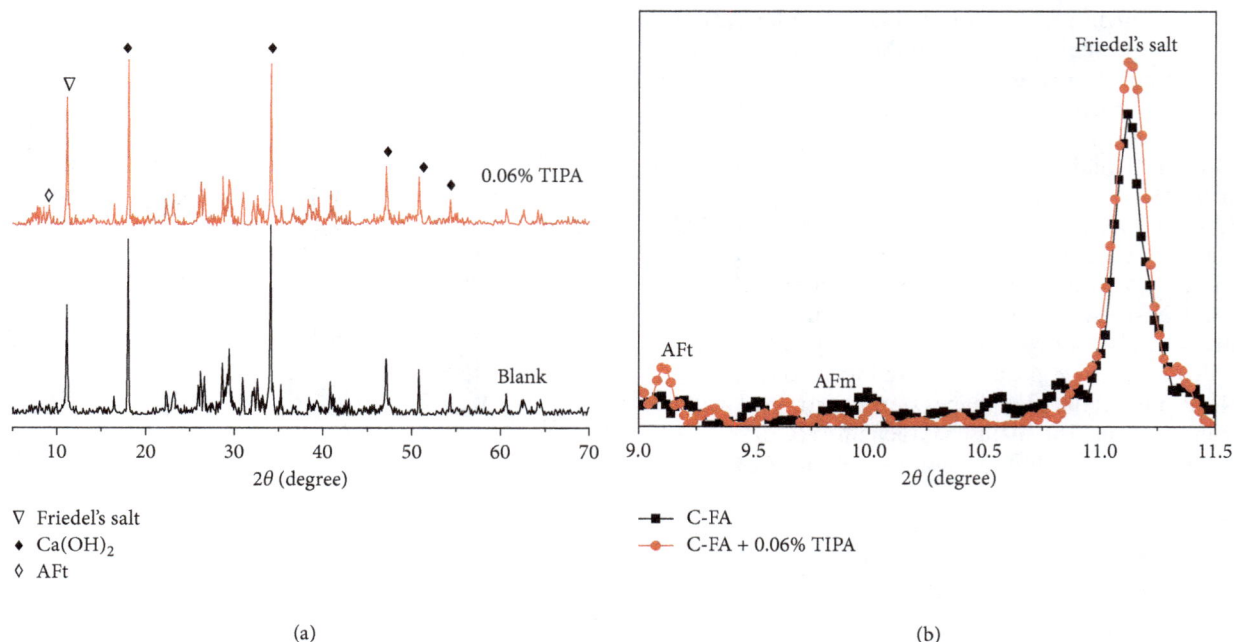

∇ Friedel's salt
♦ Ca(OH)$_2$
◊ AFt

(a)

(b)

FIGURE 3: XRD patterns of cement-FA paste hydrated at 60 days.

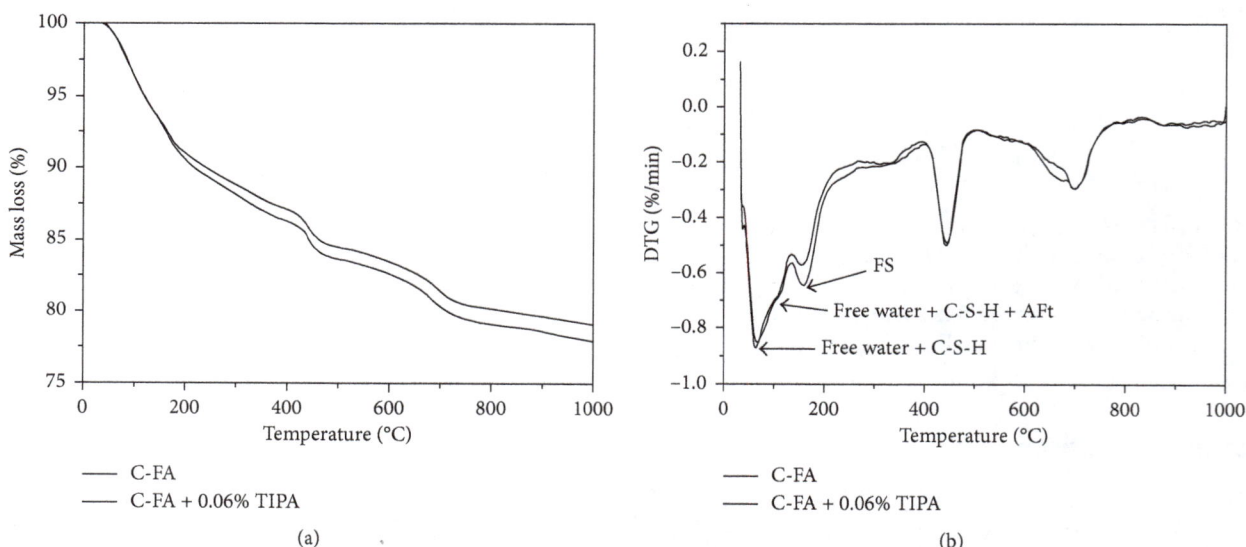

(a)

(b)

FIGURE 4: TG-DTG curve of the paste hydrated for 60 d. (a) TG; (b) DTG.

dioxide which was formed in the process of preparing for the samples [34].

The weight loss at the temperature from 50 to 200°C, mainly resulting from evaporation of free water and the decomposition of hydrate (C-S-H gel, AFt and FS), shows great interest. As reported in the literatures [35], FS would be decomposed at the temperature range of about 160°C. It is noticed that in Figure 4(b), the absorption peak, resulting from decomposition of FS, is obviously increased with addition of TIPA (0.06%), indicating more amount of FS. This result provides supplementary evidence to prove that TIPA can facilitate formation of FS, in agreement with the XRD results.

TABLE 2: Calcium hydroxide content in the cement-FA system (wt.%).

Temperature range (°C)	Blank	0.06% TIPA
400–500°C	2.66	2.55
500–750°C	4.14	4.29
CH content	17.89	17.69

However, the amount of CH in hydration products attracts more interest. As shown in Table 2, for 60 d age, the amount of CH in blank (without TIPA) is greater than that with addition of TIPA. This result indicates that TIPA can

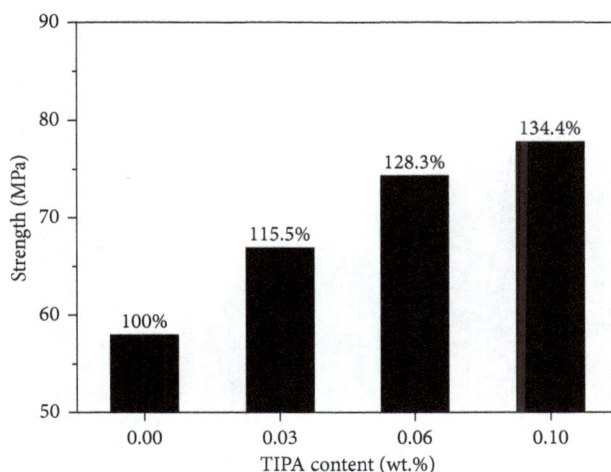

FIGURE 5: Effect of TIPA on compressive strength of 60 d.

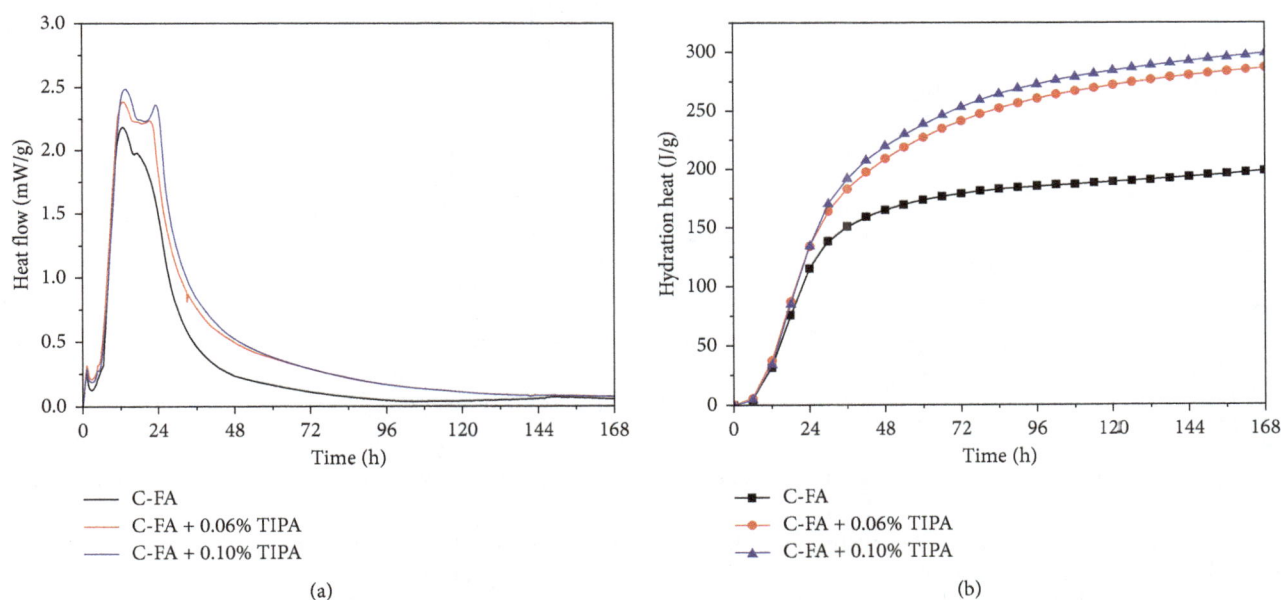

(a)

(b)

FIGURE 6: Hydration heat of cement-FA paste with TIPA. (a) Heat flow; (b) hydration heat.

reduce the amount of CH in hydration products at 60 d age. The amount of CH depends on the cement hydration and pozzolanic reaction of FA. Greater degree of cement hydration would generate more amount of CH, and greater degree of pozzolanic reaction of FA would consume more amount of CH. In comparison with the blank, the predominant aspect would determine the relative amount of CH. Obviously, the acceleration of the FA, namely, consumption of CH, should be predominated. This implies that TIPA can significantly hasten the pozzolanic reaction of FA.

Based on discussion above, pozzolanic reaction of FA and formation of FS can be hastened with addition of TIPA, and the improvement in CBR is mainly because of the accelerated formation of FS in hydrates, which would be closely related to the hydration process.

3.3. Hydration Process

3.3.1. Compressive Strength.
To further verify the effect of TIPA on hydration of cement-FA paste, the compressive strength was examined, and the results are shown in Figure 5. As can be seen from the figure, the compressive strength at the age of 60 days increases with the increasing dosage of TIPA. Compared with the sample without TIPA, 0.03% TIPA increases the strength by 15.5%; with the dosage of 0.10%, the strength reaches 77.8 MPa, with an increase by 34.4%. The increased strength implies that TIPA can also accelerate the hydration of the cement-FA system.

3.3.2. Hydration Heat.
As reported in the literatures, the hydration of cement-FA includes the initial reaction,

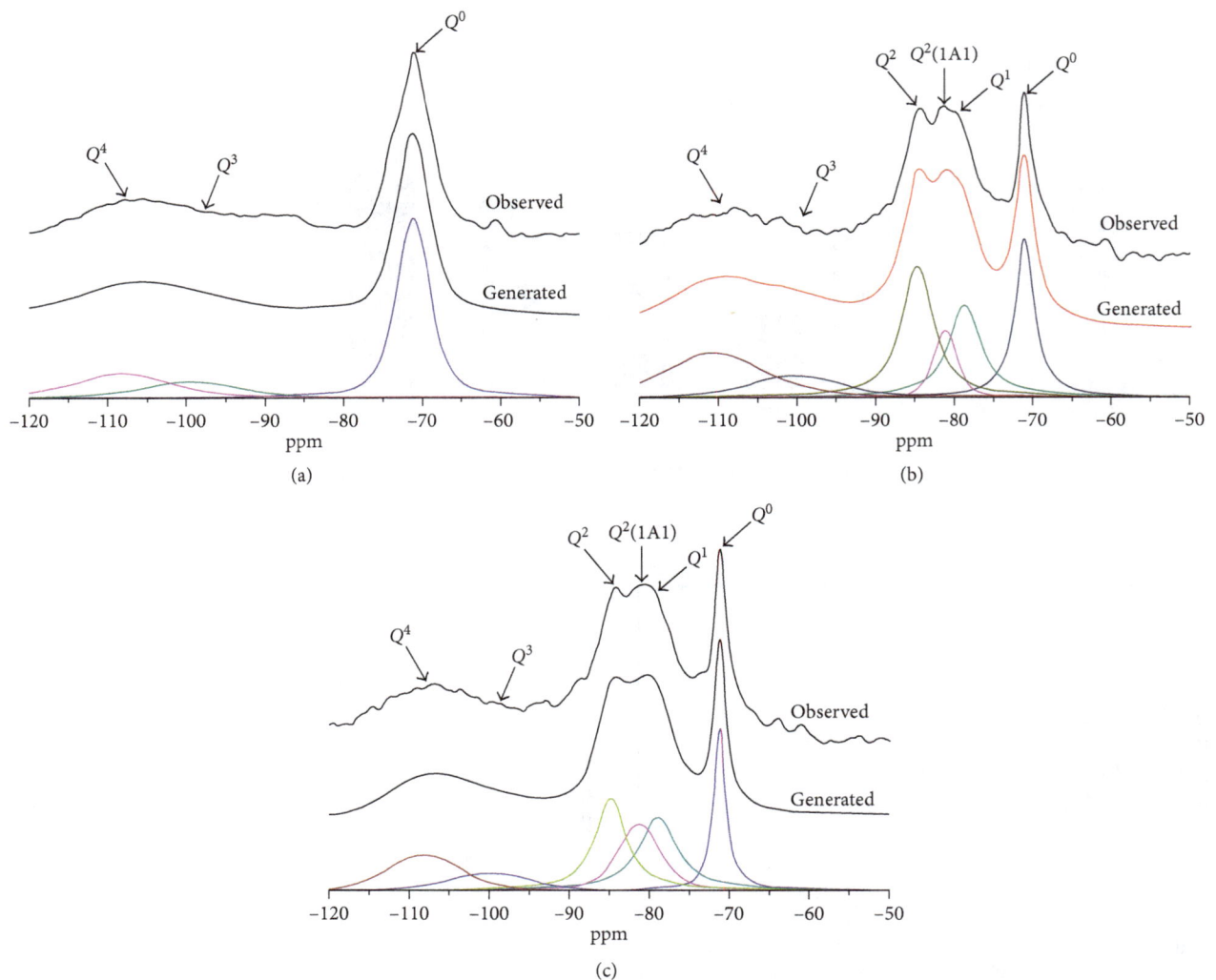

FIGURE 7: ^{29}Si-NMR spectrum of cement-FA paste. (a) Unhydrated C-FA; (b) C-FA hydrated for 60 d; (c) C-FA with 0.06% TIPA hydrated for 60 d.

a period of slow reaction, an acceleration period, and a deceleration period, as described by Taylor [36]. As shown in Figure 6(a), the addition of TIPA cannot change these four steps but can accelerate the release of hydration heat, indicating that TIPA can hasten the hydration of the cement-FA system. This can also be confirmed in Figure 6(b) that the cumulated heat is increased with addition of TIPA.

Furthermore, it is noticed that the extra peak at about 30 h can be seen clearly with addition of TIPA. As reported, the effect of TIPA on cement hydration can be divided into three phases: (1) adsorb on the surface of cement particles and slightly delay the hydration; (2) hasten the dissolution of ions, especially the ferric ion, to accelerate the formation of AFt; and (3) facilitate the conversion of AFt to AFm [37, 38]. It can be inferred that the extra peak should be related to that conversion. The reason can be revealed that at the very beginning of the hydration, AFt should be formed because of the much faster dissolution speed of gypsum to provide enough sulfates, with lower Al/S ratio. While with time going on, more amount of aluminum can increase the Al/S ratio, resulting in conversion of AFt to AFm. In cement-FA

paste with TIPA, the dissolution of aluminum in FA would be hastened, and the Al/S ratio would be much higher than that without TIPA. In this case, the conversion of AFt to AFm can be facilitated:

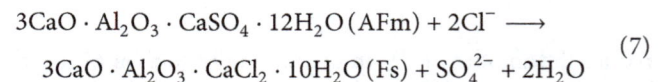

$$3CaO \cdot Al_2O_3 \cdot CaSO_4 \cdot 12H_2O\,(AFm) + 2Cl^- \longrightarrow$$
$$3CaO \cdot Al_2O_3 \cdot CaCl_2 \cdot 10H_2O\,(Fs) + SO_4^{2-} + 2H_2O \tag{7}$$

According to results in the literatures [14, 39], AFm can combine with chloride ions to form FS, and in cement-FA paste with TIPA, the conversion of AFt to AFm can be facilitated, which can further hasten the formation of FS.

3.3.3. ^{29}Si-NMR. The deconvoluted ^{29}Si MAS NMR spectra of the samples obtained from the fitting are plotted in Figure 7; reaction ratio, MCL, and Al/Si ratio were calculated, and the results are shown in Table 3. As shown in Figure 7, Q^0, Q^1, Q^2 (1Al), Q^2, Q^3, and Q^4 can be seen clearly. In comparison with the unhydrated C-FA (as shown in Figure 7(a)), the increase in peak intensity of Q^1, Q^2 (1Al),

TABLE 3: Deconvolution results of the sample.

	Q^0	Q^1	Q^2 (1Al)	Q^2	$Q^3 + Q^4$	MCL	Al/Si	Reaction ratio of cement (%)	Reaction ratio of fly ash (%)
Unhydrated C-FA	62.27	—	—	—	37.73	—	—	—	—
C-FA	19.86	24.20	9.86	22.00	24.08	5.04	0.0879	68.11	36.18
C-FA with 0.6% TIPA	16.00	21.77	18.00	22.34	21.89	6.53	0.1449	74.31	41.98

FIGURE 8: SEM images of the paste hydrated for 60 d.

and Q^2 can be found, and the decline in Q^0, Q^3, and Q^4 can also be observed. These results illustrate the hydration of cement and the pozzolanic reaction of FA. More details can be found in Table 3: in comparison with the blank sample, 0.06% TIPA obviously reduces the amount of Q^0 and slightly declines the amount of $Q^3 + Q^4$; this result demonstrates that TIPA can promote the cement hydration as well as the pozzolanic reaction of FA; by contrast, this promoting effect on cement hydration is much stronger than that of FA. Furthermore, C-FA has a MCL of 5.04, while that for C-FA-TIPA (0.06%), it was 6.53, which indicates that the addition of TIPA can increase the degree of silicate polymerization; this can also show that a higher degree of hydration has occurred. Moreover, the Al/Si ratio is also increased with 0.06% TIPA, in comparison with blank paste (C-FA). This increase definitely confirms that the incorporation of TIPA into the cement-FA system induces the substitution of Si by Al (Al [4]) into C-S-H, resulting in an increase in the length of silicate chain of C-S-H, in agreement with the results of MCL. Additionally, the reaction degree of cement without TIPA is 68.11%, while that for addition of TIPA (0.06%), it is 74.31%, which indicates the accelerated hydration of cement by TIPA. The same results can also be found in FA: the reaction degree of FA is increased from 36.18% to 41.98% with 0.06% TIPA.

According to the analysis of NMR, the addition of TIPA not only hastens the hydration of cement but also accelerates the pozzolanic reaction of FA, and it also increases the degree of silicate polymerization, the length of C-S-H, and substitution of Si by Al, with contribution to the mechanical performance. This also hastens the dissolution of aluminum of FA into liquid phase.

3.3.4. SEM.
Figure 8 shows the SEM images of the paste hydrated for 60 d in the presence and absence of TIPA.

As shown in Figure 8, more serious erosion can be seen clearly in Figure 8(b) than that in Figure 8(a), which implies that the addition of TIPA can hasten pozzolanic reaction of FA.

Based on the discussion of hydration heat, NMR, SEM, and compressive strength, it is concluded that TIPA can accelerate the pozzolanic reaction of FA and hydration of cement minerals at the age of 60 d.

3.4. Dissolution of Fly Ash

3.4.1. Ions Dissolution.
The accelerated hydration of FA by TIPA has been confirmed above, and the mechanism behind this is closely related to the dissolution of ions from FA into liquid. Accordingly, the reaction of FA in pore solution with various dosages of TIPA was investigated to further reveal the reason for the accelerated hydration of FA by TIPA.

The effect of TIPA on dissolution of FA in pore solution is shown in Figure 9. As shown in the figure, the dissolution of aluminum ions, ferric ions, and silicon ions was increased with the increasing dosage of TIPA, indicating that TIPA in pore solution can significantly hasten the dissolution of FA. TIPA (20 g/L) can make the aluminum ions about 2.5 times and silicate ions about 2 times of the blank (without TIPA) at the age of 60 days. Furthermore, as shown in Figure 9(b), very little amount of Fe can be found without TIPA, which implies that ferric in FA is not that easy to be dissolved into pore solution in hydration process. However, with addition of TIPA, surprisingly, the dissolution of ferric ions was considerably hastened, in agreement with the results in the literatures [40, 41].

The morphology of FA was observed with SEM, as a supplementary evidence to prove the dissolution of ions of FA in pore solution. As shown in Figure 10(a), at the age of 60 d, the erosion of FA in the pore solution can be seen clearly, and the amount of flocculent reaction products on

(a)

(b)

(c)

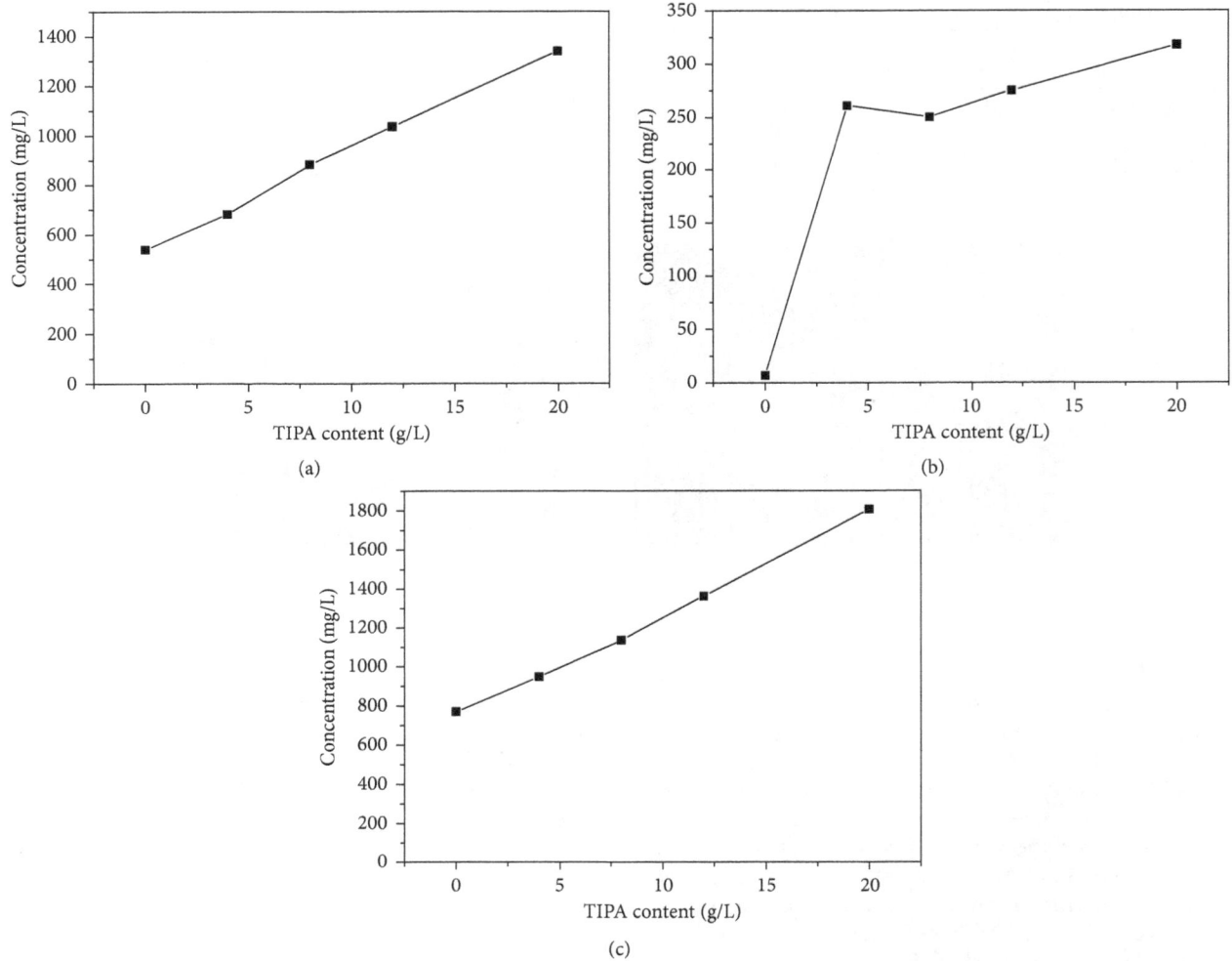

FIGURE 9: Ions dissolution of FA in the pore solution with TIPA at the age of 60 days. (a) Al; (b) Fe; (c) Si.

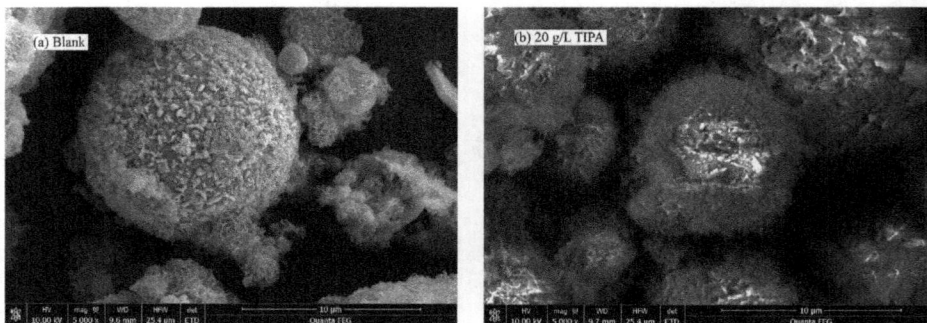

FIGURE 10: Morphology of FA immersed in pore solution for 60 days.

the surface can also be found. By contrast, as shown in Figure 10(b), with TIPA solution (20 g/L), the erosion of the surface is more seriously, indicating that dissolution of FA can be hastened at the age of 60 d, in agreement with the results of ions dissolution.

Based on the discussion above, the promoted ions dissolution of FA into the liquid phase by addition of TIPA can

be concluded, and this can benefit the improvement in CBR of the cement-FA system. The details for the reason can be summarized as follow:

On the one hand, the promotion of TIPA on cement hydration has been confirmed, which can generate more amount of CH to activate the pozzolanic reaction of FA, which means that the dissolution of FA can be hastened. On the other hand, the addition of TIPA can also accelerate dissolution of Al into the pore solution. As a result, the addition of TIPA can significantly accelerate the dissolution of FA into the liquid phase, especially the dissolution of aluminum. Furthermore, the aluminum dissolved from FA can participate in the formation of C-S-H gel as C-A-S-H gel and also increase the Al/S ratio to induce the formation of AFm from AFt. AFm can react with chloride to form FS. Additionally, the excessive can also be directly reacted with calcium and chloride to generate FS. Consequently, the formation of FS can be hastened, and the CBR of the cement-FA system can be improved.

4. Conclusion

(1) Addition of TIPA in the cement-FA system can not only hasten the hydration of cement minerals but also accelerate the pozzolanic reaction of FA, with contribution to the mechanical performance.

(2) The FS can also be accelerated with addition of TIPA at the age of 60 d, which is responsible for the improved chloride-binding capacity of the system.

(3) Dissolution of aluminum, silicate, and ferric in FA can be hastened with addition of TIPA, which attributes to the increased amount of FS and C-A-S-H gel in hydration products.

Conflicts of Interest

The authors declare that there are no conflicts of interest regarding the publication of this paper.

Acknowledgments

Financial support from the National Key R&D Program of China (2016YFC0701003-5) is gratefully acknowledged.

References

[1] X. Y. Lu, C. L. Li, and H. Zhang, "Relationship between the free and total chloride diffusivity in concrete," *Cement and Concrete Research*, vol. 30, no. 2, pp. 323–326, 2002.

[2] W. Li, B. Dong, Z. Yang et al., "Recent advances in intrinsic self-healing cementitious materials," *Advanced Materials*, vol. 30, no. 17, 2018.

[3] X. Shi, N. Xie, K. Fortune, and J. Gong, "Durability of steel reinforced concrete in chloride environments: an overview," *Construction and Building Materials*, vol. 30, pp. 125–138, 2012.

[4] B. Martın-Pereza, H. Zibarab, R. D. Hootonb, and M. D. A. Thomas, "A study of the effect of chloride binding on service life predictions," *Cement and Concrete Research*, vol. 30, no. 8, pp. 1215–1223, 2000.

[5] J. Xiao, C. Qiang, A. Nanni, and K. Zhang, "Use of sea-sand and seawater in concrete construction: current status and future opportunities," *Construction and Building Materials*, vol. 155, pp. 1101–1111, 2017.

[6] F. Zou, H. Tan, Y. Guo, B. Ma, X. He, and Y. Zhou, "Effect of sodium gluconate on dispersion of polycarboxylate superplasticizer with different grafting density in side chain," *Journal of Industrial and Engineering Chemistry*, vol. 55, pp. 91–100, 2017.

[7] H. Tan, B. Gu, Y. Guo et al., "Improvement in compatibility of polycarboxylate superplasticizer with poor-quality aggregate containing montmorillonite by incorporating polymeric ferric sulfate," *Construction and Building Materials*, vol. 162, pp. 566–575, 2018.

[8] B. Ma, Y. Peng, H. Tan et al., "Effect of hydroxypropyl-methyl cellulose ether on rheology of cement paste plasticized by polycarboxylate superplasticizer," *Construction and Building Materials*, vol. 160, pp. 341–350, 2018.

[9] F. P. Glasser, A. Kindness, and S. A. Stronach, "Stability and solubility relationships in AFm phases," *Cement and Concrete Research*, vol. 29, no. 6, pp. 861–866, 1999.

[10] S. Yoon, J. Ha, S. R. Chae et al., "Phase changes of monosulfoaluminate in NaCl aqueous solution," *Materials*, vol. 9, no. 5, p. 401, 2016.

[11] E. P. Nielsen, D. Herfort, and M. R. Geiker, "Binding of chloride and alkalis in Portland cement systems," *Cement and Concrete Research*, vol. 35, no. 1, pp. 117–123, 2005.

[12] M. Marinescu and J. Brouwers, "Chloride binding related to hydration products part I: ordinary Portland cement," *Advances in Modeling Concrete Service Life*, vol. 3, pp. 125–131, 2012.

[13] D. M. Burgos, Á. Guzmán, N. Torres, and S. Delvasto, "Chloride ion resistance of self-compacting concretes incorporating volcanic materials," *Construction and Building Materials*, vol. 156, pp. 565–573, 2017.

[14] J. Geng, D. Easterbrook, L. Y. Li, and L. W. Mo, "The stability of bound chlorides in cement paste with sulfate attack," *Cement and Concrete Research*, vol. 68, pp. 211–222, 2015.

[15] C. Alonsoa, C. Andradea, M. Castellotea, and P. Castrob, "Chloride threshold values to depassivate reinforcing bars embedded in a standardized OPC mortar," *Cement and Concrete Research*, vol. 30, no. 7, pp. 1047–1055, 2000.

[16] H. Justnes, "A review of chloride binding in cementitious systems," *Nordic Concrete Research*, vol. 21, pp. 1–6, 1998.

[17] Y. Guo, B. Ma, Z. Zhi et al., "Effect of polyacrylic acid emulsion on fluidity of cement paste," *Colloids and Surfaces A: Physicochemical and Engineering Aspects*, vol. 535, pp. 139–148, 2017.

[18] R. K. Dhir, M. A. K. El-Mohr, and T. D. Dyer, "Developing chloride resisting concrete using PFA," *Cement and Concrete Research*, vol. 17, no. 11, pp. 1633–1639, 1997.

[19] H. Zibara, R. D. Hooton, M. D. A. Thomas, and K. Stanish, "Influence of the C/S and C/A ratios of hydration products on the chloride ion binding capacity of lime-SF and lime-MK mixtures," *Cement and Concrete Research*, vol. 38, no. 3, pp. 422–426, 2008.

[20] X. Gong, Y. Wang, and T. Kuang, "ZIF-8-based membranes for carbon dioxide capture and separation," *ACS Sustainable Chemistry and Engineering*, vol. 5, no. 12, pp. 11204–11214, 2017.

[21] J. Paya, J. Monzo, M. V. Borrachero, and E. Peris-Mora, "Mechanical treatment of fly ashes. Part I: physico-chemical characterization of ground fly ashes," *Cement and Concrete Research*, vol. 25, no. 7, pp. 1469–1479, 1995.

[22] A. Dakhane, S. Tweedley, S. Kailas, R. Marzke, and N. Neithalath, "Mechanical and microstructural characterization of alkali sulfate activated high volume fly ash binders," *Materials and Design*, vol. 122, pp. 236–246, 2017.

[23] J. Mei, H. Tan, H. Li et al., "Effect of sodium sulfate and nano-SiO$_2$ on hydration and microstructure of cementitious materials containing high volume fly ash under steam curing," *Construction and Building Materials*, vol. 163, pp. 812–825, 2018.

[24] E. Gartner and D. Myers, "Influence of tertiary alkanolamines on Portland cement hydration," *Journal of the American Ceramic Society*, vol. 76, no. 6, pp. 1521–1530, 1993.

[25] Z. Xu, W. Li, J. Sun et al., "Research on cement hydration and hardening with different alkanolamines," *Construction and Building Materials*, vol. 141, pp. 296–306, 2017.

[26] H. Huang, X.-R. Li, and X.-D. Shen, "Hydration of ternary cement in the presence of triisopropanolamine," *Construction and Building Materials*, vol. 111, pp. 513–521, 2016.

[27] B. Zhang, H. Tan, B. Ma et al., "Preparation and application of fine-grinded cement in cement-based material," *Construction and Building Materials*, vol. 157, pp. 34–41, 2017.

[28] JGJ/T 322-2013, *Technical Specification for Test of Chloride Ion Content in Concrete*, China Architecture and Building Press, Beijing, China, 2013, in Chinese.

[29] U. A. Birnin-Yauri and F. P. Glasser, "Friedel's salt, Ca$_2$Al (OH)6(Cl,OH)·2H$_2$O: its solid solutions and their role in chloride binding," *Cement and Concrete Research*, vol. 28, no. 12, pp. 1713–1723, 1998.

[30] C. A. Love, I. G. Richardson, and A. R. Brough, "Composition and structure of C–S–H in white Portland cement–20% metakaolin pastes hydrated at 25°C," *Cement and Concrete Research*, vol. 37, no. 2, pp. 109–117, 2007.

[31] L. Wang and Z. He, "Quantitative of fly ash-cement hydration by 29Si MAS NMR," *Journal of the Chinese Ceramic Society*, vol. 38, no. 11, pp. 2212–2216, 2010.

[32] Z. Wang, J. Wu, P. Zhao et al., "Improving cracking resistance of cement mortar by thermo-sensitive poly N-isopropyl acrylamide (PNIPAM) gels," *Journal of Cleaner Production*, vol. 176, pp. 1292–1303, 2018.

[33] M. D. A. Thomas, R. D. Hooton, A. Scott, and H. Zibara, "The effect of supplementary cementitious materials on chloride binding in hardened cement paste," *Cement and Concrete Research*, vol. 42, no. 1, pp. 1–7, 2012.

[34] Z. Shi, M. R. Geiker, B. Lothenbach et al., "Friedel's salt profiles from thermogravimetric analysis and thermodynamic modelling of Portland cement-based mortars exposed to sodium chloride solution," *Cement and Concrete Composites*, vol. 78, pp. 73–83, 2017.

[35] J.-P. Rapina, G. Renaudinb, E. Elkaimc, and M. Francoisa, "Structural transition of Friedel's salt 3CaO·Al$_2$O$_3$·CaCl$_2$·10H$_2$O studied by synchrotron powder diffraction," *Cement and Concrete Research*, vol. 32, pp. 513–519, 2002.

[36] H. Taylor, *Cement Chemistry*, Thomas Telford, London, UK, 1997.

[37] Z. Shi, C. Shi, H. Liu, and P. Li, "Effects of triisopropanol amine, sodium chloride and limestone on the compressive strength and hydration of Portland cement," *Construction and Building Materials*, vol. 125, pp. 210–218, 2016.

[38] Y. Wang and X. Gong, "Special oleophobic and hydrophilic surfaces: approaches, mechanisms, and applications," *Journal of Materials Chemistry A*, vol. 5, no. 8, pp. 3759–3773, 2017.

[39] A. Ipavec, T. Vuk, R. Gabrovšek, and V. Kaučič, "Chloride binding into hydrated blended cements: the influence of limestone and alkalinity," *Cement and Concrete Research*, vol. 48, pp. 74–85, 2013.

[40] H. Huang, X. Shen, and J. Zheng, "Modeling, analysis of interaction effects of several chemical additives on the strength development of silicate cement," *Construction and Building Materials*, vol. 24, no. 10, pp. 1937–1943, 2010.

[41] S. Ma, W. Li, S. Zhang, Y. Hu, and X. Shen, "Study on the hydration and microstructure of Portland cement containing diethanol-isopropanolamine," *Cement and Concrete Research*, vol. 67, pp. 122–130, 2015.

15

Experimental and FEM Research on Airport Cement Concrete Direct-Thickening Double-Deck Pavement Slabs under Aircraft Single-Wheel Dynamic Loads

Qingkun Yu [ID],[1] Liangcai Cai,[1] and Jianwu Wang[2]

[1]*Air Force Engineering University, Xi'an, Shaanxi Province, 710038, China*
[2]*Air Force Service Academy, Xuzhou, Jiangsu Province, 221000, China*

Correspondence should be addressed to Qingkun Yu; 1028419220@qq.com

Academic Editor: Antonio Boccaccio

The wide-use airport cement concrete direct-thickening double-deck pavement slabs (ACCDDPS) were selected as the research object to study their mechanical properties. The airport runway simulation test station (ARSTS) was used to conduct indoor tests to demonstrate the distribution of tension stress at the bottom of slabs and slabs deflection. Furthermore, ANSYS software was applied to establish finite element model (FEM) of ACCDDPS and analyze the mechanical laws under different loads. The indoor tests results are in good agreement with the ANSYS simulation results, and some consistent conclusions can be obtained that the maximum tension stress increases with wheel load, and the slab middle of the longitudinal edge is a critical position. In addition, we studied the influence of covered layer thickness, elastic modulus, and slab size on pavement slab mechanical properties by ANSYS, and we concluded that although the structural parameters are different, the critical position of ACCDDPS is still in the middle of the longitudinal edge. However, for the covered layer and the original surface layer, the law that the tension stress values vary with the structural parameters is different, but the maximum deflection value is about 0.1.

1. Introduction

The cement concrete pavement has the advantages of large rigidity, strong bearing capacity, and good weather resistance and becomes the main form of airport pavement in many countries in the world. In the twenty-first century, there is a new question whether the pavement can meet the new aircrafts' application requirements, because many airport runways were constructed in the mid-to-late twentieth century, and most airport pavements have suffered different degrees of damage. Taking economic conditions, environmental benefits, and usage requirements into overall consideration, many countries have reinforced the existing old cement concrete pavement, which is commonly known as pavement coverings.

At present, pavement coverings can be made of asphalt and cement concrete. However, due to the asphalt pavements' short service life and poor corrosiveness to fuel oil, the cement concrete pavement, which has the advantages of good economy, high carrying capacity, and good durability, is the preferred choice for many airports. In addition, there are three kinds of structural forms for the pavement coverings, which are isolated covered layer, combined covered layer, and direct covered layer [1]. The direct covered layer is widely used because of its simple construction, low cost, and high performance. However, the traditional design theories and methods for thickening and reinforcing cement concrete pavement are out of date. Therefore, there is an urgent need to study the performance of ACCDDPS under the action of various types of aircrafts in order to develop new design theory.

The ACCDDPS mentioned in this paper means that the cement concrete thickened layer is directly laid on the clean old concrete cement pavement surface, which is characterized of a certain adhesive force and frictional resistance between the upper and lower two layers, resulting in a better overall stiffness in the upper and lower slabs. As shown in Figure 1, from the bottom to the top, the ACCDDPS

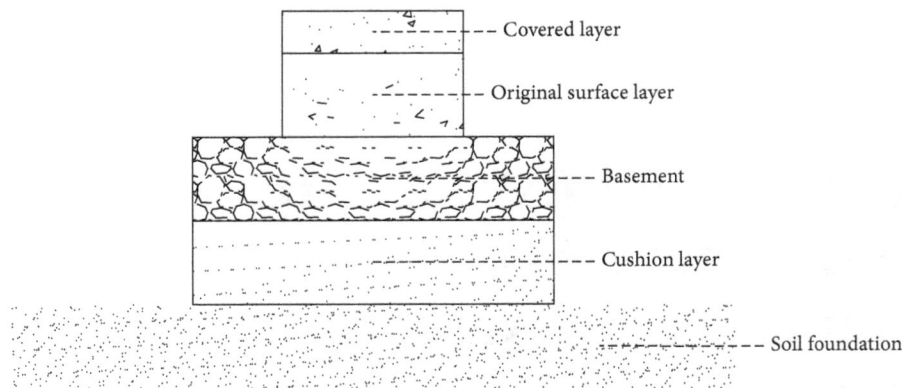

FIGURE 1: The ACCDDPS structure.

structure is composed of the natural soil foundation, the cushion (as appropriate), the compacted basement, the original surface layer, and the covered layer. When the covered layer is under construction, it is necessary to repair some functional damage that occurred in the original surface layer and to clean the entire pavement surface.

Since the first cement concrete double-deck slabs was built in the United States in 1930, the cement concrete double-deck slabs has been widely applied to the highway and airport in many countries, and the related research and practical exploration have also been carried out in full swing.

An improved solution algorithm based on Finite Element Method for dynamic analysis of rigid pavements under moving loads [2] is presented to research on the effect of soil modulus, shear modulus, pavement thickness, and the vehicle–pavement interaction.

Binchen et al. [3] established a fine airport pavement finite element model based on the Winkler foundation model to analyze the number of slabs required when the wheel loads are applied to the slab center, slab edges, and slab corners. Through analysis, he gave some suggestions: when the load is at the middle of the slab, it is recommended to establish a four-slab model; when the load is at the slab edge, two- or four-slab models are established; when the load is at the slab corner, it is recommended to establish four-slab model.

Shuyuan [4] analyzed, respectively, the load stress of isolated covered and direct covered cement concrete pavement slabs based on elastic half-space foundation and gave the moment calculation formulas of isolated covered cement concrete pavement slabs and the maximum tension stress of direct covered cement concrete pavement slab.

Chao et al. [5] applied ABAQUS software to analyze the influence of different materials at the original base (i.e., different strength) on the stress of pouring cement concrete surface after the old pavement is crushed. The analysis showed that some materials (such as the cement stabilized grit, the graded gravel, and the lime-ash soil) as the original base layer, the tension stress of the surface layer slab gradually decreases with the increase of the elastic modulus of the base material.

Haiyan et al. [6] used the theory of multilayer thin plates based on the elastic foundation model to explore the deflection and maximum stress of the double-deck pavement slab. On the basis of certain assumptions, the design methods of direct and isolated pavement structures are obtained by numerical calculation. Comparing with the analysis result of ANSYS software, it is found that the numerical calculation results can be well matched with the results of ANSYS analysis, which proves the feasibility of the method to some extent.

Xiangcheng et al. [7] firstly established a 3D finite element model to analyze the effect of fiber gratings on the maximum tension stress at the bottom of the covered layer slab. Due to the addition of the fiber grid, the friction coefficient between the covered layer and the original surface layer is changed that the tension stress gradually decreases as the friction coefficient increases, and the tension stress tends to be stable until the friction coefficient is 0.8. In addition, he adopted orthogonal test to analyze the influence of the thickness and elastic modulus of covered layer, the size and elastic modulus of the fiber grid, the elastic modulus of the original covered layer and compacted soil layer on the maximum principle stress and the defection at the bottom of the covered slab, and the tension stress at the bottom of the original surface layer.

Xianmin et al. [8] established 3D finite element models for airport runways, selected 5 types of aircrafts (A320-200, B737-800, A340-500, B777-200, and B747-300, resp., divided into 2 groups in terms of the same load level and different landing gear configuration) to analyze the influence of different landing gear configurations (including the number of wheels, landing gear spacing and landing gear layout, and other factors) on the pavement surface deflection, impact depth on the base depth, and the maximum tension stress at the bottom of surface layer. The analysis results show that with the same load, the fewer the number of wheels, the greater the deflection of the slab bottom, the greater the maximum tension stress; the greater the landing gear spacing, the smaller the impact depth on the base.

Zizheng and Hongduo [9] analyzed the structural response of the double-deck cement concrete pavement slab under the action of a large-scale aircraft with relatively complicated landing gear configuration represented by A380-800 through ABSQUS software, combined with the pavement fatigue life under different load distributions according to the space superposition effect and finally obtained the most unfavorable load combination of A380-800.

Xiang [10] established a 3-D finite element model of a composite airport pavement with nine panels and chose B777-200 aircraft as the calculation load. The bottom tension stress of the cement concrete slab is selected as the index to analyze the most unfavorable position of the composite airport pavement; the orthogonal test design is applied in studying the influence of pavement structural parameters on load stress.

The above research achievements mainly use finite element method to simulate the airport pavement slab, which inspire us to conduct ACCDDPS research by FEM. The predecessors have established the FEM under the role of the B777-200 and A380-800 models and studied the different effects of the pavement structure parameters. However, no relevant work has been carried out on ACCDDPS under the action of various main aircrafts, which is necessary to conduct this research because many airports' cement concrete pavements undertake the task of guaranteeing various types of aircrafts. Moreover, the experimental method and FEM method have not been compared in order to draw more scientific conclusions.

Taking four main aircrafts of A, B, C, and D as an example, we design the ARSTS experiment on ACCDDPS under different loads. The stress distribution and slab deflection are analyzed by the cyclic-load mechanical test on the ACCDDPS. On the basis of fully simulating the experimental conditions, we intend to establish an ACCDDPS FEM by ANSYS software. The FEM results and experimental results will be compared and analyzed, which should be compared with the existing research results to verify the credibility and accuracy of our research. Moreover, we will consider and analyze the influence of covered layer thickness, elastic modulus, and slab size on pavement mechanical properties by FEM, in order to draw some fruitful conclusions to guide the design, construction, and maintenance of the ACCDDPS.

2. Airport Runway Simulation Test Station Experimental Research on ACCDDPS under Aircraft Single-Wheel Dynamic Loads

2.1. Experimental Equipment, Materials, and Methods

2.1.1. Simplified Design of Experiment. The size of an ACCPPDS is generally more than 5 m × 5 m. Due to the limitations of the ARSTS experimental equipment, indoor experiments cannot be performed in a ratio of 1 : 1. Based on similarity theorems, the simplified designs are as follows:

(1) Ignore the influence of natural environment and other factors on the experimental conditions, including groundwater level change, temperature change and humidity change.

(2) The plane size and thickness of each pavement slab are reduced to a certain proportion according to the geometric similarity. The size of the test slab is 0.50 m × 0.50 m × 0.25 m, ignoring the lateral restraining effect.

(3) Ensure that the construction process in specimen's production part is the same as in practice, ignoring the influence of human factors during the whole process of the experiment.

(4) Suppose that the old and new surface layer, basement, and soil foundation are continuous, homogeneous, and isotropic perfectly linear elastic bodies.

2.1.2. Airport Runway Simulation Test Station. ARSTS was designed by a Changchun Experimental Instrument Company in September 2012, called Model KPD-01. This instrument simulates the movement of a wheel on a runway by regulating the size of wheel load and replacing wheels with different sizes. In addition, through the introduction of stress-strain collectors, the relevant data of pavement surface and basement can be collected, further analyzing the force conditions of the pavement.

ARSTS is mainly composed of three parts: the main part, the power system, and the human-computer interaction system, as shown in Figure 2.

2.1.3. Raw Materials and Mix Proportions. The main materials of specimens are ordinary Portland cement, medium sand, gravel, and water whose performance conforms to the provisions [1].

According to the requirements for ordinary cement concrete mix proportions [11], combined with relevant specifications on ACCDDPS, the mix proportions are shown in Table 1.

2.1.4. Equipment Preparation. The basement and soil foundation in the ARSTS box are treated as follows: fill the clay, and compact it layer by layer until it reaches a thickness of 0.70 m. Then, pave the gravel and tamp it layer by layer until the thickness is about 0.3 m, as shown in Figure 3.

In this experiment, stress gauges are placed at the bottom of the ACCDDPS specimen. Due to the symmetry of a square specimen, we only need to collect the stress values of 1/4 specimen. The layout of the measuring points is shown in Figure 4.

2.1.5. Specimens Production and Curing. The concrete steps of ACCDDPS specimens are shown below:

(1) Mold: make double-deck slab mold, size: 0.50 m × 0.50 m × 0.25 m, wood material. Considering the later stripping problem, a layer of plastic film is laid in the mold, and then a small amount of lubricating oil is brushed in the plastic film.

(2) Pouring concrete: firstly, an already accomplished single-layer cement concrete slab (0.50 m × 0.50 m × 0.25 m) is placed in the mold, and the stress gauges are pasted at 9 positions as shown in Figure 4. Secondly, make the thickening layer, uniformly pour cement concrete on the single-layer slab, tamp with a vibrator, flatten to a design thickness, and level the surface.

FIGURE 2: The ARSTS.

TABLE 1: Cement concrete material consumption per cubic meter.

Concrete materials	Cement	Water	Sand	Gravel	
				Large grain- size	Small grain-size
Weight (kg)	315	145	637	859	531

FIGURE 3: The ARSTS box preparation.

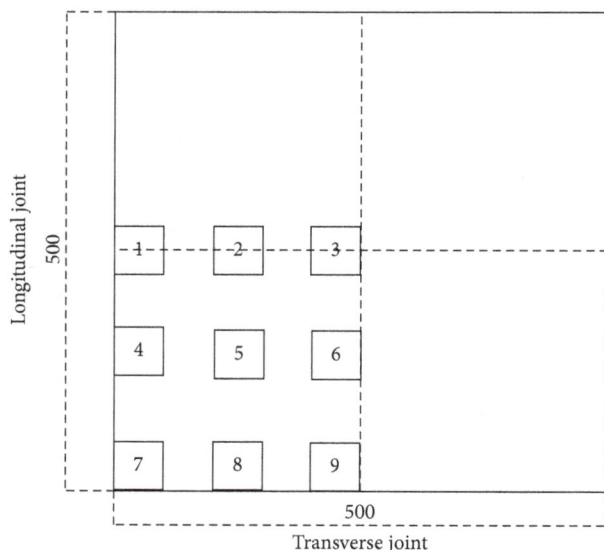

FIGURE 4: The stress gauges layout schematic diagram.

(3) Conservation: the wet geotextiles are covered on the specimens for curing. During the period, spray an appropriate amount of water daily to ensure the required moisture during the curing period.

(4) In the same steps (1) to (3), three tests are required under each aircraft load, so 12 specimens are to be poured.

The process of making ACCDDPS specimens is shown in Figure 5.

2.1.6. Experiment Procedures

(i) ARSTS Debug. Fix the cured specimens in ARSTS box, connect, and check the line. Open the experiment software system. Set the load in the vertical direction to 10 kN and slowly lower the wheel to the specimen. Observe whether the data display on the software system; if yes, continue subsequent operations; if not, recheck lines and trouble shoot problems until the data display normally.

(ii) Set Experimental Relevant Parameters. The experiment parameters are set as shown in Table 2.

(iii) Data Collection. Collect the value of the surface deflection, the maximum tension stress at the bottom of the slab, and the number of repeated action when the specimen is fractured. Considering that the maximum number of annual flight times for the four main aircrafts of A, B, C, and D is 25000, the number of repeated action is controlled to 30000 times in order to ensure the successful progress of the experiment.

(iv) Adjust the Load and Start the Experiment. The aircraft single-wheel dynamic loads are applied to the ACCDDPS

FIGURE 5: The making process of specimens. (a) Mixing. (b) Filling. (c) Leveling. (d) Tamping and mud extracting. (e) Rendering. (f) Finishing.

TABLE 2: Experiment relevant parameters.

Loading mode	Parameters						
	Control mode	Frequency (Hz)	Max/min Load (kN)	Max/min displacement (m)	Load wave	Recycle times	Protection Load (kN)
Vertical load	Displacement	1 Hz	86.64/0	Default	Sine	3000	200
Horizontal load	Displacement	1 Hz	Default	0.23 m/−0.23 m	Sine	3000	Default

specimens through the ARSTS wheel simulator, referring to partial parameters of A, B, C, and D four main aircrafts in Table 3.

Set the vertical load to 86.64 kN, and start the experiment. In the course of the experiment, pay close attention to the wheel running to ensure that the wheel does not deviate from the specimen. When the specimen is completely broken, the experiment is stopped immediately, and the relevant data are recorded. Continuing with other experiments, the vertical loads of the wheel are sequentially set to 103.07 kN, 129.16 kN, and 196.52 kN, and the other procedures are the same as above.

2.2. ARSTS Experiment Results. In order to reduce the experimental error, we choose the average value of three sets of experimental results at the same load level as the final value, and the detailed method of taking value refers to a Chinese professional standard "Specifications for Airport Cement Concrete Pavement Design"(MH/T5004-2010). The tension stress values are shown in Table 4, and the total surface deflection values are shown in Table 5.

In order to express the experimental results more directly and vividly, draw Figures 5–8 according to the above data.

As depicted in Figures 6 and 7, for both covered layer and original surface layer, the maximum tension stress is at the measuring point 1, followed by the measuring point 9, and the minimum tension stress is at the measuring point 5, which shows that the critical position of ACCDDPS is in the middle of the longitudinal edge.

As shown in Figure 8, the tension stress of covered layer increases with the increase of wheel load. When the load increases from 129.16 kN to 196.52 kN, the maximum tension stress increases to 0.26 MPa. That is, at the point of 129.16 kN, there is a clear inflection point, and after that point, the change rate of stress with load becomes larger. Similarly, the tension stress of original surface layer also increases with the increase of wheel load; however, the rate of growth has been slightly increased after the point of 129.16 kN.

Figure 9 shows that the total deflection of surface layer increases with the increase of wheel load, and the rising rate is relatively stable with no large fluctuations.

3. ACCDDPS FEM Establishment by ANSYS Software

3.1. Basic Assumptions. Before modeling the ACCDDPS, considering the differences between the ANSYS FEM and

TABLE 3: Four main Aircrafts' partial parameters.

Aircraft	Single main wheel grounding area (cm^2)	Single main wheel dynamic load (kN)	Single main wheel static load (kN)	Maximum takeoff masskg)	Landing gear Configuration	Main wheel tire pressure (MPa)
A	1110.77	86.64	72.20	6200	Double-axis double-wheel	0.78
B	1171.25	103.07	85.89	75800	Double-axis double-wheel	0.88
C	2483.85	129.16	107.63	191000	Compound	0.52
D	1284.44	196.52	157.22	34500	Single-wheel	1.53

TABLE 4: The tension stress at the slab bottom.

Gauge position	Tension stress stress (MPa)	Wheel Load (kN)			
		86.64	103.07	129.16	196.52
Covered layer bottom	1	0.1269	0.1445	0.1754	0.2601
	2	0.0832	0.1143	0.1305	0.1951
	3	0.0741	0.0995	0.1241	0.1701
	4	0.1157	0.1321	0.1543	0.2245
	5	0.0945	0.1209	0.1462	0.2069
	6	0.0842	0.1024	0.1322	0.1902
	7	0.1221	0.1401	0.1609	0.2379
	8	0.1146	0.1311	0.1507	0.2157
	9	0.1234	0.1421	0.1721	0.2542
	max	0.1269	0.1445	0.1754	0.2601
Original surface bottom	1	0.5796	0.6901	0.8015	1.2709
	2	0.5421	0.6521	0.7623	1.2461
	3	0.5013	0.6403	0.7415	1.2397
	4	0.5562	0.6709	0.7853	1.2603
	5	0.5487	0.6623	0.7779	1.2541
	6	0.5475	0.6507	0.7617	1.2405
	7	0.5694	0.6767	0.7902	1.2647
	8	0.5501	0.6715	0.7811	1.2579
	9	0.5723	0.6878	0.7963	1.2683
	max	0.5796	0.6901	0.8015	1.2709

TABLE 5: The surface defection.

The wheel Load (kN)	86.64	103.07	129.16	196.52
The surface defection (mm)	0.0816	0.0975	0.1069	0.1407

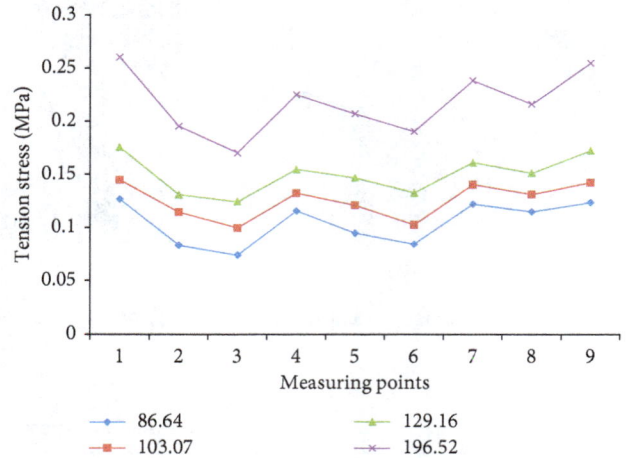

FIGURE 6: Tension stress value of each measuring point of the covered layer.

3.2. Preprocessing, Modeling, and Solving of FEM by ANSYS Software

3.2.1. Preprocessing of Contact Surface between Aircraft Wheel and Airport Pavement.
The contact surface between the wheel and the airport pavement is actually an approximate ellipse with a rectangle in the middle and a semicircle on both sides. In the design of airport pavement, the contact surface is equivalent to a circle of equal area [1], as shown in Figure 10(a). However, this conversion method is difficult to be realized in the ANSYS software because it is not convenient for the meshing, and the coupling between the wheel and the pavement is bad, resulting in the inaccuracy of calculation results. In order to facilitate the finite element calculation and further improve the accuracy of the calculation, the approximate ellipse is converted into a square of equal area, as shown in Figure 10(b), and the square size after the conversion is shown in Table 6.

3.2.2. Unit and Material Definition.
The model use SOILD65 unit commonly used in reinforced concrete engineering. The interaction between different layers is simulated by the contact unit. The rigid body target surface on the contact surface is TARGE170, and the contact surface is CONTA173 unit.

Define the thickness, elasticity modulus, Poisson's ratio, and density of different materials such as the new concrete

the entity, it is necessary to make the following basic assumptions [12]:

(1) The materials of each layer of the pavement structure are uniform, continuous, and isotropic linear elastic bodies characterized by the elastic modulus and Poisson's ratio

(2) The contact surface between the pavement slab and the basement and the basement and the soil foundation is assumed to be completely continuous with continuous contact between layers under the action of the wheel load

(3) The elastic modulus, Poisson's ratio, and other parameters of each layers' material do not change in the study

(4) Excluding the influence of the weight of each structural layer

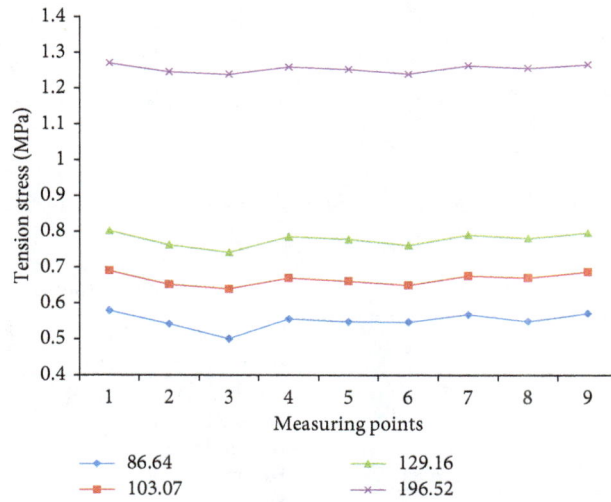

FIGURE 7: Tension stress value of each measuring point of the original surface layer.

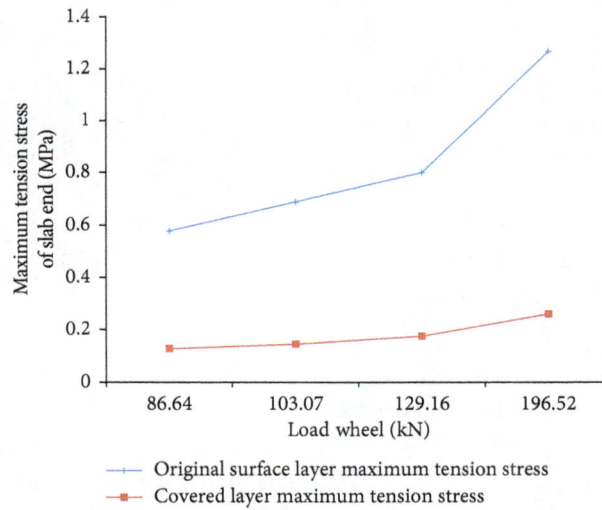

FIGURE 8: Maximum tension stress value of each measuring point.

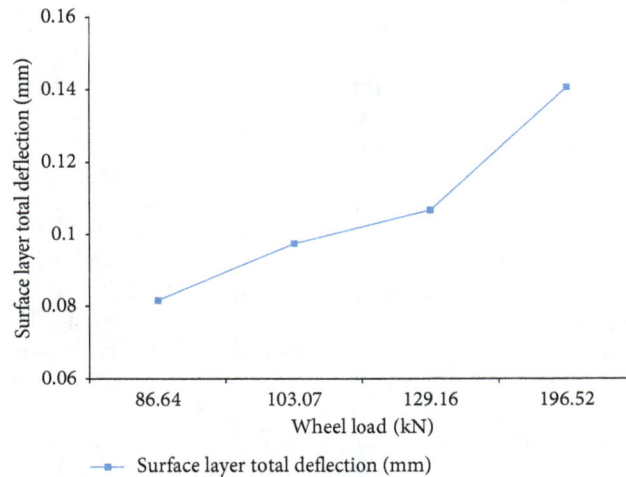

FIGURE 9: Maximum total defection values of each measuring point.

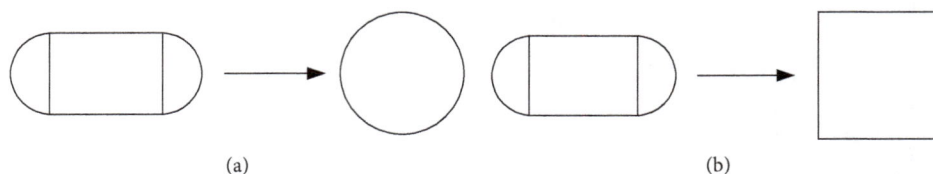

FIGURE 10: Conversion schematic diagram of the contact surface.

TABLE 6: Square size for various types od aircraft wheel.

Aircrafts	Single main wheel grounding area (cm^2)	Square size (cm)
A	1110.77	33.33
B	1171.25	34.22
C	2483.85	49.84
D	1284.44	35.84

TABLE 7: The pavement material parameters.

Structural layer	Thickness (m)	Elastic Modulus (MPa)	Poisson ratio	Density (kg/m^3)
Covered layer	0.20	36000	0.15	2400
Original surface layer	0.30	32000	0.15	2400
Basement	0.40	300	0.25	2100
Soil foundation	10	80	0.35	1800

pavement surface (the covered layer), the old concrete pavement surface (the original surface layer), the basement, and the soil foundation, as shown in Table 7.

3.2.3. Modeling. A 5 m × 5 m geometric model of a cement concrete slab is established, the length and width of which are half of the basement and the soil foundation and use the "SMART MESH" command to divide the grid to form a FEM, as shown in Figure 11.

3.2.4. Loading and Solving. Enter the solution module, impose constraints on the model, and set the analysis options. The displacement boundary is simulated with the actual consolidation of the pier bottom, and the temperature boundary is set at normal temperature (20°C). Gravity is loaded using the "ACEL" command, and concentrated force is loaded using the "F" command. A transient analysis is used to solve the dynamic effect.

3.3. FEM Results. Figures 12–15 show tension stress and maximum deflection with wheel load (each graph is composed of a two-dimensional broken line graph and a three-dimensional broken line graph).

Appendix A shows that stress analysis cloud chart under different wheel loads. As shown in Figures A1–A16, the deflection cloud diagram and the stress cloud diagram show roughly the same distribution rules that the middle of the slab longitudinal edge is the center, and the larger the radius is the smaller the deflection and stress values are, and the Y-direction maximum stress (i.e., the maximum tension stress) appears in the same position—the middle of the slab longitudinal edge, indicating that the critical position of ACCDDPS is in the middle of the longitudinal edge. We have listed Figures A1 and A3 as Figure 12(a) and Figure 12(b), respectively, and other pictures are shown in Appendix.

As depicted in Figures 13 and 14, the tension stress of covered layer increases with the increase of wheel load. When the load increases from 129.16 kN to 196.52 kN, the maximum tension stress increases to 0.257 MPa. That is, at

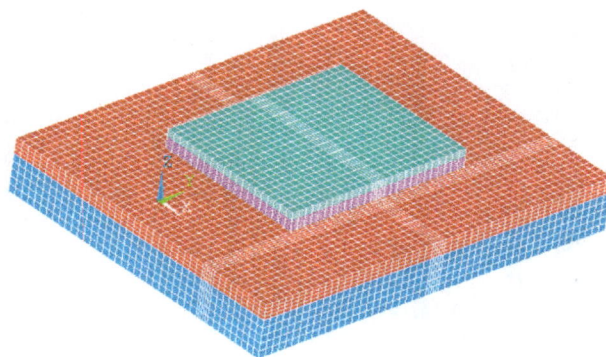

FIGURE 11: ACCDDPS FEM.

the point of 129.16 kN, there is a clear inflection point; after that point, the change rate of stress with load becomes larger. Similarly, the tension stress of original surface layer also increases with the increase of wheel load; however, at the point of 129.16 kN, the stress values of the three curves decrease slightly. Because the C-aircraft wheel load is 129.16 kN, the ground area of the wheel reaches 2483.85 cm^2 (it is nearly twice as high as the other three aircrafts) which decreases the stress. As we can see from Figures 13 and 14, for both covered layer and original surface layer, the principal stress curves almost coincide with the Y-direction stress curves, which means that the maximum principal stress of ACCSSPS is in the longitudinal direction.

Figure 15 shows that the maximum deflection values of the covered layer and the original surface layer increase with the increase of the wheel load. And, two maximum deflection curves almost coincide. This is in line with the mechanical principle of the combination method of the ACCDDPS.

3.4. FEM Verification. By comparing FEM and ARSTS experimental results, their experimental results are basically consistent, indicating that FEM is feasible for mechanical analysis of ACCDDPS. The following conclusions can be drawn:

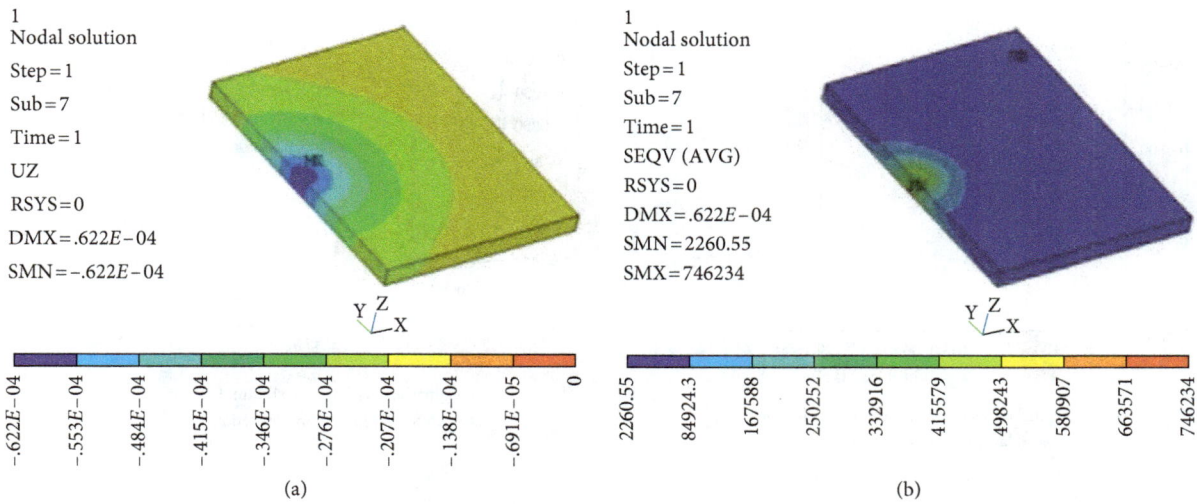

FIGURE 12: Aircraft deflection and stress cloud diagrams: (a) original surface layer deflection cloud diagram; (b) original surface layer stress cloud diagram.

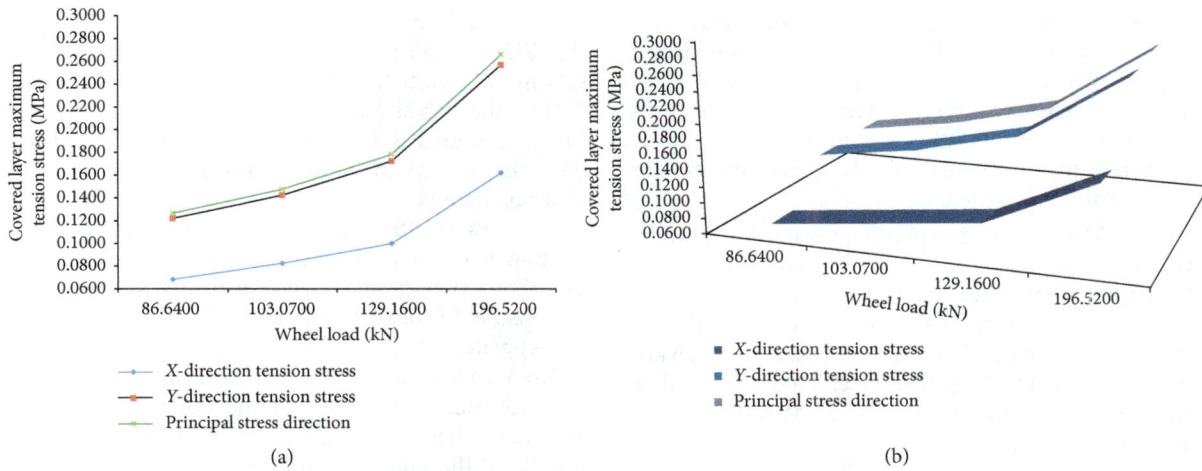

FIGURE 13: Variation of covered layer tension stress with the wheel load: (a) two-dimensional broken-line graph; (b) three-dimensional broken-line graph.

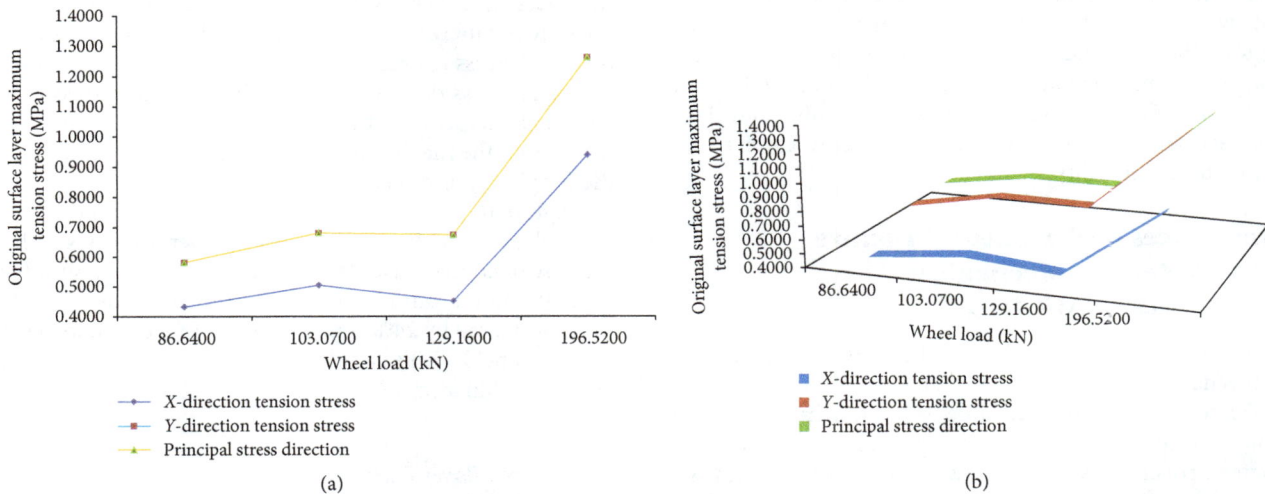

FIGURE 14: Variation of original surface layer tension stress with wheel load: (a) two-dimensional broken-line graph; (b) three-dimensional broken-line graph.

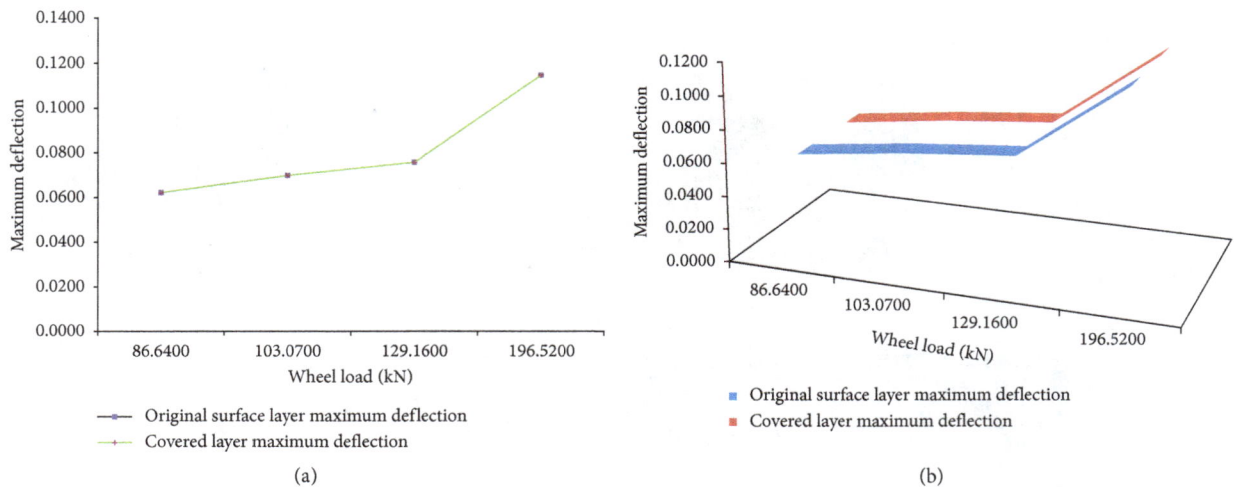

FIGURE 15: Variation of maximum deflection with wheel load: (a) two-dimensional broken-line graph; (b) three-dimensional broken-line graph.

(1) In the ANSYS stress cloud diagrams, the Y-direction maximum stress (i.e., the maximum tension stress) appears at the middle of the slab longitudinal edge. This is consistent with the results of the ARSTS experimental results—the maximum tension stress appears at the measuring point 1, namely, the middle of the slab longitudinal edge.

(2) Both FEM and ARSTS experiment results' maximum tension stress increases with the increase of wheel load, maximum deflection values of the covered layer and the original surface layer increases with the increase of the wheel load, and the point of 129.16 kN is a key point in the growth rate, which shows that ARSTS experimental results and FEM results are in good agreement.

The conclusions we obtained have a good fit with the existing results. Many experts and scholars agree that the critical load location of airport pavement and highway surface should coincide with the middle of the longitudinal joint, which coincides with our conclusion [1, 13]. Wong [14] used the theoretical formula to derive and came to such a conclusion that the maximum stress value of ACCDDPS is 0–2.35MPa. Our ARSTS experimental results and FEM results are within its range, which proves our research results have certain credibility.

4. Influences of Pavement Structure Parameters on Mechanical Properties of ACCDDPS

The above research shows that FEM is feasible and accurate for mechanical analysis of ACCDDPS. With the help of ANSYS software, we use the same FEM modeling method to further study the effect of three pavement structure parameters: covered layer thickness, elastic modulus, and slab size on mechanical properties of ACCDDPS, thus enriching and perfecting the design theory of ACCDDPS.

4.1. Covered Layer Thickness on Mechanical Properties of ACCDDPS.
The covered layer thickness is 0.13 m, 0.15 m, 0.17 m, 0.19 m, 0.21 m, 0.23 m, 0.25 m, 0.27 m, 0.29 m, and 0.31 m, the slab size is 5 m × 4 m, and the load is 196.52 kN. The action area is 1284.44 cm². The other material parameters are shown in Table 7. The modeling method is as described above.

Figures 16–18 show tension stress and maximum deflection with covered layer thickness (each graph is composed of a two-dimensional broken-line graph and a three-dimensional broken-line graph).

Appendix B shows that stress analysis cloud chart under different covered layer thickness. As shown in Figures B1–B40, the deflection cloud diagram and the stress cloud diagram also show that the critical position of ACCDDPS is in the middle of the longitudinal edge.

As depicted in Figures 16 and 17, for both covered layer and original surface layer, the tension stress decreases with the increase of covered layer thickness. However, with the thickness of cover layer decreasing, the rate of stress reduction is different. For the covered layer, the decrease of principal stress is small, but for the original surface layer, the principal stress decreases faster. The principal stress curves almost coincide with the Y-direction stress curves, which means that the maximum principal stress of ACCDDPS is in the longitudinal direction.

Figure 18 shows the maximum deflection values of the covered layer, and the original surface layer decreases with the increase of covered layer thickness, and the deflection value is about 0.1 mm. From 13 cm to 31 cm, the maximum deflection value decreased by 24%. And two maximum deflection curves almost coincide. This is in line with the mechanical principle of the combination method of the ACCDDPS.

4.2. Covered Layer Elastic Modulus on Mechanical Properties of ACCDDPS.
The covered layer elastic modulus is 24000 MPa, 26000 MPa, 28000 MPa, 30000 MPa,

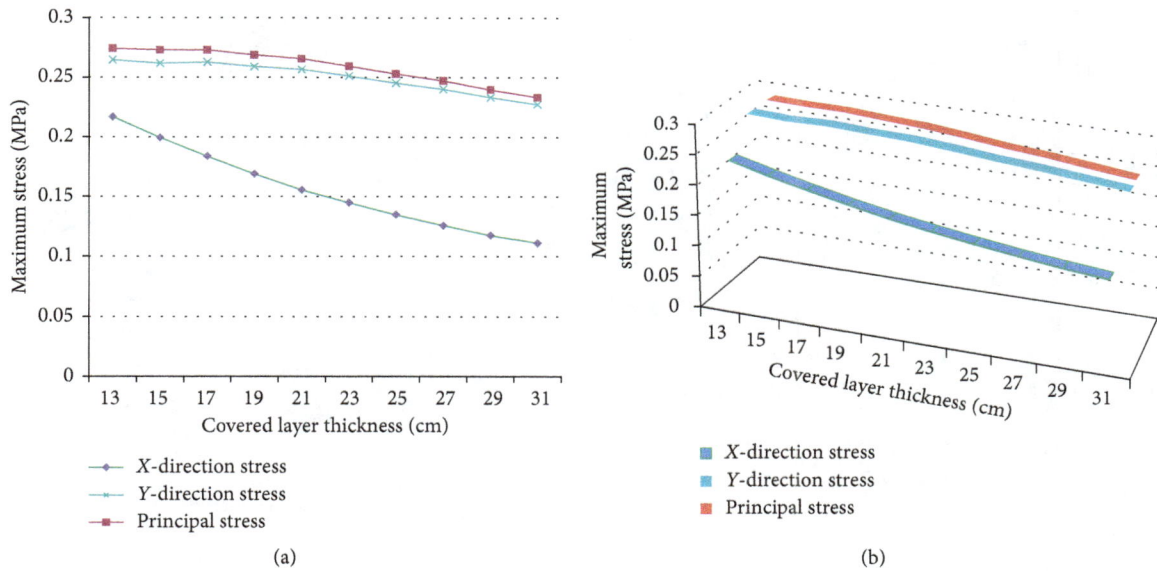

FIGURE 16: Variation of covered layer stress with covered layer thickness: (a) two-dimensional broken-line graph; (b) three-dimensional broken-line graph.

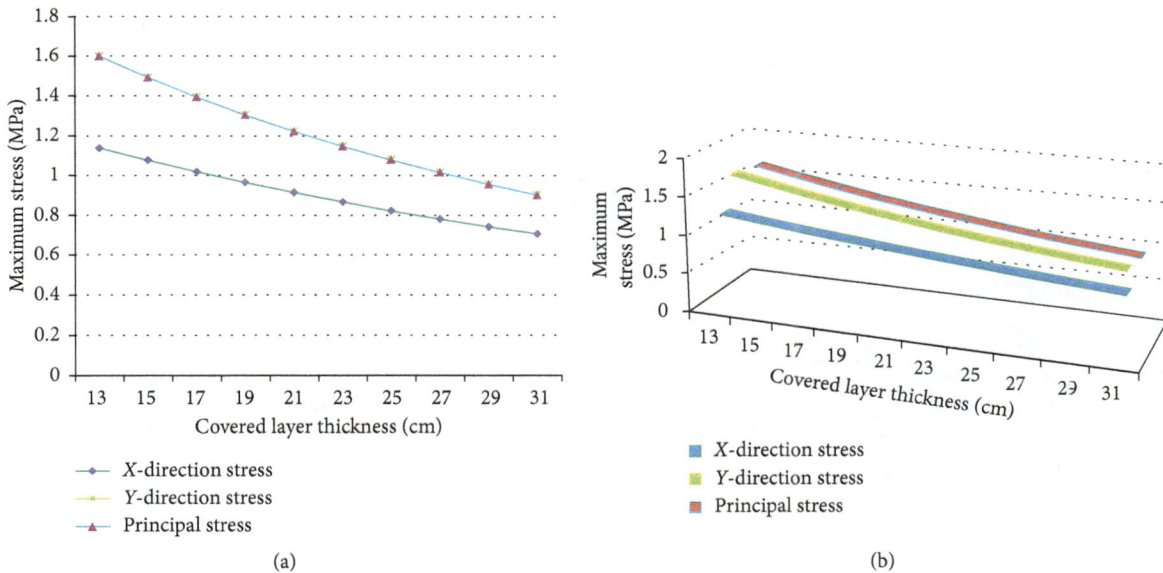

FIGURE 17: Variation of original surface layer stress with covered layer thickness: (a) two-dimensional broken-line graph; (b) three-dimensional broken-line graph.

32000 MPa, 34000 MPa, 36000 MPa, 38000 MPa, 40000 MPa, 42000 MPa, and 44000 MPa, the slab size is 5 m × 4 m, and the load is 196.52 kN. The action area is 1284.44 cm². The other material parameters are shown in Table 7. The modeling method is as described above.

Figures 19–21 show tension stress and maximum deflection with covered layer elastic modulus (each graph is composed of a two-dimensional broken-line graph and a three-dimensional broken-line graph).

Appendix C shows that stress analysis cloud chart under different covered layer thicknesses. As shown in

Figures C1–C44, the deflection cloud diagram and the stress cloud diagram also show that the critical position of ACCDDPS is in the middle of the longitudinal edge.

As depicted in Figures 19 and 20, the principal stress curves almost coincide with the Y-direction stress curves, which means that the maximum principal stress of ACCDDPS is in the longitudinal direction. For covered layer, the tension stress increases with the increase of covered layer elastic modulus. However, for original surface layer, the tension stress decreases with the increase of covered layer elastic modulus.

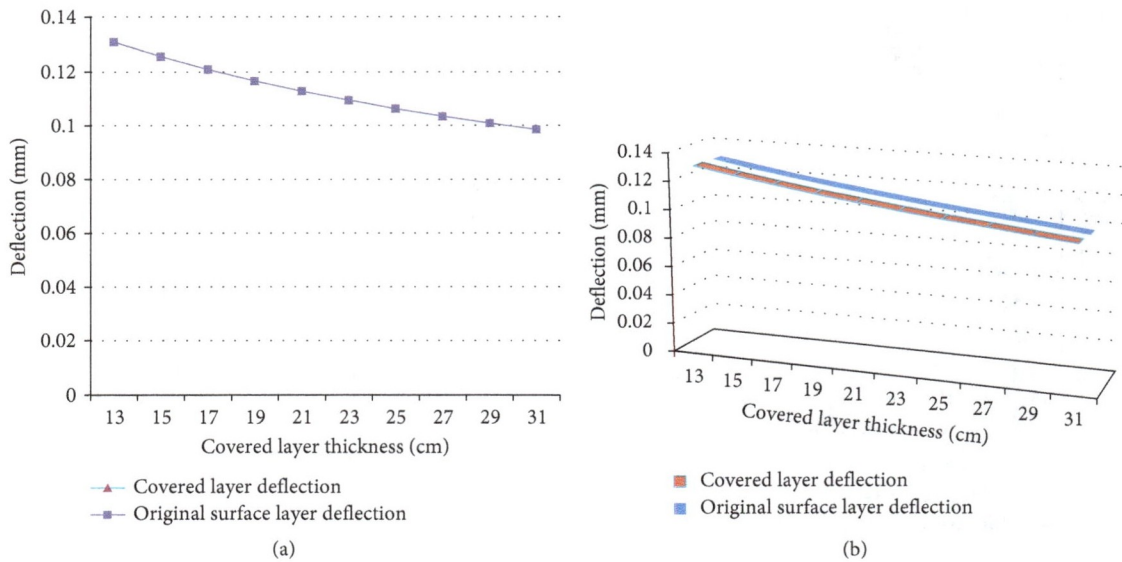

Figure 18: Variation of maximum deflection with covered layer thickness:. (a) two-dimensional broken line graph; (b) three-dimensional broken-line graph.

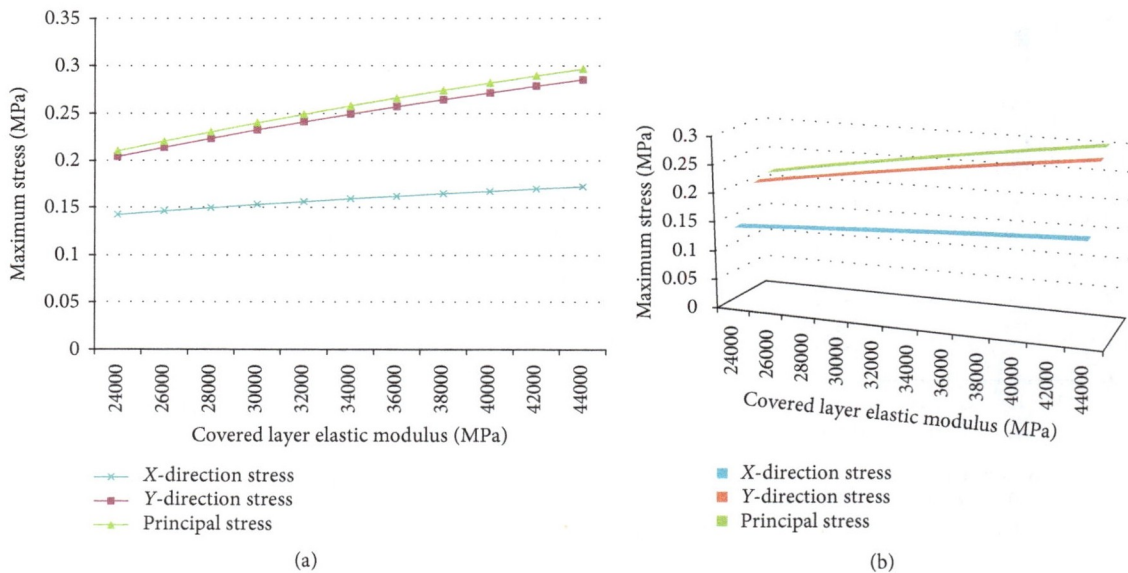

Figure 19: Variation of covered layer stress with covered layer elastic modulus: (a) two-dimensional broken-line graph; (b) three-dimensional broken-line graph.

Figure 21 shows that the maximum deflection values of the covered layer and the original surface layer decrease with the increase of covered layer elastic modulus, and the deflection value is about 0.1 mm. And, two maximum deflection curves almost coincide, which is determined by the mechanical principle of the combination method of the ACCDDPS.

4.3. Slab Size on Mechanical Properties of ACCDDPS. We select four common slab sizes (6 m × 5 m, 5 m × 5 m, 5 m × 4 m, 4 m × 4 m, resp.), and the load is 196.52 kN. The action area is 1284.44 cm². The other material parameters are shown in Table 7. The modeling method is as described above.

Figures 22–24 show tension stress and maximum deflection with slab size (each graph is composed of a two-dimensional broken-line graph and a three-dimensional broken-line graph).

Appendix D shows stress analysis cloud chart under different slab sizes. As shown in Figures D1–D16, the deflection cloud diagram and the stress cloud diagram also show that the critical position of ACCDDPS is in the middle of the longitudinal edge.

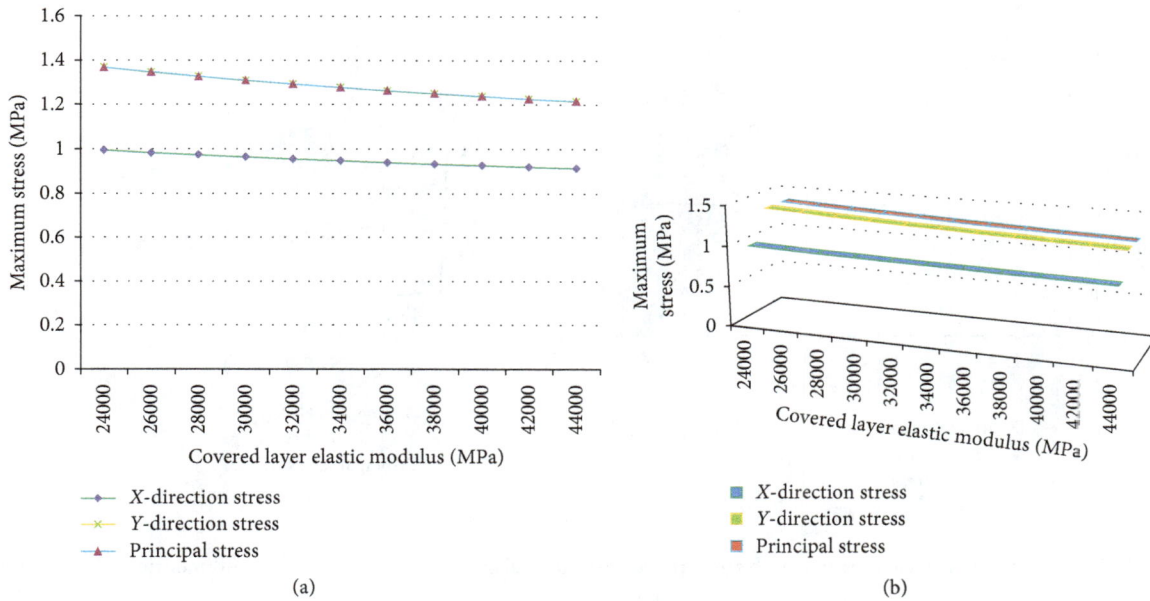

FIGURE 20: Variation of original surface layer stress with covered layer elastic modulus.

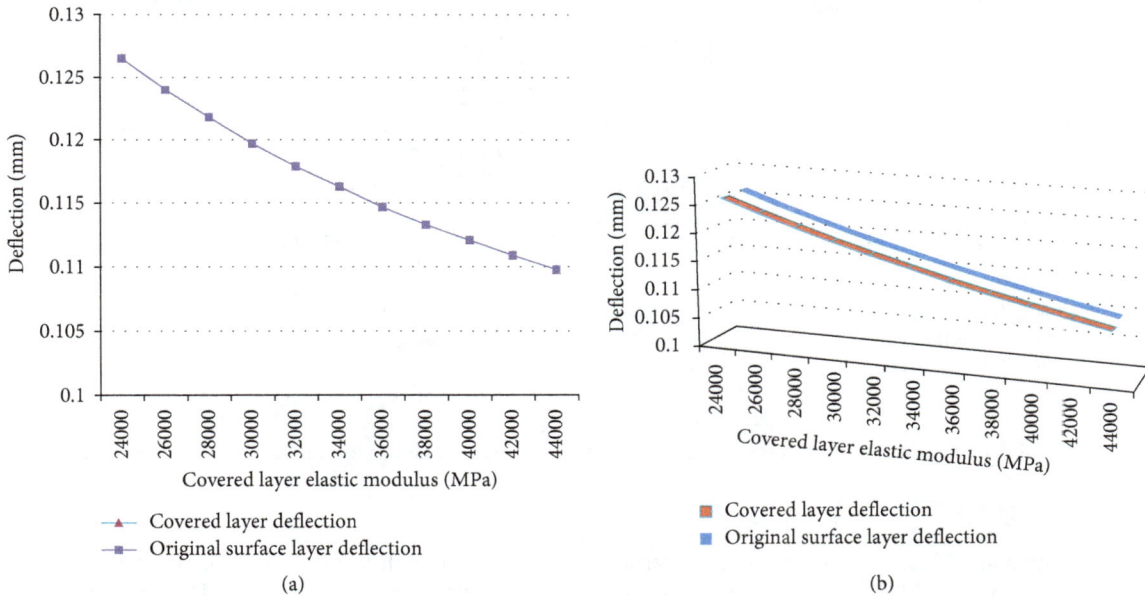

FIGURE 21: Variation of maximum deflection with covered layer elastic modulus.

As depicted in Figures 22 and 23, the principal stress curves almost coincide with the Y-direction stress curves, which means that the maximum principal stress of ACCDDPS is in the longitudinal direction. For covered layer, the tension stress decreases with the increase of slab size. However, for original surface layer, the tension stress hardly changes as the slab size increases.

Figure 24 shows the maximum deflection values of the covered layer, and the original surface layer increases with the increase of slab size, and the deflection value is about 0.1 mm. And, two maximum deflection curves almost

coincide, which is determined by the mechanical principle of the combination method of the ACCDDPS.

5. Conclusions

We used the ARSTS to study the mechanical properties of ACCDDPS under aircraft single-wheel dynamic loads. In the ANSYS software, FEM was established and the simulation experiment under the same conditions was carried out. FEM results are consistent with ARSTS results and are consistent with the existing research conclusions, which prove the

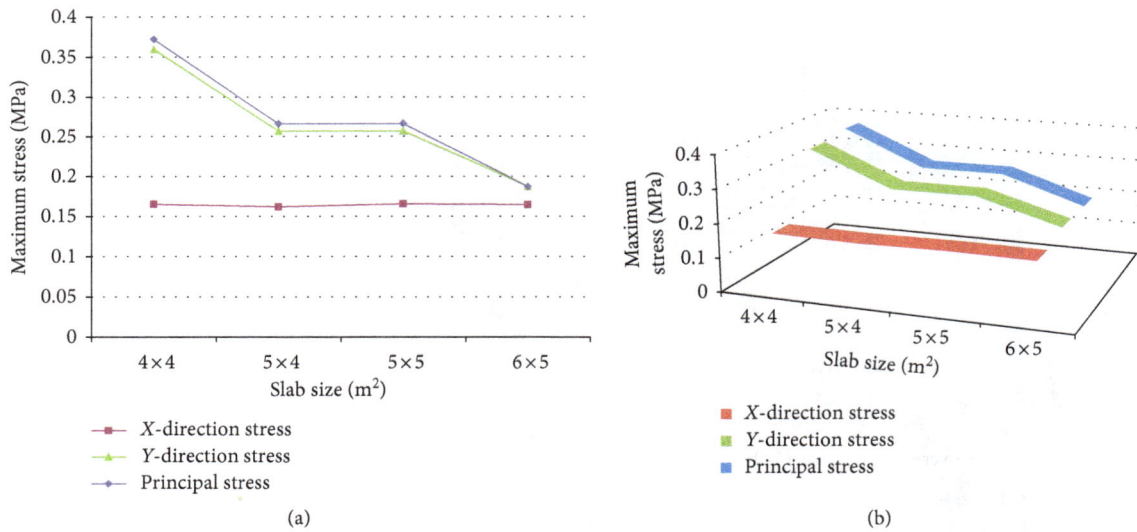

FIGURE 22: Variation of covered layer stress with slab size: (a) two-dimensional broken-line graph; (b) three-dimensional broken-line graph.

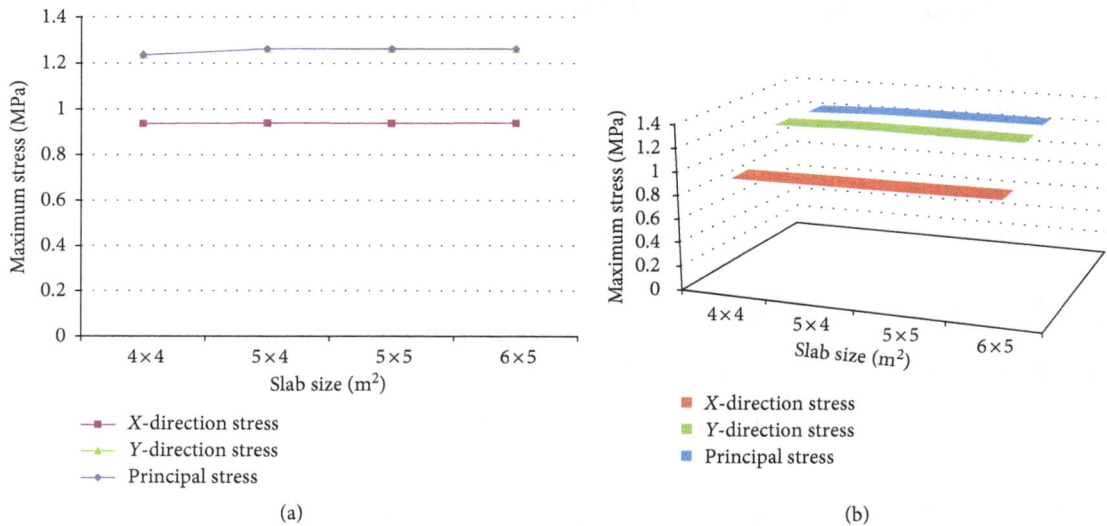

FIGURE 23: Variation of original surface layer stress with slab size: (a) two-dimensional broken-line graph; (b) three-dimensional broken-line graph.

feasibility of FEM applying ACCDDPS mechanical properties analysis. In view of this, we used FEM to analyze the influence of covered layer thickness, covered layer elastic modulus, and slab size on ACCDDPS mechanical properties. We find the following conclusions:

(1) Under the different loads, the change of the tension stress and deflection values with the load change is the same for the covered layer and the original surface layer. The critical position of ACCDDPS is in the middle of the longitudinal edge. The total deflection of surface layer increases with the increase of wheel load, and two maximum deflection curves almost coincide, which is in line with the mechanical principle of the combination method of the ACCDDPS.

(2) Although the structural parameters are different, the critical position of ACCDDPS is in the middle of the longitudinal edge, which may be determined by the slab geometry or load position. However, for the covered layer and the original surface layer, the law that the tension stress values vary with the structural parameters is different, but the maximum deflection value is about 0.1.

(a) The tension stress decreases with the increase of covered layer thickness, and the maximum deflection values decrease with the increase of covered layer thickness.

(b) For covered layer, the tension stress increases with the increase of covered layer elastic modulus. However, for original surface layer, the tension

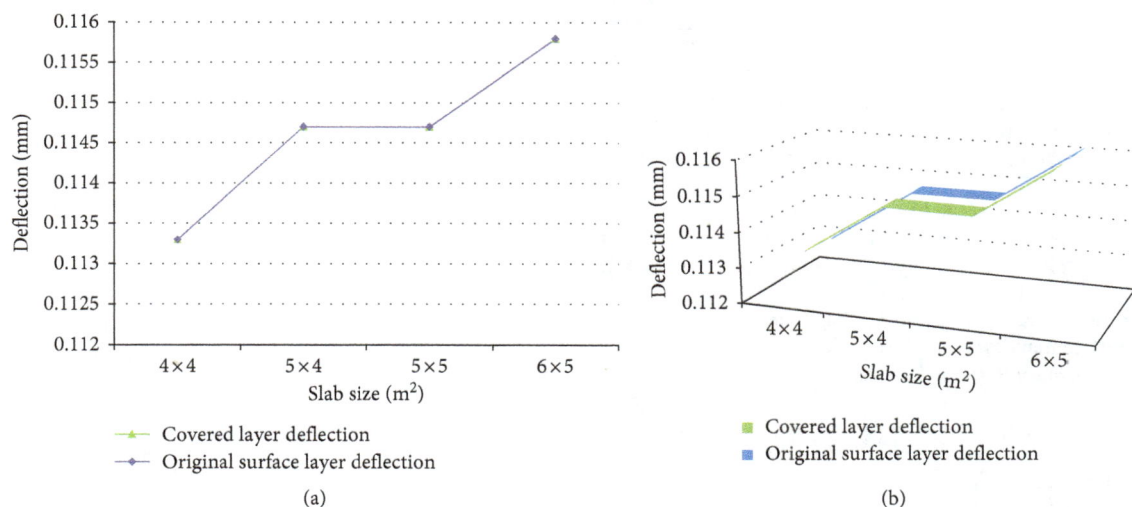

FIGURE 24: Variation of maximum deflection with slab size.

stress decreases with the increase of covered layer elastic modulus. The maximum deflection values decrease with the increase of covered layer elastic modulus.

(3) For covered layer, the tension stress decreases with the increase of slab size. However, for original surface layer, the tension stress hardly changes as the slab size increases. The maximum deflection values increase with the increase of slab size.

Conflicts of Interest

The authors declare that they have no conflicts of interest.

Authors' Contributions

Yu Qingkun designed models and wrote the manuscript. Liangcai Cai analyzed the results. Jianwu Wang conducted data collection and statistics.

Supplementary Materials

The "Appendix" file contains the cloud diagrams we got from the finite element simulation of ANSYS software, because there are 116 pictures in total, so it is placed in the appendix for the reviewers to refer to. The main contents of the Appendix are as follows: Appendix A shows stress analysis cloud chart under different wheel loads, including Figures A1–A16. Appendix B shows stress analysis cloud chart under different covered layer thicknesses, including Figures B1–B40. Appendix C shows stress analysis cloud chart under different covered layer thickness, including Figures C1–C44. Appendix D shows stress analysis cloud

chart under different slab sizes, including Figures D1–D16. (*Supplementary Materials*)

References

[1] W. Xingzhong and C. Liangcai, *Airport Pavement Surface Design*, China Communication Press, Beijing, China, 2007.

[2] V. A. Patil, V. A. Sawant, and K. Deb, "2-D finite element analysis of rigid pavement reason dynamic vehicle–pavement interaction effects," *Applied Mathematical Modeling*, vol. 37, no. 3, pp. 1282–1294, 2013.

[3] J. Binchen, Y. Jie, and Y. Ge, "Study on the method of establishing the finite element model of cement concrete pavement of airport," *Science and Technology Research*, vol. 5, no. 5, pp. 208–210, 2013.

[4] W. Shuyu, "Analysis of load stress of double-deck concrete pavement on elastic half space foundation," *Heilongjiang Traffic Science and Technology*, vol. 28, no. 2, p. 65, 2013.

[5] H. Chao, S. Jian, Y. Jianming et al., "Research on the influence of old pavement fragmentation and addition mode on cement overlay," *Chinese and Foreign Highway*, vol. 34, no. 2, pp. 109–112, 2014.

[6] Z. Haiyan, W. Xuancang, Y. Dianhai et al., "Force analysis of finite size composite pavement slab on elastic foundation," *Journal of Shenyang University of Architecture (Natural Science)*, vol. 24, no. 1, pp. 6–10, 2008.

[7] Y. Xiangcheng, W. Xingzhong, Z. Ying et al., "Force analysis on the Fiber Grill Reinforced Airport Double-deck Pavement," *Journal of Wuhan University of Technology (Transportation Science and Engineering Edition)*, vol. 36, no. 5, pp. 954–957, 2012.

[8] Z. Xianmin, D. Qian, and L. Yaozhi, "The influence of the aircraft landing-gear configuration on the mechanical response of the pavement surface," *Journal of Southwest Jiaotong University*, vol. 49, no. 4, pp. 675–681, 2014.

[9] L. Zizheng and Z. Hongduo, "Structural response analysis of composite pavement under A380-800 aircraft load," *Western Communication Technology*, vol. 79, no. 2, pp. 69–74, 2014.

[10] M. Xiang, N. Fujian, and C. Rongsheng, "Composite airport pavement surface load stress," *Journal of Chang'an University (Natural Science Edition)*, vol. 30, no. 4, pp. 23–28, 2010.

[11] S. Choi, J. Yeon, and M. C. Won, "Improvements of curing operations for Portland cement concrete pavement," *Construction and Building Materials*, vol. 35, no. 10, pp. 597–604, 2012.

[12] A. Zokaei-Ashtiani, C. Carrasco, and S. Nazarian, "Finite element modeling of slab–foundation interaction on rigid pavement applications," *Computers and Geotechnics*, vol. 62, pp. 118–127, 2014.

[13] S. Jiankang, W. X. Zhong, and Y. Xiangcheng, "Test method for airport cement concrete double deck pavement sandwich material," *Sichuan Building Science*, vol. 38, no. 1, pp. 171–174, 2012.

[14] MH/T5004-2010, *Specifications for Airport Cement Concrete Pavement Design*, Civil Aviation Administration of China, Beijing, China, 2010.

16

Influence of Cement Type and Water-to-Cement Ratio on the Formation of Thaumasite

Ailian Zhang and Linchun Zhang

Department of Civil Engineering, Sichuan College of Architectural Technology, Deyang 618000, China

Correspondence should be addressed to Ailian Zhang; zhangailian1980@163.com

Academic Editor: Xiao-Yong Wang

Cement mortar prisms were prepared with three different cement types and different water-to-cement ratios plus 30% mass of limestone filler. After 28 days of curing in water at room temperature, these samples were submerged in 2% magnesium sulfate solution at 5°C and the visual appearance and strength development for every mortar were measured at intervals up to 1 year. Samples selected from the surface of prisms after 1-year immersion were examined by X-ray diffraction (XRD) and Fourier transform infrared (FTIR) spectroscopy. The results show that mortars with sulfate resisting Portland cement (SRC) or sulphoaluminate cement (SAC) underwent weaker degradation due to the thaumasite form of sulfate attack than mortars with ordinary Portland cement (OPC). A lower water-to-cement ratio leads to better resistance to the thaumasite form of sulfate attack of the cement mortar. A great deal of thaumasite or thaumasite-containing materials formed in the OPC mortar, and a trace of thaumasite can also be detected in SRC and SAC mortars. Therefore, the thaumasite form of sulfate attack can be alleviated but cannot be avoided by the use of SAC or SRC.

1. Introduction

In recent years, it has become a common practice to incorporate fine limestone powder as an additional constituent in the cement production [1]. A high volume of limestone filler is also used frequently to increase the content of fine particles and optimize the particle packing in self-compacting concrete (SCC) mixes [2]. The use of limestone in cement or concrete seems to have many benefits, such as reducing water demand, improving strength development, and being economical [3, 4]. It was also reported that the addition of a finely ground limestone filler has a positive effect on the behavior of mortars exposed to magnesium sulfate solution due to the improved compactness [5]. However, it has been widely reported that cement and concrete containing limestone are subject to a special type of sulfate attack, attributed to the formation of thaumasite ($CaSiO_3 \cdot CaCO_3 \cdot CaSO_4 \cdot 15H_2O$) at low temperatures (lower than 15°C) [6, 7]. Since the formation of thaumasite involves the reaction of C-S-H in the cement paste with carbonate and sulfate ions [8], it results in severer and quicker decomposition of cement and concrete than conventional sulfate attack attributed to ettringite

($3CaO \cdot Al_2O_3 \cdot 3CaSO_4 \cdot 31H_2O$). And the use of sulfate resisting Portland cement with low C_3A content becomes ineffective to preserve cementitious materials against the thaumasite form of sulfate attack [9]. Sulphoaluminate cement, being completely different from Portland cement in mineral composition, has been used increasingly in concrete structures constructed in winter, especially in hydraulic engineering and structure remedial engineering [10–12], and there are few published studies concerning the thaumasite form of sulfate attack of this cement. Therefore, this paper presents experimental results related to the effect of cement type and water-to-cement ratio on the resistance to thaumasite sulfate attack of mortars containing a limestone filler.

2. Experimental

Three types of cement were used: ordinary Portland cement (OPC), sulfate resisting Portland cement (SRC), and sulphoaluminate cement (SAC). All of them have the same strength grade of 42.5 according to Chinese standards. The chemical compositions of these cement types are shown in Table 1. The ground limestone filler has a specific surface area of

TABLE 1: Chemical compositions of cement (wt.%).

Number	CaO	SiO$_2$	Al$_2$O$_3$	Fe$_2$O$_3$	MgO	SO$_3$	R$_2$O	IL
OPC	61.27	21.04	6.94	2.36	1.32	1.94	0.97	3.76
SRC	63.52	22.75	4.12	4.37	2.19	2.01	0.68	0.33
SAC	41.53	8.10	30.32	3.41	3.60	11.93	—	0.65

TABLE 2: Mixing proportions of mortar.

Number	OPC	SRC	SAC	Sand	Limestone	Water
OPC (0.6)	1.0	0	0	2.2	0.3	0.6
SRC	0	1.0	0	2.2	0.3	0.6
SAC	0	0	1.0	2.2	0.3	0.6
SAC50	0.5	0	0.5	2.2	0.3	0.6
0.5	1.0	0	0	2.2	0.3	0.5
0.4	1.0	0	0	2.2	0.3	0.4

420 m^2/kg (Blaine). The fine aggregate was quartz sand with a fineness modulus of 2.5, an apparent density of 2.65 g/cm^3, and a bulk density of 1.60 g/cm^3. The sulfate solutions were prepared by synthetic MgSO$_4$.

Mortar prisms (40 × 40 × 160 mm) were cast according to the mixture proportions presented in Table 2. The prepared specimens were kept in a moist cabinet and cured for 24 hours at room temperature. Then, all the specimens were removed from the mould and cured in water. After 27 days of curing, all the specimens were immersed into 2%-by-weight MgSO$_4$ solution at 5°C. After every 2 months, the sulfate solution was replaced keeping the volume ratio of MgSO$_4$ solution to prepared specimens at 2:1. For every mixture specimen, the visual appearance and strength were measured after immersion treatment in MgSO$_4$ solution at regular intervals up to 1 year. Samples were selected from the surface of prisms after strength testing at different immersion ages. X-ray diffraction (XRD) and Fourier transform infrared (FTIR) spectroscopy were used to analyze these samples and distinguish the degraded products of cement mortar after sulfate exposure at lower temperature.

X-ray diffraction (XRD) analysis was conducted using a diffractometer (D/MAX-IIIA, Rigaku, Japan) with Cu-Kα radiation to identify the phase formed before and after immersion into the sulfate solution. The diffraction data was collected for each specimen under the following conditions: 2θ range of 5 to 60°, a count time of 0.6 s per step, and sampling width of 0.02°. Fourier transform infrared (FTIR) spectroscopy was carried out to identify the chemical and structural changes using a Nicolet 60 SXB FTIR Spectropho-tometer. A mixture of synthesized powder and spectroscopic grade KBr was ground and then pressed at 2000 psi for 5 min to obtain a pellet, and the corresponding wavenumber ranged from 400 cm^{-1} to 4000 cm^{-1}.

3. Results and Discussion

3.1. Visual Appearance. Visual appearances of the tested mortar specimens were detected for every mixture monthly.

It can be found that there is a little of a white precipitate on the surface of the mortar specimens after the initial 27-day curing in water for all the mortars except for SAC mixture, being mainly attributed to the CH leaching out from the mortar and a little of carbonate in water. During the first several months of immersion in the sulfate solution, such white mass exudation continued and it seemed to have no negative influence on the mortar mechanical performance. There were firstly some signs of deterioration on the surfaces and edges of the samples after 4 months for the OPC mortar but a longer time of 6 months for other mortars.

Visual appearances of the tested specimens were detected for every mixture after 1-year exposure to sulfate solution at 5°C. The results for typical mixtures are presented in Figure 1. It can be found that there are obvious signs of deterioration on the specimen surfaces for every specimen. And a white soft substance was formed on the container's bottom resulting from the spalling specimen surfaces. From the visual appearance, it was found that the OPC mortar exhibited the worst damage, sulfate resisting Portland cement (SRC) mortar showed a limited improvement, and the mortars containing sulphoaluminate cement (SAC and SAC50) underwent much weaker damage. Therefore, both SRC and SAC showed improved resistance to the thaumasite form of sulfate attack, and the latter is better. With the decreasing of the water-to-cement ratio from 0.6 to 0.4, the mortar presented less appearance damage.

3.2. Strength Development. Figure 2 presents the strength development of mortars with different types of cement immersed in sulfate solution. No significant strength reduction occurred on the OPC mortar during the beginning of the 3 months, and a little strength increase happened on other mortars. With the continued exposure to sulfate attack, the strength of every mortar began to decrease and the strength loss increased evidently with the increasing of immersion time. After 1-year immersion in sulfate solution, OPC, SRC, SAC, and SAC50 mortars showed 72.8%, 53.6%, 35.6%, and 25.6% compressive strength loss and 45.6%, 35.1%, 25.0%,

FIGURE 1: Visual appearance of mortars after 1-year exposure to sulfate: (a) OPC (0.6), (b) SRC, (c) SAC, (d) SAC50, (e) 0.5, and (f) 0.4.

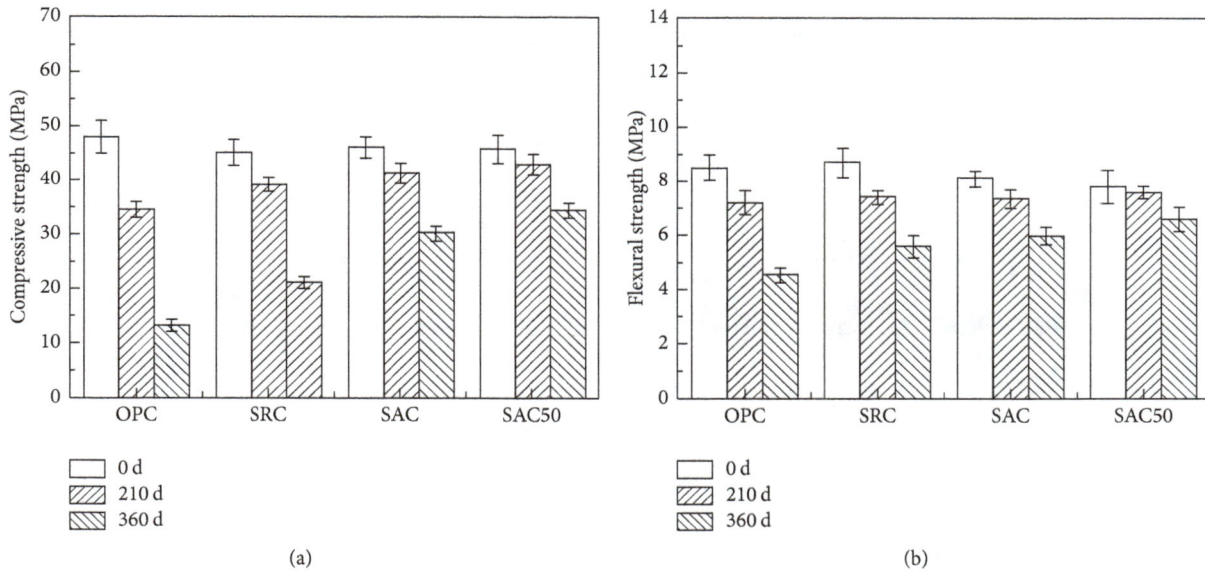

FIGURE 2: Strength development of mortars with different types of cement with immersion time: (a) compressive strength, (b) flexural strength.

and 15.6% flexural strength loss, respectively. Though sulfate resisting Portland cement mortar shows a less strength loss than OPC mortar, it does not behave as good as expected for the traditional sulfate attack. Sulphoaluminate cement shows much better resistance to sulfate attack at lower temperature than OPC mortar. These results agree with those of the above visual inspection.

Figure 3 presents the strength development of mortars with different water-to-cement ratios immersed in a sulfate solution. After the first 3 months of immersion, just a very little decrease in compressive strength was found for the control mortar with water-to-cement ratio of 0.6, and the other two mortars with lower water-to-cement ratios of 0.5 and 0.4 showed an obvious strength increase. With the continued exposure to sulfate attack, the strength of every

mortar began to decrease and the strength loss increased with the increasing of immersion time. After 1-year immersion in sulfate solution, mortars with water-to-cement ratios of 0.6, 0.5, and 0.4 showed 72.8%, 26.9%, and 15.0% compressive strength loss and 45.6%, 9.8%, and 3.1% flexural strength loss, respectively. This weaker damage is mostly attributed to the improved porosity and impermeability of mortar with a lower water-to-cement ratio. Therefore, a lower water-to-cement ratio is very effective for improving the resistance to the thaumasite form of sulfate attack.

3.3. Mineralogy. The samples were selected from mortars after the initial 27-day curing in water and XRD patterns of them are shown in Figure 4. For all the samples, strong peaks corresponding to quartz (SiO_2) from the sand and calcite

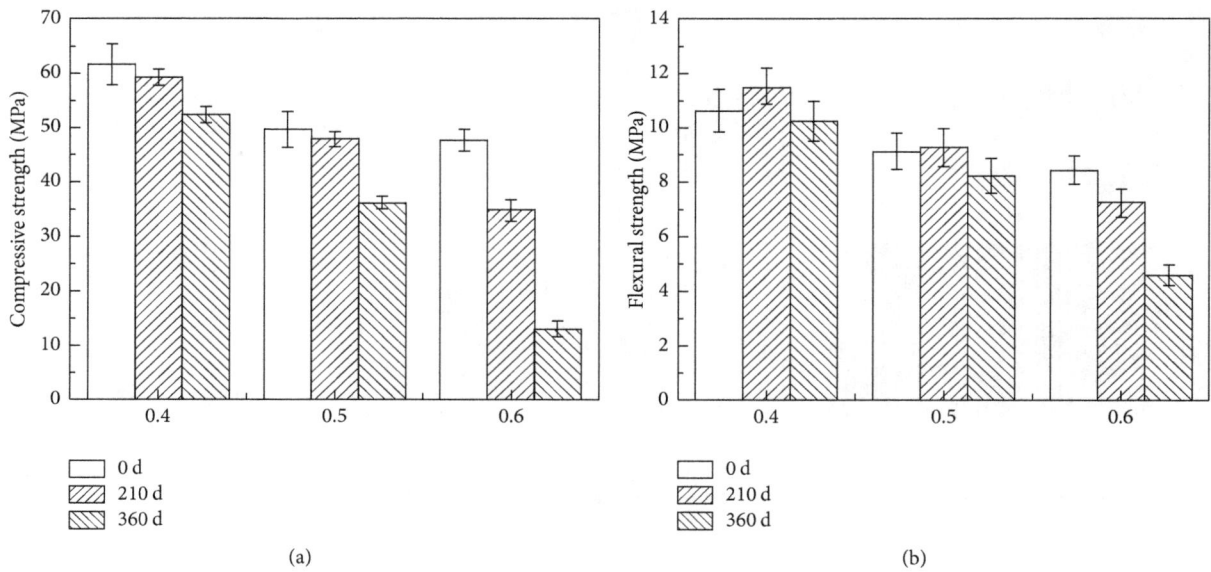

FIGURE 3: Strength development of mortars with different water-to-cement ratios with immersion time: (a) compressive strength, (b) flexural strength.

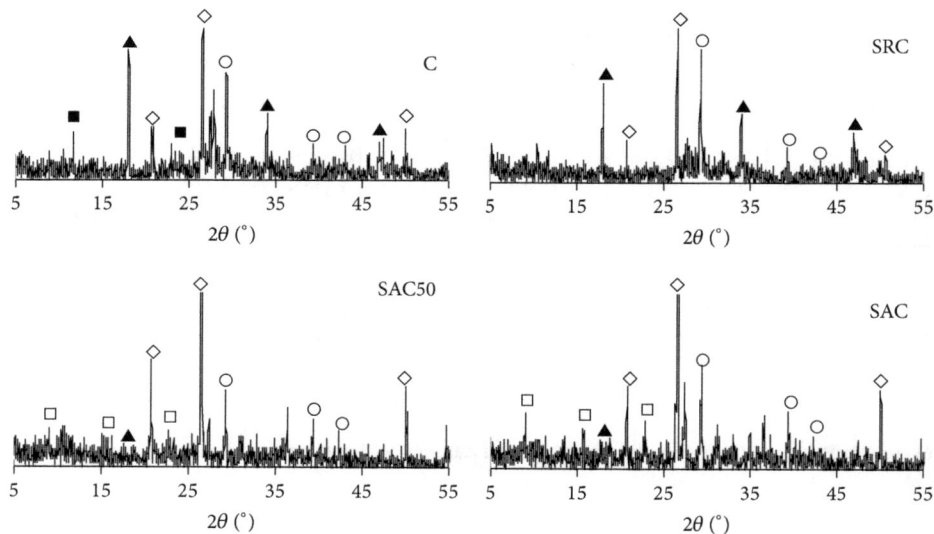

FIGURE 4: XRD patterns of mortars before immersion into sulfate solution. $\diamond$: quartz; $\circ$: calcite; $\blacktriangle$: portlandite; $\blacksquare$: monocarboaluminate; $\square$: ettringite.

($CaCO_3$) from the limestone filler were found. As expected, a mass of portlandite ($Ca(OH)_2$) formed in OPC and SRC mortars. Monocarboaluminate ($3CaO \cdot Al_2O_3 \cdot CaCO_3 \cdot 11H_2O$) formed as one of the hydration products of C_3A from OPC and limestone filler. When the mortar is immersed into the sulfate solution, sulfate ions penetrate into the mortar and react with portlandite, monosulfoaluminate, CSH gel to form ettringite, gypsum, and thaumasite [13, 14]. The formation of these products leads to the swelling, cracking, and finally decomposition of the OPC mortar. No detectable

monocarboaluminate was found in the sulfate resisting Portland cement mortar containing much less C_3A, and the mortar suffered from a slighter attack attributed to ettringite formation. Other mineral compositions in SRC are similar to OPC, so the SRC mortar also suffered from the formation of gypsum and thaumasite and it showed limitedly better resistance to sulfate attack at low temperature than the OPC mortar. The main minerals of sulphoaluminate cement are $C_4A_3\check{S}$, C_2S, and $C_{12}A_7$, and their hydration products mainly contain ettringite, CSH gel, and $Al(OH)_3$ [15]. Very strong

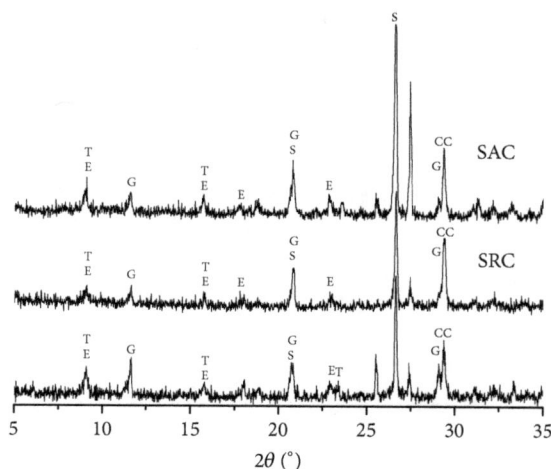

FIGURE 5: XRD patterns of mortars after 1-year immersion in sulfate solution. S: quartz; CC: calcite; E: ettringite; T: thaumasite; G: gypsum.

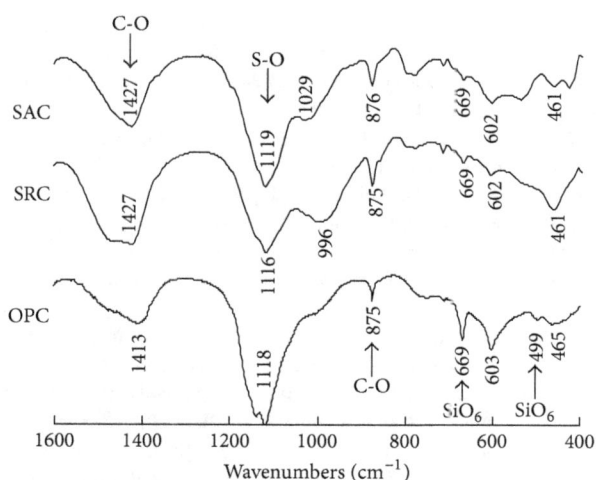

FIGURE 6: FTIR spectra of mortars with different types of cement after 1-year exposure to sulfate solution.

peaks corresponding to ettringite and weak peaks attributed to portlandite can be found in the XRD pattern of the SAC mortar. On the one hand, there is no enough portlandite or unstable aluminates such as monocarboaluminate or monosulfoaluminate to form gypsum and ettringite. On the other hand, there is less CSH gel for thaumasite sulfate attack in the SAC paste than in the OPC paste. Therefore, the SAC mortar behaves much better than the OPC mortar.

Figure 5 presents the XRD patterns of samples selected from the surfaces of mortars with different types of cement after 1-year immersion. In all samples, even no detectable trace of portlandite can be seen. And a great deal of sulfate-bearing substances including ettringite/thaumasite and gypsum formed in the samples. According to the relative intensities of major peaks at around $9.1°$ and $11.6°$ 2θ, there are more sulfate-bearing substances formed in OPC mortar than in SRC and SAC mortars which suffered from slighter deterioration.

It is difficult to distinguish ettringite and thaumasite from XRD patterns when only small amounts are present in a sample being attributed to their very similar crystal structures [16]. So, spectra of samples were further analyzed by FTIR spectra as shown in Figure 6. For all the samples, there are strong peaks at around 1110 cm^{-1} corresponding to S-O, showing a large number of sulfate-bearing substances [17]. And the OPC mortar shows a stronger peak, attributed to more sulfate-bearing substances, than the other two samples. The C-O peaks at 875 cm^{-1} and around 1400 cm^{-1} are occurring in all the samples as expected, and they are attributable to the presence of carbonates. The obvious peaks at 499 cm^{-1} and 669 cm^{-1}, being assigned to the presence of SiO_6 bonds [18], indicate a mass of thaumasite or thaumasite-containing solid solution formed in the OPC mortar. Weak peaks of SiO_6 bonds in the other two samples and the XRD patterns mentioned above show that SRC and SAC postpone the thaumasite formation.

4. Conclusions

(i) Based on the mechanical performance degradation, the relative resistance to the thaumasite form of sulfate attack of cement is outlined below, from the best to the worst: mixture of sulphoaluminate cement and OPC, sulphoaluminate cement, sulfate resisting Portland cement, and OPC.

(ii) A lower water-to-cement ratio leads to better resistance to the thaumasite form of sulfate attack of cement mortar.

(iii) After 1 year of exposure to sulfate solution, a great deal of thaumasite or thaumasite-containing materials formed in the OPC mortar, and a trace of thaumasite can also be detected in SRC and SAC mortars. Therefore, the thaumasite form of sulfate attack can be alleviated but cannot be avoided by the use of SAC or SRC.

Conflicts of Interest

The authors declare that they have no conflicts of interest.

References

[1] H. F. Sun, B. Hohl, Y. Z. Cao et al., "Jet mill grinding of portland cement, limestone, and fly ash: impact on particle size, hydration rate, and strength," *Cement and Concrete Composites*, vol. 44, pp. 41–49, 2013.

[2] M. Valcuende, E. Marco, C. Parra, and P. Serna, "Influence of limestone filler and viscosity-modifying admixture on the shrinkage of self-compacting concrete," *Cement and Concrete Research*, vol. 42, no. 4, pp. 583–592, 2012.

[3] S. Tsivilis, J. Tsantilas, G. Kakali, E. Chaniotakis, and A. Sakellariou, "The permeability of Portland limestone cement concrete," *Cement and Concrete Research*, vol. 33, no. 9, pp. 1465–1471, 2003.

[4] S. Tsivilis, E. Chaniotakis, E. Badogiannis, G. Pahoulas, and A. Ilias, "A study on the parameters affecting the properties of Portland limestone cements," *Cement and Concrete Composites*, vol. 21, no. 2, pp. 107–116, 1999.

[5] N. Saca and M. Georgescu, "Behaviour of Portland limestone cement mortars in magnesium sulfate solution," *Romanian Journal of Materials*, vol. 44, no. 1, pp. 5–16, 2014.

[6] A. P. Barker and D. W. Hobbs, "Performance of Portland limestone cements in mortar prisms immersed in sulfate solutions at 5°C," *Cement and Concrete Composites*, vol. 21, no. 2, pp. 129–137, 1999.

[7] A. Skaropoulou, G. Kakali, and S. Tsivilis, "Thaumasite form of sulfate attack in limestone cement concrete: the effect of cement composition, sand type and exposure temperature," *Construction and Building Materials*, vol. 36, pp. 527–533, 2012.

[8] D. W. Hobbs and M. G. Taylor, "Nature of the thaumasite sulfate attack mechanism in field concrete," *Cement and Concrete Research*, vol. 30, no. 4, pp. 529–533, 2000.

[9] P. Brown and R. D. Hooton, "Ettringite and thaumasite formation in laboratory concretes prepared using sulfate-resisting cements," *Cement and Concrete Composites*, vol. 24, no. 3-4, pp. 361–370, 2002.

[10] L. Pelletier-Chaignat, F. Winnefeld, B. Lothenbach, and C. J. Müller, "Beneficial use of limestone filler with calcium sulphoaluminate cement," *Construction and Building Materials*, vol. 26, no. 1, pp. 619–627, 2012.

[11] J. Zhao, G. Cai, D. Gao, and S. Zhao, "Influences of freeze-thaw cycle and curing time on chloride ion penetration resistance of Sulphoaluminate cement concrete," *Construction and Building Materials*, vol. 53, pp. 305–311, 2014.

[12] H. Wang, X. Gao, and R. Wang, "The influence of rheological parameters of cement paste on the dispersion of carbon nanofibers and self-sensing performance," *Construction and Building Materials*, vol. 134, pp. 673–683, 2017.

[13] M. E. Gaze and N. J. Crammond, "Formation of thaumasite in a cement:lime:sand mortar exposed to cold magnesium and potassium sulfate solutions," *Cement and Concrete Composites*, vol. 22, no. 3, pp. 209–222, 2000.

[14] G. Qu and A. Zhang, "Influence of temperature on the resistance to sulfate attack of limestone filler concrete," *Romanian Journal of Materials*, vol. 42, no. 4, pp. 381–386, 2012.

[15] X. Fu, C. Yang, Z. Liu, W. Tao, W. Hou, and X. Wu, "Studies on effects of activators on properties and mechanism of hydration of sulphoaluminate cement," *Cement and Concrete Research*, vol. 33, no. 3, pp. 317–324, 2003.

[16] S. M. Torres, J. H. Sharp, R. N. Swamy, C. J. Lynsdale, and S. A. Huntley, "Long term durability of Portland-limestone cement mortars exposed to magnesium sulfate attack," *Cement and Concrete Composites*, vol. 25, no. 8, pp. 947–954, 2003.

[17] X. Gao, B. Ma, Y. Yang, and A. Su, "Sulfate attack of cement-based material with limestone filler exposed to different environments," *Journal of Materials Engineering and Performance*, vol. 17, no. 4, pp. 543–549, 2008.

[18] J. Bensted and S. P. Varma, "Studies of thaumasite-part II," *Silictes Industrials*, vol. 39, no. 1, pp. 11–19, 1974.

17

Effect of TiO$_2$ Nanoparticles on Physical and Mechanical Properties of Cement at Low Temperatures

Li Wang, Hongliang Zhang ⓘ, and Yang Gao

Key Laboratory for Special Area Highway Engineering of Ministry of Education, Chang'an University, Xi'an, Shaanxi, China

Correspondence should be addressed to Hongliang Zhang; zhliang0105@163.com

Academic Editor: Ali Nazari

Low temperature negatively affects the engineering performance of cementitious materials and hinders the construction productivity. Previous studies have already demonstrated that TiO$_2$ nanoparticles can accelerate cement hydration and enhance the strength development of cementitious materials at room temperature. However, the performance of cementitious materials containing TiO$_2$ nanoparticles at low temperatures is still unknown. In this study, specimens were prepared through the replacement of cement with 1 wt.%, 2 wt.%, 3 wt.%, 4 wt.%, and 5 wt.% TiO$_2$ nanoparticles and cured under temperatures of 0°C, 5°C, 10°C, and 20°C for specific ages. Physical and mechanical properties of the specimens were evaluated through the setting time test, compressive strength test, flexural strength test, hydration degree test, mercury intrusion porosimetry (MIP), X-ray diffraction (XRD) analysis, thermal gravimetric analysis (TGA), and scanning electron microscopy (SEM) in order to examine the performance of cementitious materials with and without TiO$_2$ nanoparticles at various curing temperatures. It was found that low temperature delayed the process of cement hydration while TiO$_2$ nanoparticles had a positive effect on accelerating the cement hydration and reducing the setting time in terms of the results of the setting time test, hydration degree test, and strength test, and the specimen with the addition of 2 wt.% TiO$_2$ nanoparticles showed the superior performance. Refined pore structure in the MIP tests, more mass loss of CH in TGA, intense peak appearance associated with the hydration products in XRD analysis, and denser microstructure in SEM demonstrated that the specimen with 2 wt.% TiO$_2$ nanoparticles exhibited preferable physical and mechanical properties compared with that without TiO$_2$ nanoparticles under various curing temperatures.

1. Introduction

Pavement structures, owing to their nature of closely contacted with the environment, are highly influenced by environmental factors such as low temperatures in cold regions, specifically. Cement, as one of the most widely used pavement materials, is widely used in pavement engineering because of its low price and easy availability. However, previous studies have demonstrated that low temperature prolongs the setting time of cement and reduces the strength of cement, resulting in lower construction productivity in cold regions [1].

To minimize these negative impacts associated with low temperature, accelerators of cement hydration are used as additives in concrete to reduce the setting time and enhance the strength development [2]. Commonly, chloride salts are the most widely used and cost-effective accelerators [3].

However, in some areas which suffered from chemical attacks such as chloride-induced corrosion, addition of these chloride accelerators should be avoided to reduce the occurrence of corrosion of steel reinforcement and strength loss of concrete. Hence, numerous studies have been carried out to develop nonchloride accelerator instead to remove the possibility of corrosion, especially in aggressive environments. Calcium nitrate [4, 5], aluminate [6], and lithium salts [7] can change the setting time and strength of cement through their chemical properties. In addition, limestone powder [1, 8] and silica fume [9] can be used as physical accelerators to modify the cement at low temperatures.

Nanoparticles, owing to its ultrafine size and high surface area to volume ratio, can accelerate the cement hydration process and enable the formation of a denser microstructure, thereby improving the durability and mechanical properties of cementitious materials [10, 11]. Therefore, nanoparticles could be considered as an alternative choice to serve as accelerators for cementitious materials under low temperatures. Numerous studies have been conducted on the utilization of nanoparticles, including nano-SiO$_2$ [12–22], nano-CaCO$_3$ [17, 23, 24], nano-Al$_2$O$_3$ [16, 20, 25], nano-TiO$_2$ [14, 16, 26–38], nano-Fe$_2$O$_3$ [10, 13], and nano-CuO [13, 39], in cementitious materials to improve their durability and mechanical properties. Nano-SiO$_2$ is the most commonly used nanoparticle, while nano-TiO$_2$ is comparatively favorable for pavement structures in terms of air-cleaning property. With respect to TiO$_2$ nanoparticles, the study by Nazari et al. showed that partial replacement of cement with TiO$_2$ nanoparticles reduced the setting time, enhanced the hydration process, and increased the flexural strength of concrete at the temperature of 20°C, while the flexural strength reduced when the content of TiO$_2$ nanoparticles was more than 4% [26–29]. Lee et al. reported that, through the addition of TiO$_2$ nanoparticles, the early-age hydration of cement was accelerated, the setting time was reduced, and the compressive strength was increased at 23 ± 2°C [31, 32]. Pimenta Teixeira et al. indicated that high temperature (40°C and 60°C) facilitated the hydration of cement paste containing TiO$_2$ nanoparticles and improved the early-age compressive strength [34]. Mohseni et al. found that the incorporation of TiO$_2$ nanoparticles enhanced the compressive strength of self-compacting mortar at ambient temperature and with the increase in the content of TiO$_2$, the compressive strength also exhibited an upward trend [16]. However, Zhang et al. examined that, at 20 ± 3°C, the compressive strength and flexural strength of concrete decreased with the increasing dosage of TiO$_2$ nanoparticles [33]. Feng et al. discovered that, at room temperature, the microstructure of cement paste became denser with the incorporation of TiO$_2$ nanoparticles and a new type of hydration product was examined [30]. In addition, due to photocatalytic action of TiO$_2$ nanoparticles, these particles are commonly mixed into construction materials to obtain the property of air cleaning [36–38].

Most of these studies were carried out under ambient temperature, and very few studies have investigated the influence of SiO$_2$ nanoparticles and Al$_2$O$_3$ nanoparticles on properties of cementitious materials under low temperatures [18–20]. The effect of TiO$_2$ nanoparticles on the performance of cementitious materials under low temperatures was still undiscovered, and whether TiO$_2$ nanoparticles still possess its characteristics under low temperatures was still unknown. Better understanding the knowledge of nanoparticles under low temperatures can promote the practical application of nanoparticles in cold regions. The aim of the current study was to evaluate the effect of TiO$_2$ nanoparticles on the physical and mechanical properties of cement under low temperatures, and low temperatures of 0°C, 5°C, and 10°C were selected in this paper with the ambient temperature 20°C as a reference temperature. Furthermore, the influences of TiO$_2$ on the pore size distribution, microstructure, and thermal property were studied when the specimens were cured under varying temperatures at specific ages.

2. Materials and Tests

2.1. Materials and Specimen Proportions. Raw materials used in this study include cement, TiO$_2$ nanoparticles, and natural river sand. Type I ordinary Portland cement (P.O) 42.5 conforming to ASTM C150 specification [40] and TiO$_2$ nanoparticles with the average particle size of 15 nm were used. The chemical and physical properties of cement and TiO$_2$ nanoparticles, provided by the suppliers, are listed in Table 1 and Table 2, respectively. The particle size distribution of the sand used was 0.63–1.25 mm, 0.315–0.63 mm, and 0.16–0.315 mm, with a sand grading of 2 : 4 : 2. Polycarboxylic-type water-reducing agent complied with ASTM C494-type F [41] was used to keep a favorable workability.

In this study, the laboratory experiments were accomplished in two steps. In the first step, effects of TiO$_2$ nanoparticles of different dosages of 1 wt.%, 2 wt.%, 3 wt.%, 4 wt.%, and 5 wt.% on the physical and mechanical properties of the specimens cured under temperatures of 0°C, 5°C, 10°C, and 20°C were evaluated. The optimum dosage of TiO$_2$ nanoparticles that exhibited the superior performance in terms of strength, setting time, and hydration degree was selected as a representative to investigate the reason why TiO$_2$ nanoparticles could accelerate cement hydration. In the second step, the influence of optimum TiO$_2$ nanoparticles on the pore structure, hydration products, and microstructure of cement pastes under different curing temperatures was studied.

Cement mortar specimens were prepared for the test of mechanical strength. Mortar specimens included the ordinary mortar specimen without TiO$_2$ nanoparticles and mortar specimens with the cement replacement of 1 wt.%, 2 wt.%, 3 wt.%, 4 wt.%, and 5 wt.% TiO$_2$ nanoparticles, respectively. 50 × 50 × 50 mm cube specimens for the compressive strength test and 40 × 40 × 160 mm prism specimens for the flexural strength test were prepared.

TiO$_2$ nanoparticles were uniformly dispersed in 30% of water using ultrasonication firstly. Then, cement and sand were mixed together for about 1 minute, and after that, the well-dispersed TiO$_2$ nanoparticles were added and mixed for another 1 minute before the rest of the water was added. Mortar was then placed into molds and cured under temperatures of 0°C, 5°C, 10°C, and 20°C with a constant relative humidity of 95% until testing. The water to binder ratio of specimens was 0.5. With regard to the control of curing temperature, automated control chambers with an adjustable temperature range of 0–50°C were employed in the tests. It should be emphasized that the raw materials and molds used were precooled first in order to achieve the temperatures close to the target curing temperature after the mixing process.

In addition, cement paste specimens with the same water to binder ratio of mortar specimens were also prepared and cured under the same curing conditions, and the small pieces cut from the 50 × 50 × 50 mm cube paste specimen were used in the tests of hydration degree, MIP, TGA, XRD, and SEM analysis.

It should be noted that the test process was carried out at room temperature. When the specimens reached the testing

TABLE 1: Chemical and physical properties of ordinary Portland cement.

Chemical composition	Content (%)
CaO	59.48
SiO_2	18.73
Al_2O_3	5.12
Fe_2O_3	3.45
MgO	4.02
SO_3	2.83
LOI	2.51
Specific gravity(g/cm^3)	3.15
Specific surface area (m^2/g)	0.39

TABLE 2: Properties of TiO_2 nanoparticles.

	Diameter (nm)	Surface volume ratio (m^2/g)	Purity (%)
TiO_2	10–25	200	99.8

age, they were taken out of the temperature-controlled curing chambers for testing, and the test process was completed within 30 minutes.

2.2. Physical and Mechanical Property Tests. Initial and final setting times of ordinary cement pastes and cement pastes with different additions of TiO_2 nanoparticles were measured in accordance with ASTM C191 [42] with a manually operated Vicat needle apparatus under varying curing temperatures. The initial setting time was determined as the time between the initial contact of cement and water and the time when the penetration depth was measured to be 25 mm, and the final setting time was calculated as the time between initial contact of cement and water and the time when the needle did not leave a complete circular impression on the surface. Repeated penetration tests were conducted on paste specimens by allowing the Vicat needle to penetrate into the paste. Cement paste used for this test should be prepared to satisfy a standard normal consistency according to ASTM C187 [43], and the cement paste should be of standard consistency when the measuring needle settles to a point 10 ± 1 mm below the original surface within 30 s after being released. Fresh cement pastes with various amounts of water were made until the standard consistency was achieved.

The compressive strength test was conducted on ordinary cement mortars and cement mortars containing different additions of TiO_2 nanoparticles according to ASTM C109 [44] by operating a hydraulic testing machine under a controlled rate of 1350 N/s, and flexural strength was determined in accordance with ASTM C293 [45]. These tests were carried out, respectively, at 3, 7, 28, and 56 days of curing under curing temperatures of 0°C, 5°C, 10°C, and 20°C, and the strength was determined by the average value of triplicate specimens.

To evaluate the hydration process of ordinary cement pastes and cement pastes with various TiO_2 nanoparticles under different curing temperatures, the nonevaporable water chemically bounded in hydration products was calculated to determine the degree of cement hydration at 3, 7, 28, and 56 days of curing [12, 15]. A piece of cement paste was chosen and grinded into powder, and then, 2 g of this powder was dried at 105°C for 3 h until a constant weight was achieved; after that, the powder was ignited at 1000°C for 3 h, and then, the degree of hydration at a certain age can be calculated as the ratio of the non-evaporable water for paste cured at that age to the non-evaporable water for paste fully hydrated, as indicated in the following equation:

$$\alpha = \frac{m_{105} - m_{1000}}{0.25 \cdot m_{105}}, \tag{1}$$

where α is the hydration degree at a certain curing age, m_{105} and m_{1000} are the masses of the specimens after drying at 105°C and then heating at 1000°C, respectively, and $0.25m_{105}$ is considered as the amount of chemically bounded water for a fully hydrated cement paste.

2.3. MIP. In order to observe the influence of low temperature on the pore structure of cement pastes in the hydration process, based on the results of physical and mechanical tests already conducted in this study, the ordinary cement paste and cement paste containing the optimum dosage of TiO_2 nanoparticles were selected to measure the pore size distribution with mercury intrusion porosimetry (MIP) referring to ISO 15901-1 [46]. After 28 days of curing under different temperatures, cement paste specimens were broken into small square fragments with an approximate size of $5 \times 5 \times 5$ mm with the knife, and hardened cement slurry pieces chosen from the middle of the specimen were employed to measure the pore size distribution in order to ensure accuracy. The hardened cement slurry pieces should be soaked in ethanol to stop hydration. After soaking in ethanol for 24 h, the pieces were taken out from ethanol to fully evaporate the ethanol and dried at 105°C for 3 h in an oven to remove the water from the pores and ensure a stable weight before the mercury test. The pore pressure was in a range from 0.2 to 220 MPa, the density of mercury was 13.5335 g/mL, the angle of contact was 130°, and the surface tension was 0.485 N/m.

2.4. Thermal Gravimetric Analysis (TGA). The mass variation of a material exposed to steadily growing heating temperature can be obtained using thermogravimetric analysis. Because the decomposition temperatures of different constituents in a material are different, the content of the corresponding constituents can therefore be determined. In this study, to quantify the CH content of the ordinary cement paste and cement paste modified with optimum dosage of TiO_2 nanoparticles under different curing temperatures, corresponding specimens at 28 days of age were grinded into powder and 2 g of powder was heated with the temperature rising from 20°C to 1000°C at a rate of 10°C/min under a nitrogen environment.

2.5. X-Ray Diffraction (XRD). X-ray diffraction can be utilized to identify the main mineral phases of the cementitious material during hydration. The phases of ordinary cement paste and cement paste with optimum dosage of TiO_2 nanoparticles under varying curing temperatures were analyzed at 28 days of age. The XRD test was conducted to scan the specimens between 5° and 90° (2θ) with a step of 0.02° (2θ) using Cu Ka radiation, at a scan rate of 1° per minute.

2.6. SEM. Microstructural images of ordinary cement paste and cement paste containing the optimum dosage of TiO_2 nanoparticles under various curing temperatures were acquired at 28 days of age by scanning electron microscopy (SEM). After 28 days, the specimen was broken into pieces and the middle part was used in this test. This part was immersed into ethanol to stop hydration, and then, it was dried and sprayed with gold to be conductive prior to SEM testing.

3. Results and Discussion

3.1. Physical and Mechanical Properties

3.1.1. The Setting Times. The initial and final setting times of ordinary cement paste and pastes with different proportions of TiO_2 nanoparticles at various curing temperatures are shown in Figure 1. It is examined from the results that both the initial setting time and the final setting time were extended as the curing temperature reduced. The setting times of cement pastes cured at 0°C were notably longer than those cured at ambient temperature regardless of the content of TiO_2 nanoparticles. With the increase in the curing temperature, the setting time showed an obvious downward trend. The initial setting time of pastes cured at 0°C was three times longer than that of pastes cured at 5°C, nearly five times in comparison with that of pastes cured at 20°C, while pastes cured at 10°C had the initial setting time 9–20% longer compared to that cured at 20°C, almost the same. Similar trends were observed in the final setting time. This is because low temperature restrains the pace of the cement hydration and cement pastes require much longer time to set especially at 0°C. Although the hydration process is demonstrated to proceed at a slower rate in previous study [1, 8], low temperature still has a significant impact on the dissolution of clinkers and their reaction with water, resulting in delay in setting time.

Another phenomenon can also be noticed that, with the increasing content of TiO_2 nanoparticles, the setting time of cement pastes was shortened steadily despite the variation in curing temperature. This is due to the fact that, in addition to the filler effect, nanoparticles itself can act as activator to promote the hydration of cement due to its high activity. TiO_2 nanoparticles, despite chemically inert with cement constituents, can provide extra space for hydrated products to precipitation due to its large surface area to volume ratio, thereby accelerating the cement hydration and shortening the setting time.

3.1.2. Degree of Hydration. Hydration degree of ordinary cement paste and pastes with varying addition of TiO_2 nanoparticles determined by chemically bounded water at different curing temperatures is shown in Figure 2. From the results, it can be seen that the hydration degree of cement was improved with the increase in the curing age because the hydration products could be produced with time growing. It is also worth noting that the hydration degree of cement increased gradually at the early age, while the hydration of cement increased at a relatively slower rate after 7 days.

This phenomenon is mainly attributed to the fact that low temperature generally suppresses the hydration process, causing the slow rate of hydrated product generation. At low temperatures, the dissolution of clinkers will be slowed down and the lower temperature indicates the slower rate of dissolution. Also, low temperature impedes the mutual movement of molecular reaction and accordingly delays the process of the hydration reaction, resulting in poor development of hydration products, which can be reflected in strength gain and microstructure development. It is generally considered that a hydrated cement particle is composed of an anhydrous cement core and a hydrated cement shell, and the core turns smaller with the shell turning thicker as the hydration proceeds. The reason why the hydration degree showed a slower increase after 7 days might be that the diffusion rate of Ca^{2+} and OH^- through the hydrated cement shell decreases as more hydration products will be produced with time and therefore delays the increase of the hydration degree.

On the contrary, with the increase of TiO_2 nanoparticles, the hydration degree of cement pastes was improved firstly, then the improvement was not very evident, and this trend was similar regardless of the curing temperature. TiO_2 nanoparticles, owing to its large surface area to volume ratio, can provide an extra space for the precipitation of hydration products; therefore, the hydrated shell will not get thicker and thicker with time compared to that without TiO_2 nanoparticles, since the thicker the hydrated cement shell reflects the lower the diffusion rate of ions through the shell; finally, the hydration process can be accelerated. With the increasing content of TiO_2 nanoparticles, the nucleation site provided by the total surface area for growth of hydration products also increases, therefore more hydration products will be deposit on the surface of TiO_2 nanoparticles, and the hydration degree will be improved.

3.1.3. Mechanical Strength. The compressive strength and flexural strength of ordinary cement mortar and mortars with different proportions of TiO_2 nanoparticles at various curing temperatures are shown in Figure 3. It is noted that both the two strengths showed a downward trend as the curing temperature decreased and low temperature had an undesirable impact on the strengths especially at 3 days. On the contrary, the strength was steadily enhanced with the growth of the curing time regardless of the variation in temperature. The strength of the specimens increased fast

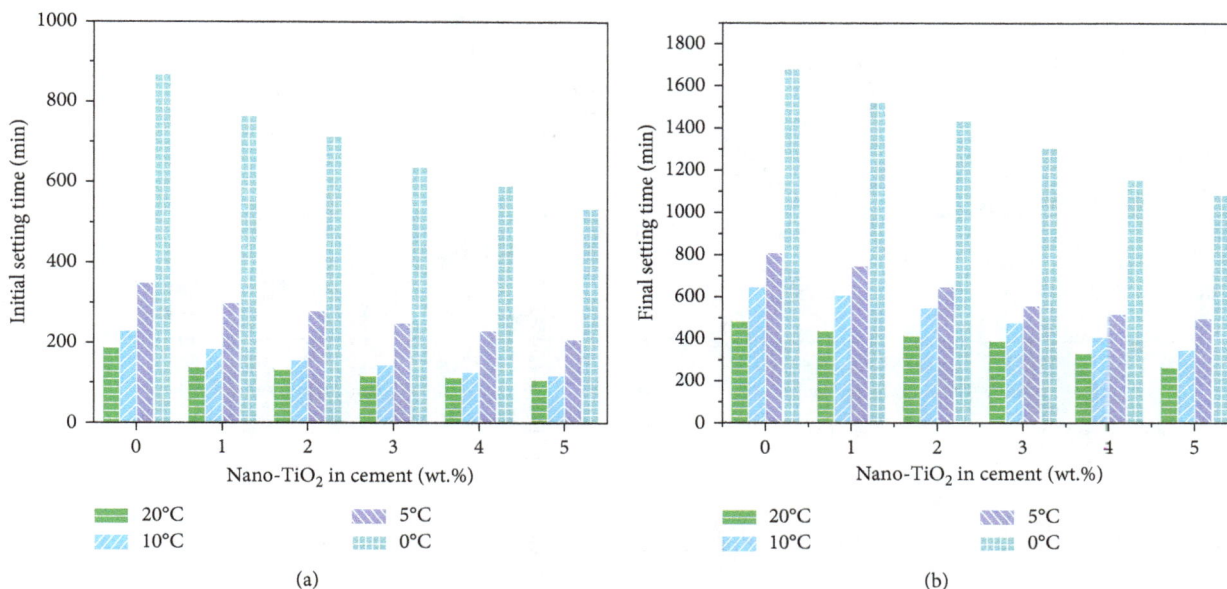

FIGURE 1: (a) Initial setting time of cement pastes; (b) final setting time of cement pastes.

before 7 days while after 7 days the increase rate slowed down, and finally at 56 days, the maximum strength was exhibited. Meanwhile, it is discovered that, with the incorporation of TiO_2 nanoparticles, the strength firstly showed a fast increase compared with the ordinary mortar until the dosage of TiO_2 nanoparticles reached up to 2 wt.% and then the rate of this increase slowed down.

The strength of cement mortar is mainly attributed to the uniform distribution and density of the hydrated product, mainly gels of calcium silicate hydrate (C-S-H), from the hydration of the tricalcium silicate (C_3S) and dicalcium silicate (C_2S). Nevertheless, the hydration rate at low temperatures is lower, therefore insufficient hydration products will be generated to shorten the distance between them and fill the pores in the matrix, and accordingly much more time will be needed for the formation of a dense microstructure compared to the time required at ambient temperature. Moreover, the diffusion of Ca^{2+} and OH^- through the hydrated shell during the hydration process is slowed down at low temperatures, which in turn delays the strength gain. This may be the reason that, at 3 days, although the cement hydration was still carried on at 0°C, the retardation in hydration resulted in the generation of few hydrated products, leading to approximately no strength gain.

The effect of TiO_2 nanoparticles on the strength of mortars in this study was evident. It is clear that the strength of mortars was improved rapidly until the content of TiO_2 nanoparticles is up to 2 wt.%, and this improvement was more remarkable in flexural strength. As is mentioned above, the strength of the cement mortar is closely related to the amount of ettringite and C-S-H gels, and the existence of nanoparticles facilitates the cement hydration, thereby producing more hydration products. In addition to the filler property of nanoparticles to fill the pores in C-S-H gels, it is well known that nanoparticles

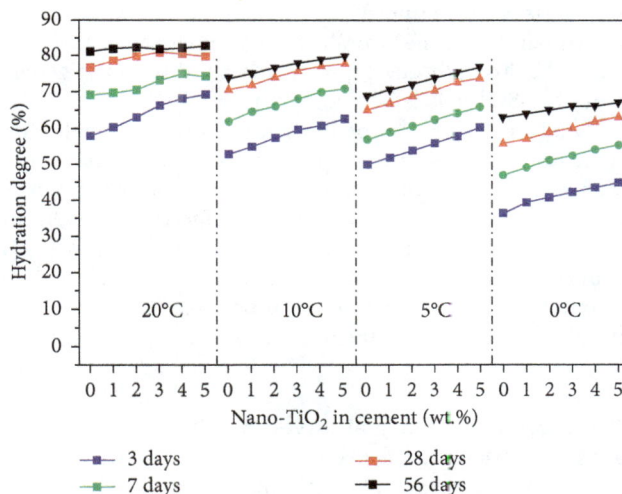

FIGURE 2: Hydration degree of cement pastes.

have a large surface area to volume ratio, and hence, the additional surface area turns to be an appropriate place for hydration products to precipitate. Besides, nanoparticles enable the formation of a bond between itself and C-S-H gels, as a result the strength can be accordingly improved. However, there is also an undesirable effect due to the large ratio of surface area to volume, since nanoparticles are liable to glue together and many nanoparticle clusters which are very weak in strength will be generated, leading to a heterogeneous microstructure [47]. This may be the reason why the hydration degree was still increased slightly while a limited increase in the strength of mortar was observed when the dosage of TiO_2 nanoparticles was over 2 wt.% in this study. This phenomenon may also be attributed to the fact that TiO_2 nanoparticles are enough to

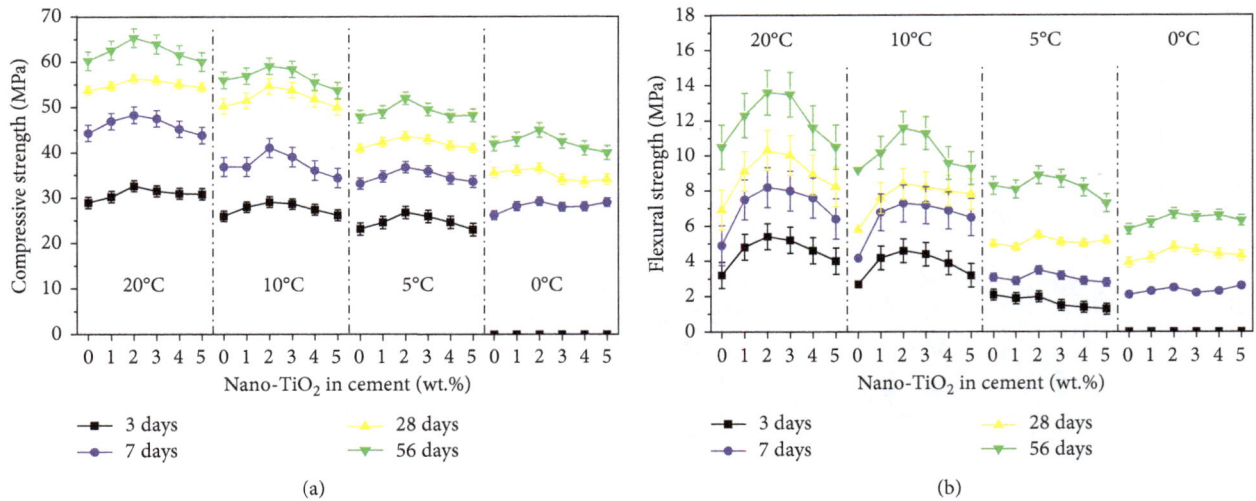

FIGURE 3: (a) Compressive strength of cement mortars; (b) flexural strength of cement mortars.

combine with the free lime, resulting in a loss in strength due to leaching of excessive silica [28].

It should be worthy to note that, although the strength was enhanced by TiO$_2$ nanoparticles, the enhancement was still not very pronounced at low temperatures. Because at low temperatures, the overall rate of cement hydration reduced, the strength development mainly due to the generation of C-S-H gels is accordingly postponed.

Based on the results of the setting time test, hydration degree test, and mechanical strength test, the specimen with 2 wt.% TiO$_2$ nanoparticles showed the superior performance, therefore the optimum dosage was considered as 2 wt.%. In the following experiments, the specimens with 2 wt.% TiO$_2$ nanoparticles and the ordinary specimen were selected to illustrate the mechanism why TiO$_2$ nanoparticles can accelerate the hydration of cement under low temperatures.

3.2. Pore Structure. Integral curves and differential curves of pore size distribution of specimens without and with 2 wt.% TiO$_2$ nanoparticles at different temperatures are shown in Figure 4. These two curves can be utilized to illustrate the pore distribution in specimens. The highest point in the integral curves of pore distribution is the total specific pore volume of the specimen, and the smaller the total pore volume indicates the lower porosity of the specimen. The pore diameter corresponding to the peak in the differential curves of pore distribution is considered as the most probable pore size, which accounts for the largest proportion of the pore size in the specimen compared to the rest of pore sizes, and the larger the most probable pore size indicates the bigger mean pore size [33].

From the test results in Figure 4(a), it can be seen obviously that specimens with TiO$_2$ nanoparticles had smaller pore volume size in comparison with those without nanoparticles, and with 2 wt.% TiO$_2$ nanoparticles, the total pore volume decreased by 12.1%, 11.5%, 10.1%, and 7.3%, at 20°C, 10°C, 5°C, and 0°C, respectively. Also, we noticed that, with 2 wt.% TiO$_2$ nanoparticles in the specimen, the volume of

harmless pore (<20 nm) and few-harm pore (20–50 nm) increased, while the volume of harmful pore (50–200 nm) and multiharm pore (>200 nm) reduced, indicating that TiO$_2$ nanoparticles could refine the pore structure and reduce the porosity [48]. On the contrary, it was observed that, at low temperatures, the harmful and multiharmful pores increased while the harmless and few-harm pores decreased, which demonstrated that low temperature had an adverse effect on the development of the pore structure. From Figure 4(b), the most probable pore sizes in specimen containing TiO$_2$ nanoparticles were also smaller compared to the ordinary specimen, indicating that the size of gel pores was reduced due to the addition of TiO$_2$ nanoparticles [49]. Viewed from these two indexes, it can be found that the incorporation of 2 wt.% TiO$_2$ nanoparticles optimized the pore distribution in specimens, although the overall effectiveness was inhibited by low temperatures. This may be the reason why specimens modified with 2 wt.% TiO$_2$ nanoparticles showed superior performance in mechanical strength, since the strength of the specimen is closely associated with its porosity and microstructure.

For specimens without TiO$_2$ nanoparticles, the total specific pore volume became larger as the curing temperature declined and the most probable pore size also became bigger, indicating that low temperature deteriorated pore structures of specimens and increased their porosity. Therefore, the strength of the specimen was damaged by low temperatures. TiO$_2$ nanoparticles, due to its specialty, can fill the pores in the cement specimen and act as a core for hydration products to precipitate, thus making a homogeneous and dense structure, which may explain the corresponding variation in strength.

However, even with the addition of TiO$_2$ nanoparticles, the pore structure of the cement specimen can hardly be improved at 0°C. The main reason for this phenomenon is that low temperature suppresses the hydration reaction of each constituent of cement and reduces the mutual movement between molecules, especially at 0°C, resulting in the

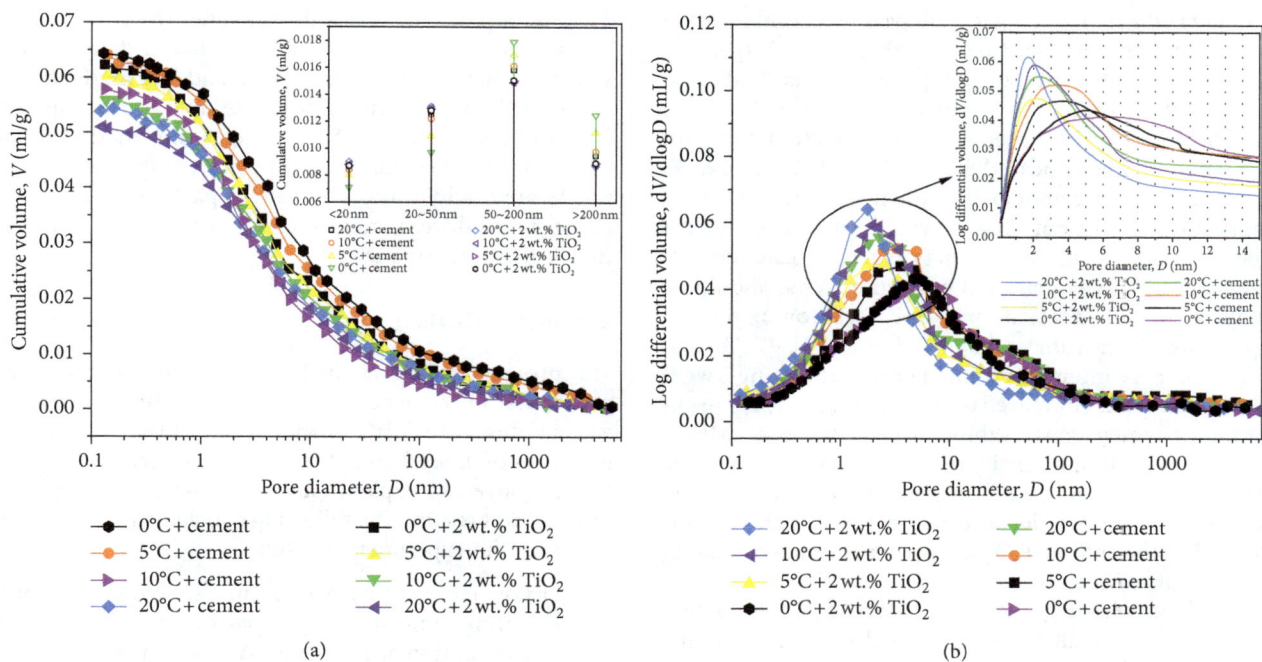

FIGURE 4: (a) Integral curves of pore size distribution of specimens; (b) differential curves of pore size distribution of specimens.

overall reduction of the hydration reaction and ultimately leading to poor development of the pore structure.

3.3. *Thermal Gravimetric Analysis.* According to findings of previous researchers, it is generally considered that, between 30 and 120°C, the evaporable water in hardened cement paste is eliminated, the decomposition of calcium hydroxide (CH) takes place between 400 and 550°C, and the decarbonation of calcium carbonate ($CaCO_3$) appears between 700 and 900°C [50]. Based on this theory, thermal gravimetric analysis was conducted on ordinary cement paste and paste with 2 wt.% TiO_2 nanoparticles cured at different curing temperatures for 28 days to examine the mass loss of the specimens subjected to an elevated temperature, and CH content are presented in Figure 5. To be exact, a portion of CH may participate in the process of carbonation along with C3S, C2S, and C-S-H [51]. However, it is quite difficult to quantify the mass loss of CH during the carbonation process, and as a result, CH content calculated from the mass loss in the temperature domain of 400–550°C is generally selected as an index to describe the hydration degree of pastes. The content of CH can be determined from the following equation:

$$W_{CH} = (W_1 - W_2) \frac{M_{CH}}{M_H}, \quad (2)$$

where W_{CH} is the percentage content of CH, W_1 and W_2 are the mass losses of specimens at 400°C and 550°C, respectively, and M_{CH} and M_H are the molar masses of CH and water, which are 74.10 g/mol and 18.02 g/mol, respectively.

The results indicated that the content of CH in cement pastes was steadily increased as the curing temperature went up from 0°C to 20°C regardless of with or without TiO_2 nanoparticles, demonstrating that low temperature inhibited

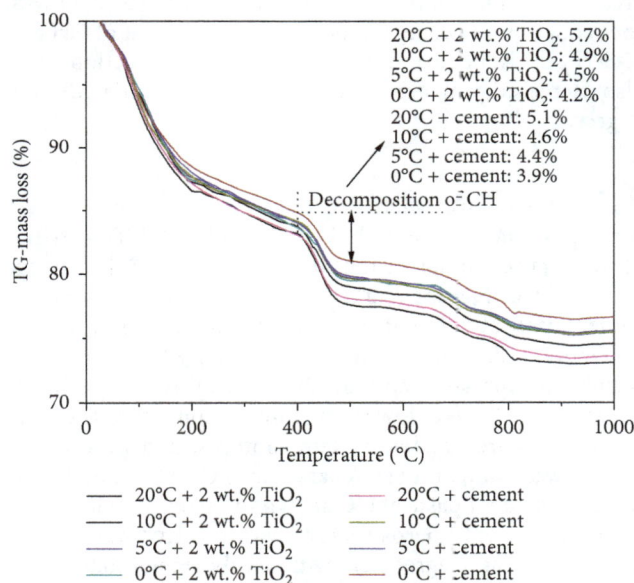

FIGURE 5: TG mass loss in percentage for cement pastes cured at various temperatures.

the process of cement hydration. On the contrary, cement pastes with TiO_2 nanoparticles had more CH in content compared to the ordinary cement pastes because TiO_2 nanoparticles could enhance the hydration of cement due to its characteristic of high ratio of surface area to volume. The thermal gravimetric test results conformed to the strength test very well, since the formation of more CH due to the hydration reaction of C_3S and C_2S directly illustrated more generation of C-S-H gels, leading to the improvement in strength.

3.4. X-Ray Diffraction. The XRD analysis was conducted on ordinary cement paste and paste with 2 wt.% TiO₂ nanoparticles under different curing temperatures at 28 days of age, and the results are shown in Figure 6. From Figure 6(b), it can be observed that diffraction peaks of TiO₂ nanoparticles appeared at 25.33°, 36.25°, and 42.33°, indicating that TiO₂ nanoparticles did not participate in the hydration of cement but only provided extra space for the attachment of hydration products, such as calcium hydroxide (CH), C-S-H gels, and ettringite. What should be worthy to note is that, despite the variation in curing temperature, diffraction peaks of CH at 29.63°, 47.73°, and 48.77° were more intense in cement paste containing 2 wt.% TiO₂ nanoparticles compared to those corresponding peaks in ordinary cement paste exhibited in Figure 6(a). Meanwhile, the diffraction intensity of C-S-H gels and ettringite at 23.22°, 39.20°, and 39.65° in cement paste containing 2 wt.% TiO₂ nanoparticles also became stronger, demonstrating that the role of TiO₂ nanoparticles in accelerating the hydration of cement.

Combined with Figures 2 and 5, the XRD diffraction agreed with the results of hydration degree and thermal gravimetric analysis. The hydration degree was enhanced with 2 wt.% TiO₂ nanoparticles, as more CH and C-S-H gels were generated during the hydration process with the increase in diffraction peaks of CH and C-S-H gels. And with the reduction in curing temperature, the amount of CH and C-S-H gels declined and the corresponding diffraction peaks also decreased, which in turn verified the results of hydration degree.

3.5. SEM Analysis. Microstructural images of ordinary cement paste and paste with 2 wt.% addition of TiO₂ nanoparticles cured under various temperatures at 28 days of age are shown in Figures 7 and 8. Microstructures of ordinary cement paste at different curing temperatures can be noticed in Figure 7. In light of mechanical strength and porosity results, compressive and flexural strengths of specimens cured at 0°C were lower in comparison with those of specimens cured at higher temperatures, and pore distribution was comparatively coarser at 0°C. Accordingly, the image of cement paste at 0°C shown in Figure 7(a) indicated a loosely packed microstructure with less hydration products, which was in agreement with the abovementioned tests. With the increase in the curing temperature, more hydrated products containing CH crystal and C-S-H gels were generated as indicated in Figures 7(b)–7(d), more C-S-H gels were found in SEM images, and the microstructures turned to be much denser.

With the incorporation of 2 wt.% TiO₂ nanoparticles, the hydration process was accelerated, and more hydration products were produced as seen in Figure 8, and as the additional space for precipitation of hydration products, it is not difficult to find that C-S-H gels and ettringite were generated on the surface of TiO₂ nanoparticles with the increasing curing age, thereby promoting the formation of a denser and closely bonded microstructure. However, the overall improvement by TiO₂ nanoparticles was still weakened by low temperatures, and the pastes cured at 20°C had a denser microstructure with more C-S-H gels and ettringite bonded together than those cured at low temperatures, especially 0°C. These evolution of SEM images indicated that TiO₂ nanoparticles, despite chemically inert with cement, could accelerate the hydration of cement paste, thus improving the strength development under low temperatures due to its special performance.

4. Conclusions

The present study evaluated the physical and mechanical properties of cementitious materials without and with various dosages of TiO₂ nanoparticles under curing temperatures of 0, 5, 10, and 20°C, and microstructural analysis was employed to explain and verify the physical and mechanical test results. The following conclusions can be drawn based on the experimental results:

(1) Low temperature had an undesirable impact on the setting time of cement pastes while with the incorporation of TiO₂ nanoparticles, the setting time of pastes remarkably reduced, and the more addition of TiO₂ nanoparticles, the shorter the setting time.

(2) It was discovered that the hydration degree was enhanced with the increase in curing time and curing temperature, and TiO₂ nanoparticles had the ability to accelerate the hydration of the cement.

(3) To increase the dosage of TiO₂ nanoparticles not necessarily enhanced the strength development of cement mortar, and the optimal addition was 2 wt.% in this study, and the strength showed a downward trend when the dosage of TiO₂ nanoparticles was over 2 wt.%.

(4) The results of MIP indicated that, with the incorporation of 2 wt.% TiO₂ nanoparticles, the pore volume and most probable pore size turned smaller, demonstrating a much denser microstructure of cement.

(5) TiO₂ nanoparticles, although inert with cement, could accelerate the hydration of cement, thereby promoting the formation of more hydration products, which could be confirmed by the XRD test conducted on cement paste at 28 days of age.

(6) With the addition of TiO₂ nanoparticles, no new hydrate was generated in the cement pastes, but compared with ordinary cement pastes, the diffraction peaks of hydration products were more intense. Low temperatures decelerated cement hydration with weak diffraction peaks in all the cement pastes.

(7) Morphology obtained by the SEM test on ordinary cement paste and paste with 2 wt.% TiO₂ nanoparticles cured under different temperatures at 28 days of age explained the experimental results. The specimen with TiO₂ nanoparticles showed a comparatively denser microstructure with the closely bonded C-S-H gels and ettringite.

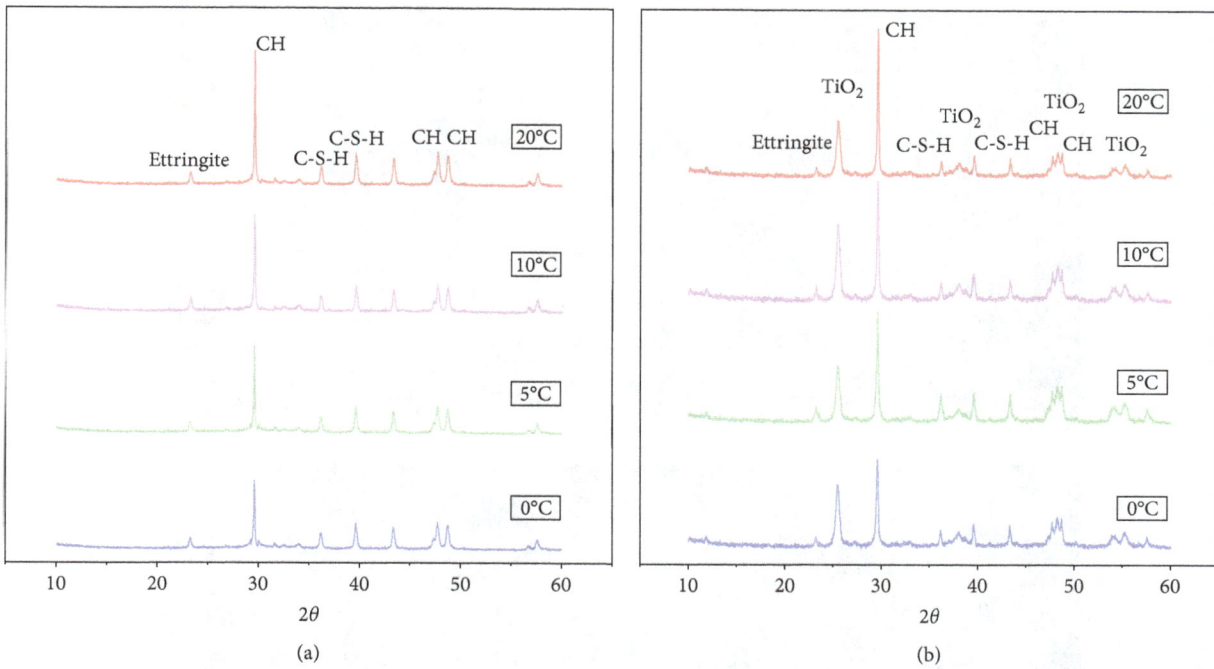

FIGURE 6: (a) XRD patterns of ordinary cement pastes at various temperatures; (b) XRD patterns of cement pastes with 2 wt.% TiO$_2$ nanoparticles at various temperatures.

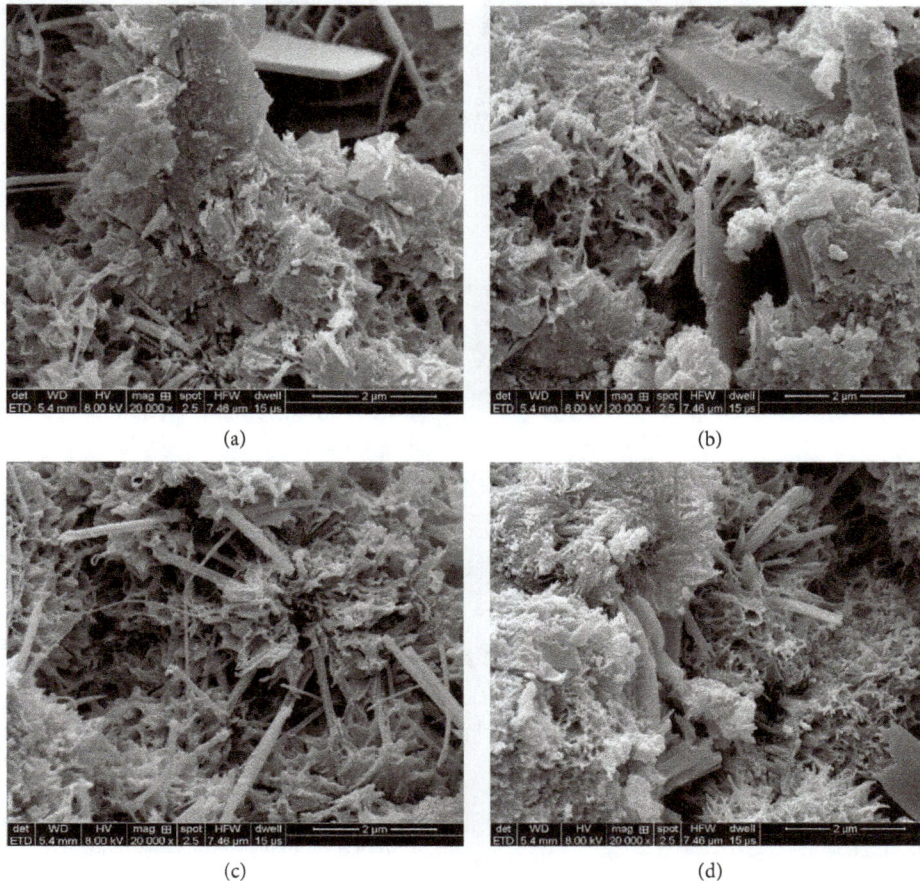

FIGURE 7: SEM images of ordinary cement pastes cured at (a) 0°C, (b) 5°C, (c) 10°C, and (d) 20°C at 28 days.

Figure 8: SEM images of cement pastes containing 2 wt.% TiO$_2$ nanoparticles cured at (a) 0°C, (b) 5°C, (c) 10°C, and (d) 20°C at 28 days.

Conflicts of Interest

The authors declare that there are no conflicts of interest.

Acknowledgments

This research was supported by the Fundamental Research Funds for the Central Universities of Chang'an University (no. 300102218523). Some experimental studies such as SEM analysis and thermal gravimetric analysis were conducted in Xi'an Jiaotong University, and this is hereby acknowledged.

References

[1] D. P. Bentz, P. E. Stutzman, and F. Zunino, "Low-temperature curing strength enhancement in cement-based materials containing limestone powder," *Materials and structures*, vol. 50, no. 3, p. 173, 2017.

[2] J. Cheung, A. Jeknavorian, L. Roberts, and D. Silva, "Impact of admixtures on the hydration kinetics of Portland cement," *Cement and Concrete Research*, vol. 41, no. 12, pp. 1289–1309, 2011.

[3] P. A. Rosskopf, F. J. Linton, and R. B. Peppler, "Effect of various accelerating chemical admixtures on setting and strength development of concrete," *Journal of Testing and Evaluation*, vol. 3, no. 4, pp. 322–330, 1975.

[4] H. Justnes and E. C. Nygaard, "Technical calcium nitrate as set accelerator for cement at low temperatures," *Cement and Concrete Research*, vol. 25, no. 8, pp. 1766–1774, 1995.

[5] F. Karagöl, R. Demirboğa, M. A. Kaygusuz, M. M. Yadollahi, and R. Polat, "The influence of calcium nitrate as antifreeze admixture on the compressive strength of concrete exposed to low temperatures," *Cold Regions Science and Technology*, vol. 89, pp. 30–35, 2013.

[6] J. Han, K. Wang, J. Shi, and Y. Wang, "Influence of sodium aluminate on cement hydration and concrete properties," *Construction and Building Materials*, vol. 64, pp. 342–349, 2014.

[7] T. Matusinović and D. Curlin, "Lithium salts as set accelerators for high alumina cement," *Cement and Concrete Research*, vol. 23, no. 4, pp. 885–895, 1993.

[8] D. P. Bentz, F. Zunino, and D. Lootens, "Chemical vs. physical acceleration of cement hydration," *Concrete international: design and construction*, vol. 38, pp. 37–44, 2016.

[9] J. Zelić, D. Rušić, D. Veza, and R. Krstulović, "The role of silica fume in the kinetics and mechanisms during the early stage of

cement hydration," *Cement and Concrete Research*, vol. 30, no. 10, pp. 1655–1662, 2000.

[10] H. Li, H. G. Xiao, J. Yuan, and J. Ou, "Microstructure of cement mortar with nanoparticles," *Composites Part B: Engineering*, vol. 35, no. 2, pp. 185–189, 2004.

[11] V. Vishwakarma and D. Ramachandran, "Green concrete mix using solid waste and nanoparticles as alternatives–a review," *Construction and Building Materials*, vol. 162, pp. 96–103, 2018.

[12] R. Yu, P. Spiesz, and H. J. H. Brouwers, "Effect of nano-silica on the hydration and microstructure development of ultra-high performance concrete (UHPC) with a low binder amount," *Construction and Building Materials*, vol. 65, pp. 140–150, 2014.

[13] R. Madandoust, E. Mohseni, S. Y. Mousavi, and M. Namnevis, "An experimental investigation on the durability of self-compacting mortar containing nano-SiO$_2$, nano-Fe$_2$O$_3$ and nano-CuO," *Construction and Building Materials*, vol. 86, pp. 44–50, 2015.

[14] H. Li, M. H. Zhang, and J. P. Ou, "Flexural fatigue performance of concrete containing nano-particles for pavement," *International Journal of fatigue*, vol. 29, pp. 1292–1301, 2007.

[15] S. Bahafid, S. Ghabezloo, M. Duc, P. Faure, and J. Sulem, "Effect of the hydration temperature on the microstructure of Class G cement: CSH composition and density," *Cement and Concrete Research*, vol. 95, pp. 270–281, 2017.

[16] E. Mohseni, B. M. Miyandehi, J. Yang, and M. A. Yazdi, "Single and combined effects of nano-SiO$_2$, nano-Al$_2$O$_3$ and nano-TiO2 on the mechanical, rheological and durability properties of self-compacting mortar containing fly ash," *Construction and Building Materials*, vol. 84, pp. 331–340, 2015.

[17] F. U. A. Shaikh and S. W. Supit, "Chloride induced corrosion durability of high volume fly ash concretes containing nano particles," *Construction and Building Materials*, vol. 99, pp. 208–225, 2015.

[18] M. Li, S. Deng, F. Meng, J. Hao, and X. Guo, "Effect of nanosilica on the mechanical properties of oil well cement at low temperature," *Magazine of Concrete Research*, vol. 69, no. 10, pp. 493–501, 2017.

[19] X. Pang, P. J. Boul, and W. Cuello Jimenez, "Nanosilicas as accelerators in oil well cementing at low temperatures," *SPE Drilling and Completion*, vol. 29, no. 1, pp. 98–105, 2014.

[20] H. Kazempour, M. T. Bassuoni, and F. Hashemian, "Masonry mortar with nanoparticles at a low temperature," *Proceedings of the Institution of Civil Engineers-Construction Materials*, vol. 170, no. 6, pp. 297–308, 2016.

[21] M. H. Zhang and J. Islam, "Use of nano-silica to reduce setting time and increase early strength of concretes with high volumes of fly ash or slag," *Construction and Building Materials*, vol. 29, pp. 573–580, 2012.

[22] M. H. Zhang, J. Islam, and S. Peethamparan, "Use of nano-silica to increase early strength and reduce setting time of concretes with high volumes of slag," *Cement and Concrete Composites*, vol. 34, no. 5, pp. 650–662, 2012.

[23] S. W. Supit and F. U. Shaikh, "Effect of nano-CaCO$_3$ on compressive strength development of high volume fly ash mortars and concretes," *Journal of Advanced Concrete Technology*, vol. 12, no. 6, pp. 178–186, 2014.

[24] T. Sato and J. J. Beaudoin, "Effect of nano-CaCO$_3$ on hydration of cement containing supplementary cementitious materials," *Advances in Cement Research*, vol. 23, no. 1, pp. 33–43, 2011.

[25] E. Mohseni and K. D. Tsavdaridis, "Effect of nano-alumina on pore structure and durability of Class F fly ash self-compacting mortar," *American Journal of Engineering and Applied Sciences*, vol. 9, no. 2, pp. 323–333, 2016.

[26] A. Nazari and S. Riahi, "The effects of TiO$_2$ nanoparticles on physical, thermal and mechanical properties of concrete using ground granulated blast furnace slag as binder," *Materials Science and Engineering: A*, vol. 528, pp. 2085–2092, 2011.

[27] A. Nazari and S. Riahi, "The effects of TiO$_2$ nanoparticles on properties of binary blended concrete," *Journal of Composite Materials*, vol. 45, pp. 1181–1188, 2011.

[28] A. Nazari and S. Riahi, "The effects of TiO$_2$ nanoparticles on flexural damage of self-compacting concrete," *International Journal of Damage Mechanics*, vol. 20, no. 7, pp. 1049–1072, 2011.

[29] A. Nazari, S. Riahi, S. Riahi, S. F. Shamekhi, and A. Khademno, "Improvement the mechanical properties of the cementitious composite by using TiO$_2$ nanoparticles," *Journal of American Science*, vol. 6, pp. 98–101, 2010.

[30] D. Feng, N. Xie, C. Gong et al., "Portland cement paste modified by TiO$_2$ nanoparticles: a microstructure perspective," *Industrial and Engineering Chemistry Research*, vol. 52, no. 33, pp. 11575–11582, 2013.

[31] B. Y. Lee, A. R. Jayapalan, and K. E. Kurtis, "Effects of nano-TiO$_2$ on properties of cement-based materials," *Magazine of Concrete Research*, vol. 65, pp. 1293–1302, 2013.

[32] A. Jayapalan, B. Lee, S. Fredrich, and K. Kurtis, "Influence of additions of anatase TiO$_2$ nanoparticles on early-age properties of cement-based materials," *Transportation Research Record: Journal of the Transportation Research Board*, vol. 2141, no. 1, pp. 41–46, 2010.

[33] M. H. Zhang and H. Li, "Pore structure and chloride permeability of concrete containing nano-particles for pavement," *Construction and Building Materials*, vol. 25, no. 2, pp. 608–616, 2011.

[34] K. Pimenta Teixeira, I. Perdigão Rocha, L. De Sá Carneiro, J. Flores, E. A. Dauer, and A. Ghahremaninezhad, "The effect of curing temperature on the properties of cement pastes modified with TiO$_2$ nanoparticles," *Materials*, vol. 9, no. 11, p. 952, 2016.

[35] E. Mohseni, M. M. Ranjbar, and K. D. Tsavdaridis, "Durability properties of high-performance concrete incorporating nano-TiO$_2$ and fly ash," *American Journal of Engineering and Applied Sciences*, vol. 8, no. 4, pp. 519–526, 2015.

[36] R. Khataee, V. Heydari, L. Moradkhannejhad, M. Safarpour, and S. W. Joo, "Self-cleaning and mechanical properties of modified white cement with nanostructured TiO$_2$," *Journal of nanoscience and nanotechnology*, vol. 13, no. 7, pp. 5109–5114, 2013.

[37] V. Binas, D. Venieri, D. Kotzias, and G. Kiriakidis, "Modified TiO$_2$ based photocatalysts for improved air and health quality," *Journal of Materiomics*, vol. 3, no. 1, pp. 3–16, 2017.

[38] M. M. Hassan, H. Dylla, L. N. Mohammad, and T. Rupnow, "Evaluation of the durability of titanium dioxide photocatalyst coating for concrete pavement," *Construction and building materials*, vol. 24, no. 8, pp. 1456–1461, 2010.

[39] M. M. Khotbehsara, E. Mohseni, M. A. Yazdi, P. Sarker, and M. M. Ranjbar, "Effect of nano-CuO and fly ash on the properties of self-compacting mortar," *Construction and Building Materials*, vol. 94, pp. 758–766, 2015.

[40] ASTM (American Society for Testing and Materials) ASTM, "C150:standard specification for Portland cement," in *Annual Book of ASTM*, ASTM, West Conshohocken, PA, USA, 2001.

[41] ASTM (American Society for Testing and Materials) ASTM, *C494: TYPE F: Standard Specification for Chemical Admixtures for Concrete*, ASTM International, West Conshohocken, PA, USA, 2001.

[42] ASTM C 191-08, *Standard Test Method for Time of Setting of Hydraulic Cement by Vicat Needle*, ASTM International, West Conshohocken, PA, USA, 2008.

[43] ASTM C 187-16, *Standard Test Method for Amount of Water Required for Normal Consistency of Hydraulic Cement Paste*, ASTM International, West Conshohocken, PA, USA, 2016.

[44] ASTM, *ASTM C109-93: Standard Specification for Compressive Strength of Mortars*, American Society for Testing and Materials, West Conshohocken, PA, USA, 2007.

[45] ASTM, *ASTM C293/C293M-10: Standard Test Method for Flexural Strength of Concrete*, American Society for Testing and Materials, West Conshohocken, PA, USA, 2007.

[46] B. Standard and B. ISO, *Pore size Distribution and Porosity of Solid Materials by Mercury Porosimetry and Gas Adsorption*, BS ISO, London, UK, 2005.

[47] M. H. Beigi, J. Berenjian, O. L. Omran, A. S. Nik, and I. M. Nikbin, "An experimental survey on combined effects of fibers and nanosilica on the mechanical, rheological, and durability properties of self-compacting concrete," *Materials and Design*, vol. 50, pp. 1019–1029, 2013.

[48] A. Nazari and S. Riahi, "Al_2O_3 nanoparticles in concrete and different curing media," *Energy and Buildings*, vol. 43, no. 6, pp. 1480–1488, 2011.

[49] L. Cui and J. H. Cahyadi, "Permeability and pore structure of OPC paste," *Cement and Concrete Research*, vol. 31, no. 2, pp. 277–282, 2001.

[50] L. Alarconruiz, G. Platret, E. Massieu, and A. Ehrlacher, "The use of thermal analysis in assessing the effect of temperature on a cement paste," *Cement and Concrete Research*, vol. 35, pp. 609–613, 2005.

[51] V. G. Papadakis, C. G. Vayenas, and M. N. Fardis, "Fundamental modeling and experimental investigation of concrete carbonation," *ACI Materials Journal*, vol. 88, no. 4, pp. 363–373, 1991.

18

Viability Study of a Safe Method for Health to Prepare Cement Pastes with Simultaneous Nanometric Functional Additions

M. A. de la Rubia ⓘ,[1] E. de Lucas-Gil ⓘ,[2] E. Reyes,[1] F. Rubio-Marcos ⓘ,[2]
M. Torres-Carrasco,[2] J. F. Fernández,[2] and A. Moragues[1]

[1]Department of Civil Engineering-Construction, School of Civil Engineering, Polytechnic University of Madrid,
  C/Professor Aranguren s/n, 28040 Madrid, Spain
[2]Electroceramic Department, Instituto de Cerámica y Vidrio (CSIC), C/Kelsen 5, 28049 Madrid, Spain

Correspondence should be addressed to M. A. de la Rubia; ma.rubia@caminos.upm.es

Academic Editor: Antonio Gilson Barbosa de Lima

The use of a mixing method based on a novel dry dispersion procedure that enables a proper mixing of simultaneous nanometric functional additions while avoiding the health risks derived from the exposure to nanoparticles is reported and compared with a common manual mixing in this work. Such a dry dispersion method allows a greater workability by avoiding problems associated with the dispersion of the particles. The two mixing methods have been used to prepare Portland cement CEM I 52.5R pastes with additions of nano-ZnO with bactericide properties and micro- or nanopozzolanic SiO_2. The hydration process performed by both mixing methods is compared in order to determine the efficiency of using the method. The hydration analysis of these cement pastes is carried out at different ages (from one to twenty-eight days) by means of differential thermal analysis and thermogravimetry (DTA-TG), X-ray diffraction (XRD), scanning electron microscopy (SEM), and Fourier transform infrared spectroscopy (FTIR) analyses. Regardless of composition, all the mixtures of cement pastes obtained by the novel dispersion method showed a higher retardation of cement hydration at intermediate ages which did not occur at higher ages. In agreement with the resulting hydration behaviour, the use of this new dispersion method makes it possible to prepare homogeneous cement pastes with simultaneous functional nanoparticles which are physically supported on the larger particles of cement, avoiding exposure to the nanoparticles and therefore minimizing health risks. Manual mixing of cement-based materials with simultaneous nanometric functional nanoparticles on a large scale would make it difficult to obtain a homogenous material together with the health risks derived from the handling of nanoparticles.

1. Introduction

Cement is the most used binder material in construction and building, with it mainly being a fundamental component of mortar and concrete. The microstructure and functional behaviour of cement-based materials strongly depend on the chemical composition and additions, water-cement ratio, and rate of hydration. The incorporation of additions to modify and improve its fresh properties and its physical, mechanical, and durability properties, as well as to obtain new functional properties such as photocatalysis [1, 2], antibacterial effects (Sikora et al. [3, 4]), hydrophobicity (Tittarelli [5] and Nunes and Slizkova [6]), and fungicide effects (De Muynck et al. [7]) among others is currently an important research topic. In recent years, nanotechnology applied to cement-based materials has been growing since the use of additions of nanometric size significantly increases the effects on their mechanical and functional properties, and the combination of several additions might be very promising (Sikora et al. [4], León et al. [8], and Mohseni et al. [9]). However, the simultaneous addition of several functional nanoparticles simultaneously is an even greater challenge due to their tendency to agglomeration and by the risk to health during its manipulation (Vishwakarma et al. [10], Albrecht et al. [11], and Mueller and Nowack [12]).

In the case of the addition of silica, the pozzolanic activity and filling capacity are the two main characteristics that significantly influence the cement-based material properties (Haruehansapong et al. [13], Senff et al. [14], and Lu and Poon [15]). Pozzolanic materials provide an increase in the amount of calcium-silicate-hydrate (CSH) gels, as well as a denser and more discontinuous and tortuous microstructure via pozzolanic reactions. As a result, an improvement of the mechanical and durable properties of the cement-based materials occurs (Massana et al. [16] and Ramezanianpour and Moeini [17]). A significant factor is the distribution particle size of silica, related with the specific surface. Silica fume is a by-product and generally shows a wide particle size range from submicronic particles to a few tens of microns. In contrast, in the case of nanosilica, the particle size is lesser. Smaller sized particles provide nucleating sites to hydration products since the high surface area, particularly in the case of nanosilica, allows the formation of calcium-silicate-hydrate (CSH) gel seeds on its surface formed by an early pozzolanic reaction that accelerate silicate hydration (Cheng-yi and Feldman [18], Larbi et al. [19], Thomas et al. [20], and Land and Stephan [21]). The surface area of silica becomes higher so does the acceleration of hydration (Land and Stephan [21]).

Zinc oxide, ZnO, can present three crystalline structures, with the most stable being the hexagonal structure, wurtzite. ZnO is an n-type semiconductor with a band gap of 3.2 eV at room temperature RT. ZnO is produced directly by oxidizing zinc metal, a reduction of an ore to zinc metal followed by re-oxidation and to a lesser extent, by precipitation of oxide or carbonate from aqueous solution and a final thermal treatment (Moezzi et al. [22]). This oxide has a large number of applications, with a vulcanization activator in the rubber industry being the main function. ZnO is a highly important raw material in ceramics (glazes and enamels in tiles and sanitary ware), in electroceramics such as overvoltage protection devices (varistors) (Sendi [23] and Xu [24]), and in optoelectronics (Kahouli et al. [25], Hussein et al. [26], and Torchynska et al. [27]). It is currently being researched as a photocatalyst and novel antifungal precursor. The morphology of the ZnO particles mainly depends on the synthesis techniques, precursors, process conditions, and pH (Moezzi et al. [22]).

According to Klingshirn, the use of ZnO particles in concrete increases the setting time and, therefore, the workability for early ages, and provides an improvement in its resistance against water (Klingshirn [28]). The main effects of zinc oxide additions to the Portland cement have been known for some time: retardation of setting and hardening (reducing the rate of heat evolution) leads to an improvement in whiteness and final strength (Ramachandran [29]). Recently, authors such as Nivethitha et al. have suggested that the addition of nano-ZnO particles may also improve the mechanical properties of ordinary Portland cement mortar (Nivethitha and Dharmar [30]).

Several authors have studied the influence of ZnO on the hydration of the Portland cement. Arliguie showed that the delay in hydration is due to the precipitation of an amorphous $Zn(OH)_2$ layer around the anhydrous C_3S grains which inhibits cement dissolution (Arliguie and Grandet [31]). The hydration of C_3A in the presence of ZnO occurs too, when there is a significant presence of SO_3. When the pore concentration of Ca^{2+} and OH^- is enough, the hydration reaction starts again and the amorphous $Zn(OH)_2$ layer transforms into crystalline calcium zinc hydroxide $CaZn_2(OH)_6 \cdot 2H_2O$ (CZ) (Arliguie and Grandet [31]). Yousuf et al. identified a calcium zinc hydroxide crystalline phase by means of Fourier transform infrared spectroscopy (FTIR) in paste samples with ZnO cured for 28 days (Yousuf et al. [32]). Some authors corroborate the formation of this calcium hydroxyzincate dihydrate $CaZn_2(OH)_6 \cdot 2H_2O$ (CZ), suggesting that the formation of this phase controls Zn solubility in cementitious materials (Cocke et al. [33]). Johnson and Kersten showed that solid solution is a possible binding mechanism for Zn(II) in the CSH gel, indicating that at least 10% of Zn may be incorporated in the CSH structure (Johnson and Kersten [34]). In such a way, Lieber and Gebauer confirmed the formation of crystalline calcium hydroxyzincate during the retardation period as an intermediate product (though after some days of reaction, it is no longer detectable) (Lieber and Gebauer [35]). In a recent study, Ataie et al. [36] in contrast with Arliguie, indicated that the mechanism by which ZnO retards hydration reaction could be nucleation and/or growth poisoning of CSH and suggested that ZnO does not inhibit cement dissolution. Ataie showed that ZnO strongly delays the cement hydration and the retardation increases as the ZnO quantity in the cement rises, so the amount of ZnO added is directly correlated with such an increase in the retardation period. Zn ions hinder nucleation and growth of hydrated products poisoning the hydration product nuclei. When there are not enough dissolved Zn ions, retardation ends and a greater acceleration of hydration occurs. In the same sense, various authors suggest that the CSH gel is the most probable site for metal fixation in cement and Zn^{2+} is incorporated in the interlayer CSH (Stumm et al. [37], Ouki and Hills [38], Ziegler et al. [39], and Ziegler and Johnson [40]).

Therefore, calcium hydroxyzincate dihydrate $CaZn_2(OH)_6 \cdot 2H_2O$ (CZ) has an important influence in the retardation of the Portland cement hydration (Arliguie and Grandet [31]) and in the prevention from corrosion of the galvanized steel in concrete when this hydroxyzincate phase forms a coating on the rebars (Tittarelli and Bellezze [41]). Protection against corrosion is due to the formation of a compact and protective layer of this phase at high pH values between 12.2 and 13.3.

ZnO and compounds such as $CaZn_2(OH)_6 \cdot 2H_2O$ are being researched as antifungal, bactericide, and photocatalyst materials (Xie et al. [42], Sirelkhatim et al. [43], Madhusudhana et al. [44], Gomez-Ortiz et al. [45], and Hernandez Sierra et al. [46]). Gomez-Ortiz et al. used CZ as the antifungal precursor of protective coatings for marble and limestone (Gomez-Ortiz et al. [45]). A drawback when using both ZnO and SiO_2 nanoparticles as an addition in the Portland cement may involve the requirement of a greater amount of mixing water due to a greater amount of fines, which increases the water-cement ratio. The use of superplasticizers is common in order to improve the workability.

Nanoparticles, due to their high specific surface area, have a high tendency to form agglomerates. Such agglomeration drastically reduces the effectiveness and modifies the properties regarding the dispersed state of nanoparticles avoiding obtaining homogeneous mixtures. Therefore, agglomeration should be avoided in order to achieve homogeneous admixtures using a low amount of nanoparticles and minimizing the health risks derived from its management. The difficulty in avoiding the agglomeration of functional nanoparticles and the difficulty in obtaining a material with high homogeneity increase in manual mixing when trying to introduce several additions simultaneously in cement pastes. Manual mixing to achieve homogeneous admixtures implies an increase in exposure time to nanoparticles and therefore an increase in health risks.

Nanotechnology is increasingly used in production processes. Currently, it is difficult to determine the risks to health and the environment during handling in the preparation of composites. Because of this, it is infeasible to prepare cement-based materials with functional nano-additions (silica, zinc oxide, and titania, among others) at an industrial level by a manual mixing method to achieve a homogeneous dispersion of functional nanoparticles, given that they have a high health risk due to a small particle size. Therefore, in order to minimize health risks derived from the use of nanoparticles, it is necessary to minimize exposure time and handling. The patented low-energy dry dispersion method (Fernandez Lozano et al. [47]) allows the nanoparticles to be incorporated in an effective way, by simplifying the steps, avoiding exposure, and handling nanoparticles, as the dispersion of the nanoparticles takes place on dry larger particles (cement particles). These nanoparticles are supported on the cement particles by cohesive forces, thus avoiding or decreasing spreading. A consistent powder with a particle distribution consisting of nanoparticles dispersed and anchored by short-range forces on particles of different morphologies is obtained. Such short-range forces avoid the presence of free nanoparticles when the composite powder is handled or used, decreasing the health risk. This process is in the absence of solvents (dry process) and does not require high shear rates. Implementation of such a process would be viable at an industrial level in a cement production plant.

The object of this study, therefore, is to assess the feasibility of preparing homogeneous functional cement-based powders with simultaneous functional nanoadditions (SiO_2 and ZnO) by means of an easy-to-use dry dispersion method minimizing health risks and avoiding exposure to nanoparticles during the preparation of cement pastes.

2. Materials and Methods

2.1. Materials. The cement paste is a 52.5R Type I commercial Portland cement with a particle size d_{50} of 6.3 μm and specific surface of 1.45 m^2/g. The nanometric ZnO was obtained as a product of hydrozincite ($Zn_5(CO_3)_2(OH)_6$) calcination at 500°C for five minutes in air. The XRD characterization indicates that the phase matches the hexagonal wurtzite structure ZnO (JCPDS card number 36-1451), without peaks, of other phases. The nanometric ZnO shows a specific surface area of 18.59 m^2/g determined by the Brunauer–Emmett–Teller (BET) method and a monomodal distribution with an average particle size of 53 nm from field emission scanning electron microscopy FE-SEM micrographs. The micrometric

SiO_2 (MS) used is silica fume produced in the manufacturing of ferro-silicon alloy. It is noncrystalline silica with wide distribution particle sizes with a d_{50} of 45.2 μm and a specific surface area of 28.52 m^2/g. Microsilica contains 91 wt.% of SiO_2 and 5.3 wt.% of Fe_2O_3.

The nanometric SiO_2 (NS) Sipernat® 22S from Evonik Industries [48] is manufactured hydrophilic silica in a liquid-phase precipitation process that obtains aggregates with a micrometer size of 14 μm composed of tiny primary particles (5–100 nm). The advantages of using Sipernat 22S involve an amorphous structure and a high surface area of 180 m^2/g. Sipernat 22S increases the early strength and consistency of concrete, reduces bleeding and segregation in self-compacting concrete, and has a hydrophobic effect that improves the water resistance of concrete (Aerosil, Aeroxide, and Sipernat products, Evonik Industries, Technical Information). Due to the difference between the specific surface area of microsilica (MS) and nanosilica (NS), the pozzolanic reactivity of nanosilica Sipernat is much higher. Both have an acidic character.

2.2. Sample Preparation. Three cement paste mixtures were studied, including 52.5R Type I Portland cement and different combinations of additions of nano-ZnO (NZ), nanosilica, and microsilica as a substitute of anhydrous cement: one with 2 wt.% of nano-ZnO (NZ), another with 2 wt.% of nano-ZnO (NZ) and 10 wt.% of nanosilica (NS), and the last one with 2 wt.% of nano-ZnO (NZ) and 10 wt.% of microsilica (MS). The mixtures were prepared by two mixing methods. A first mixing method involved a common manual mixing. The other mixing method used was the dispersion of the different functional particles (nano-ZnO, microsilica, and nanosilica) over the 52.5R Type I Portland cement particles, using the aforementioned low-energy dry dispersion (low shear rate) novel method (Fernandez Lozano et al. [47]). In this method of dry dispersion, the cement particles as the different additions were dried and then mixed in a Turbula-type shaker with 15 mm diameter alumina balls (Lorite et al. [49] and Alonso-Domínguez et al. [50]). The water-cement ratio was fixed at 0.35 for the admixtures except for that which had 2 wt.% nano-ZnO (NZ) + 10 wt.% nanosilica with a ratio fixed at 0 6 since nanosilica has a high specific surface and demands more water to obtain complete hydration of the mixture.

In order to obtain good rheological properties such as flowability but without segregation, a policarboxilate-based superplasticizer (Sika ViscoCrete® 5720) was used. The superplasticizer quantities depend on the wt.% of cement replaced, as well as the morphological characteristic of the functional particles. Ghafari et al. pointed out that cement paste composites containing 0.4 wt.% ZnO nanoparticles (specific surface area of 54 ± 20 m^2/g) or lower showed excellent workability retention compared to the reference mix and a poor workability retention was observed at higher dosage (Ghafari et al. [51]). Once the dosages have been prepared by both mixing methods, they are placed in a rotary mixer that contains the required water and the superplasticizer (following with a low-speed knead for 90 seconds).

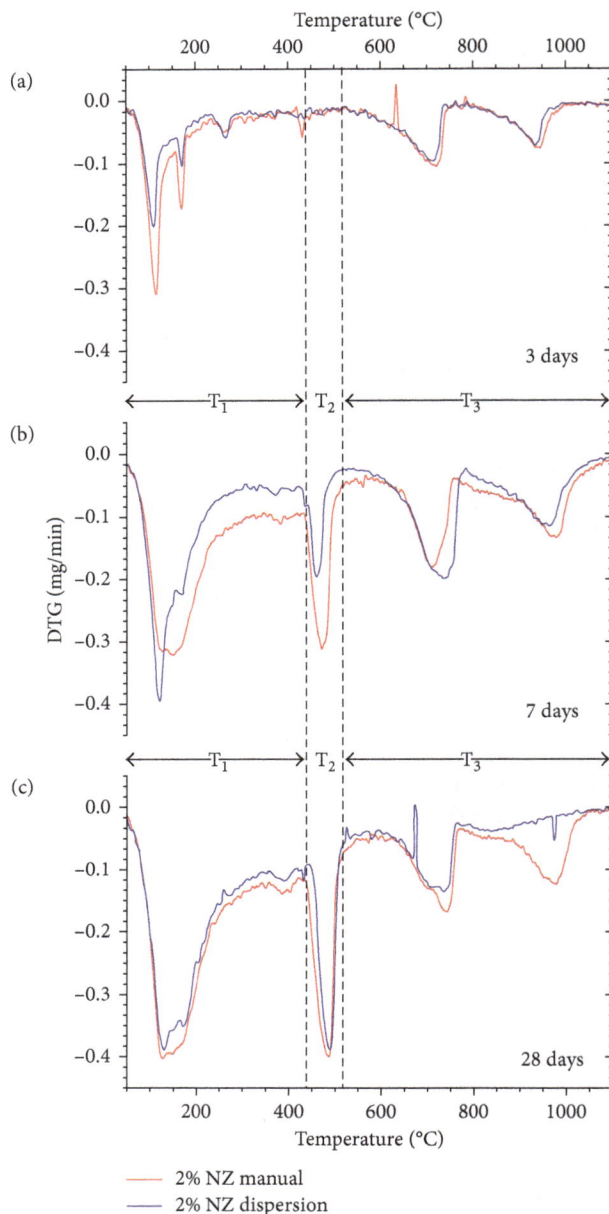

FIGURE 1: DTG-T^a analysis for cement paste with 2 wt.% of nanozinc (NZ) for (a) 3, (b) 7, and (c) 28 days.

After 30 seconds of rest, the mixture was mixed again for 90 seconds at a low speed. After mixing, the cement pastes were shaped as a prismatic bar in molds and covered with a plastic bag and placed in a curing chamber with the temperature and humidity controlled (20 ± 2°C and >95%, resp.). As zinc oxide is a cement hydration retarder, samples were demolded after 72 hours and cured in water inside the chamber. In order to stop the hydration of the samples at the ages analyzed, the cement paste was placed in vacuum for 30 minutes and then in acetone for two hours, following 24 hours in a stove at 60°C. The sample was stored in a stove at 40°C until its structural and microstructural characterization had been obtained. Cement pastes were cured for one, two, three, seven, and

twenty-eight days in water and characterized by means of DTA-TG, SEM, FTIR, and XRD.

2.3. Characterization of Cement Pastes with Additions. The differential thermal analysis and thermogravimetry (DTA-TG) of the cement pastes were performed by using LABSYS evo equipment provided by SETARAM Instrumentation, in the range of 25°C to 1100°C with a heating rate of 10°/min in N_2 atmosphere in alumina crucibles. The cement pastes were microstructurally characterized by scanning electron microscopy (SEM) with a JEOL JSM 6335F microscope, and the energy dispersive spectroscopy (EDS) analysis was carried out with a detector provided by Oxford Instruments, the X-Max[N] of 80 mm^2 with a resolution between 127 eV and 5.9 KeV. In addition, X-ray diffraction (XRD) was carried out by means of a powder diffractometer Bruker D8 AD-VANCE with Cu Kα radiation, with a high-speed detector (Lynxeye). The identification of the crystalline phases was done by comparison with the Joint Committee on Powder Diffraction Standards (JCPDS) guidelines. FTIR analysis of the cement pastes shaped in the KBr pellet method was performed by a PerkinElmer Spectrum 100 spectrophotometer from 400 to 4000 cm^{-1}.

3. Results and Discussion

The thermogravimetric derivative (DTG) versus temperature plot represents the mass change rate with temperature (T^e) and allows a better resolution of the complex thermogravimetric curves. By quantifying the water related with the CSH gel and the portlandite phase (CH), it is possible to analyze the hydration grade. The DTG versus T^e plot of cement pastes shows the following characteristic bands: first, the free water loss occurs approximately at 100°C, then the interlayer water loss occurs at around 120°C, and finally, from 140°C until 430°C, the water of gel (water combined in the CSH gel, T1) loss occurs. The portlandite content is related with the water weight loss that takes place in the temperature range between 430°C and 515°C (labeled as T2). Due to the possible portlandite carbonation, it is necessary to adjust the portlandite quantity and include it in the temperature range where the carbonate loss takes place (515°C–1100°C). Villain et al. [52] indicated that the calcium carbonate ($CaCO_3$) ensuing of the CSH carbonation dissociates in a lower temperature range than the $CaCO_3$ ensuing of the portlandite carbonation allowing it to differentiate both carbonations thermally. The authors consider that the water contained in the carbonates is 40 wt.% of the total weight loss in this temperature range. The 40 wt.% weight loss in this temperature range (515°C–1100°C) will be termed T3.

Figures 1(a)–1(c) and Table 1 show the plot of DTG versus temperature, such as mass losses in the temperature ranges previously examined (T1, T2, and T3, resp.) for cement pastes with a substitution of 2 wt.% of nano-ZnO (NZ), obtained by both mixing methods at cured ages of three, seven, and twenty-eight days, in comparison with a cement paste without addition as reference. In the DTG plot of three-day samples (Figure 1(a)), there appear peaks corresponding to weight loss of water consequence of

TABLE 1: Water losses associated with different phases in cement pastes with 2 wt.% of nanozinc (NZ).

	2% NZ manual mixing			2% NZ low shear rate dispersion			52.5R		
	3 d	7 d	28 d	3 d	7 d	28 d	3 d	7 d	28 d
T1 (140°C–430°C) wt.%	2.0	7.3	9.2	1.5	4.6	8.6	6.6	7.0	7.8
T2 (430°C–515°C) wt.%	0.3	2.7	3.5	0.2	1.4	3.2	3.4	3.5	3.5
T3 (515°C–1100°C) wt.%	1.3	3.0	2.7	1.2	2.9	1.8	2.6	2.8	2.6
Total combined water (T1 +T2 + T3) wt.%	3.6	13.0	15.4	2.9	8.9	13.6	12.6	13.3	13.9
Equivalent calcium (T2 + T3) wt.%	1.6	5.7	6.2	1.4	4.3	5.0	6.0	6.3	6.1
CSH gel (T1) wt.%/total portlandite (T2 + T3) wt.%	1.25	1.28	1.48	1.07	1.07	1.72	1.10	1.11	1.28

hydration of tricalcium aluminates C_3A (170°C and 265°C) corresponding to the initial formation of hexagonal phases. Subsequently, these will be converted to a C_3AH_6 cubic phase according to the literature and weight loss of carbonates as consequence carbonation of the samples (Ramachandran et al. [53]). The DTG of hydrated calcium hydroxyzincate $(CaZn_2(OH)_6 \cdot 2H_2O)$ synthesized (Mellado et al. [54]) shows three peaks, the first two overlapped in a temperature range between 140°C and 220°C corresponding to the two first stages' decomposition of $CaZn_2(OH)_6 \cdot 2H_2O$. At the lowest temperature, the dehydration of $CaZn_2(OH)_6 \cdot 2H_2O$ takes place, and between 160°C and 220°C in the DTG, a wide peak appears corresponding to the decomposition of $CaZn_2(OH)_6$ in ZnO and $Ca(OH)_2$. Finally, at temperature around 500°C, the $Ca(OH)_2$ decomposition takes place. In Portland cement samples with 0.1% of Zn(II), Mellado et al. indicated the presence of two peaks in DTG at 180 and between 270°C and 290°C corresponding to the decomposition of hydrated calcium hydroxyzincate $CaZn_2(OH)_6 \cdot 2H_2O$ and hydrozincite $Zn_5(CO_3)_2(OH)_6$, respectively.

In Figures 1(a)–1(c), two peaks appear before 200°C and 300°C, respectively, that could correspond to the decomposition of hydrated calcium hydroxyzincate $(CaZn_2(OH)_6 \cdot 2H_2O)$ and hydrozincite according to Mellado et al. [54]. In Figure 1(a), these peaks can appear overlapping with the hydration of tricalcium aluminates and therefore have a greater intensity.

Figure 1(a) and Table 1 data indicate that the hydration of calcium silicates has not taken place at this age, confirming the hydration retarder behaviour of ZnO on cement pastes in agreement with Mellado et al. that confirmed the absence of portlandite decomposition in cement pastes containing 1% of Zn(II) at 3 days due to the delay in the setting of the cement. Pastes at seven days show that hydration is taking place according to Figure 1(b) and Table 1 to a greater extent for cement paste prepared by a manual mixing method since the water of the CSH gel combined with portlandite for manual mixing cement pastes is higher than that obtained by the low shear dispersion method. For 28 days (Figure 1(c)), the hydration process for cement pastes obtained by both mixing methods is very similar. Only small differences can be observed in regard to the total combined water, in particular, due to the water contained in carbonates (Table 1, T3). According to Villain et al. [52], calcium carbonate $CaCO_3$ ensuing form the degradation of the CSH gel gives place to vaterite and/or aragonite whose dissociation takes place between 650°C and 800°C and $CaCO_3$ calcite ensuing from portlandite $Ca(OH)_2$ carbonation dissociated between 850°C

and 1000°C according to Figures 1(a)–1(c). For 28 days, the water loss combined with portlandite indicates that about 24% of total loss is due to the portlandite, with the portlandite content for samples obtained by both mixing methods being about 14 wt.%. The CSH gel-portlandite ratio is higher for samples with nanozinc oxide than for the reference cement paste at 28 days for both experimental methods of obtaining the cement pastes. This result is a consequence of a higher content of water (higher combined water) in the gel (T1), though this does not mean that there is more CSH gel since the portlandite content (equivalent calcium) for all the samples with and without nano-ZnO is similar. This higher water content for samples with nano-ZnO could be due to the water bound to zinc in the calcium hydroxyzincate, given that, in these samples, there is no silica and the pozzolanic reaction cannot take place to form a secondary CSH gel. The hydration process in the first ages is different for each cement paste with nanozinc oxide depending on its mixing method, although in both cases hydration is later, with it being higher for cement pastes obtained at low shear rate dispersion. ZnO is not involved in the hydration process in cement pastes prepared by this method in the first ages, and the water bonded is lower than that in pastes prepared by manual mixing. For 28 days, hydration is similar for cement pastes obtained by both mixing methods. Cement pastes obtained by the novel dispersion method show a higher CSH gel-to-portlandite ratio than the manual mixing ones, indicating a higher Ca/Si ratio and denser CSH gels.

The Ca/Si ratio in a CSH gel of cement pastes obtained by both mixing methods was determined by SEM-EDX analysis at a different point of the surface of each sample at the age of 28 days. The Ca/Si ratio for both mixing methods was similar, with the values of 2.44 and 2.52 for manual and dispersing methods, respectively. These Ca/Si ratios are slightly higher than those for 52.5R cement pastes with a Ca/Si ratio of 2.28. Accordingly, the ZnO addition increases the Ca/Si ratio in the CSH gel.

FTIR analysis for cement pastes with nano-ZnO additions (Figure 2) is in agreement with the DTA-TG characterization. At three days, the hydration of calcium silicates has not started, and only the hydration of the calcium aluminates can be observed in the cement pastes and the mixing methods studied. FTIR analysis identified alite (A) as unhydrated cement particles and an ettringite (E) phase, as a consequence of calcium aluminate hydration. The carbonate peaks as a consequence of paste carbonation and peaks corresponding to bending water or in capillaries are

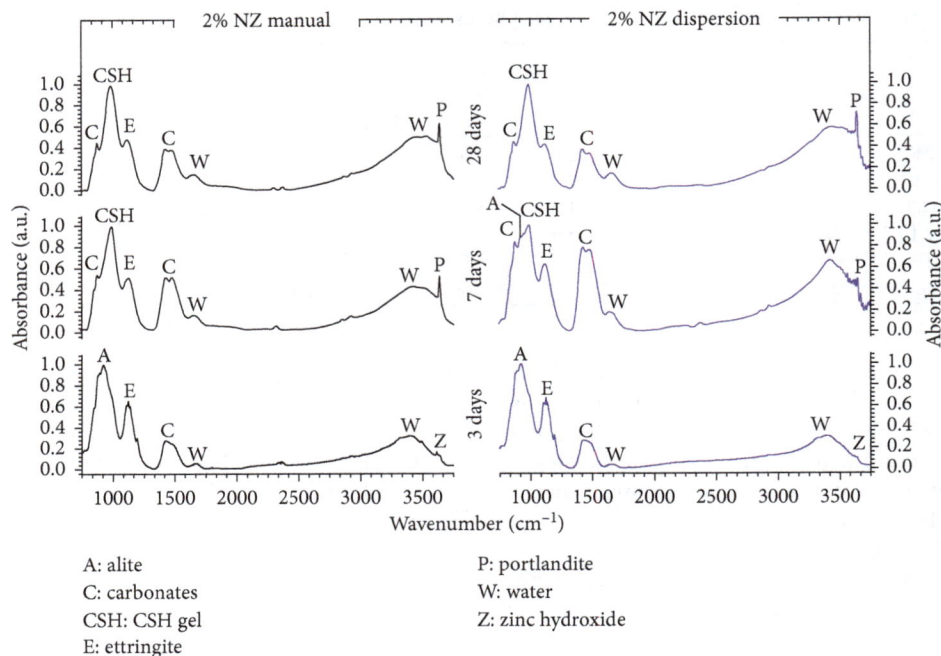

FIGURE 2: FTIR analysis for cement paste with 2 wt.% of nanozinc (NZ).

identified too. The point to the presence at three days of the cement paste by mixing oxide of a weak peak at $3615 \, \text{cm}^{-1}$ that could be attributed to OH groups joined to Zn ions (zinc hydroxide) (Mollah et al. [55] and Trezza [56]). These results are in agreement with Arliguie, who pointed out that the formation of an amorphous layer of zinc hydroxide on the surface of anhydrous cement particles inhibits cement hydration (Arliguie and Grandet [31]). For cured cement pastes at seven days, there is agreement with the DTA-TG analysis, with the hydration of manual mixing cement pastes being almost complete since the displacement of the peak corresponding to the unhydrated alite phase from 920 to $985 \, \text{cm}^{-1}$ and corresponding to the C_3S hydration product (formation of the CSH gel) is clearly identified. In this sample, a sharp peak at $3645 \, \text{cm}^{-1}$ corresponding to the O-H stretching of portlandite $Ca(OH)_2$ formed together with the CSH gel in the hydration of calcium silicates is identified. For dispersed samples at seven days, portlandite identification is not possible, although the formation of the CSH gel takes place according to the incipient peak at $985 \, \text{cm}^{-1}$ such as a decrease in the alite phase, confirming the progress of hydration. In agreement with the literature (Vázquez Moreno [57]), the intensity of the band corresponding to ettringite decreases with its transformation in monosulfoaluminates. In this case, for pastes with NZ, this change takes place when the age increases from three to seven days. For 28 days, FTIR analysis would indicate a slightly greater advance of hydration of cement pastes obtained by a manual mixing method as the greater intensity of the portlandite and CSH gel peaks, together with alite, is not identified.

X-ray diffraction characterization (Figure 3) is in agreement with DTA-TG and FTIR analyses for cement pastes with nanozinc oxide additions for both mixing methods. For samples at three days, calcium silicate hydration has not

started (alite and belite identification), with there even being aluminates identified. For pastes at seven days, hydration is at a greater point of advance in cement pastes obtained by manual mixing than for low shear rate dispersion for seven days. At this age, manual mixing sample characterization shows that the peaks corresponding to the alite phase (A) have a lower intensity than those corresponding to low shear rate dispersion ones. For 28 days, according to XRD analysis, there are no hydration differences between cement pastes obtained by both mixing methods. At this age, by means of this characterization technique, it is possible to identify traces of unhydrated alite. It should be noted that it is possible to identify the crystalline calcium hydroxyzincate dihydrate $CaZn_2(OH)_6 \cdot 2H_2O$ (CZ) phase at any cured age. The identification by the FTIR of the amorphous zinc hydroxide $Zn(OH)_2$ phase at the earliest age (three days) and crystalline calcium hydroxyzincate dihydrate $CaZn_2(OH)_6 \cdot 2H_2O$ (CZ) for all ages in XRD analysis would agree with the Arliguie and Grandet proposal [31] where these authors indicated that the delay in hydration is due to the precipitation of an amorphous $Zn(OH)_2$ layer around the anhydrous C_3S grains which inhibits cement dissolution. When the hydration reaction starts again, the amorphous $Zn(OH)_2$ layer transforms into a crystalline calcium hydroxyzincate $CaZn_2(OH)_6 \cdot 2H_2O$ (CZ). In a recent published study, Andrade et al. studied cement pastes with zinc oxide addition for electric arc furnace dust (EAFD) recycling (Andrade Brehm et al. [58]). They proposed a model to explain hydration reactions of cement admixed with ZnO. In this work, the authors identified the $CaZn_2(OH)_6 \cdot 2H_2O$ (CZ) phase at two days and suggested the starting of the hydration reactions in these cement pastes with ZnO in agreement with Arliguie and stated that $Zn(OH)_2$ is not the responsible phase of retardation of setting time since, according to the pH diagram at the pH of the cement pastes

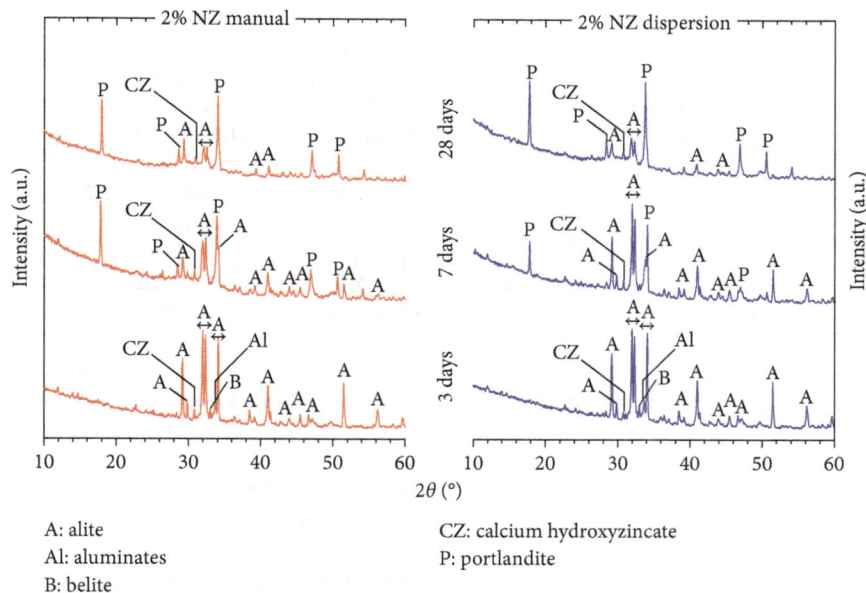

A: alite
Al: aluminates
B: belite

CZ: calcium hydroxyzincate
P: portlandite

FIGURE 3: XRD characterization for cement paste with 2 wt.% of nanozinc (NZ).

($\approx$12), there would be a small amount of this phase. The most important result, regardless of the characterization technique, is that there are no significant differences in hydration of cement pastes at 28 days with a substitution of 2 wt.% of nano-ZnO obtained by both mixing methods.

Figures 4(a)–4(c) and Table 2 show the DTG (mass change speed) versus temperature as mass loss from the TG data for the interest temperature ranges defined previously (T1, T2, and T3) and associated with water loss of a different nature for cement pastes with a cement substitution of 2 wt.% nano-ZnO and 10 wt.% microsilica (MS) obtained for both mixing methods at ages of three, seven, and twenty-eight days, respectively.

DTG analysis and water losses associated with different phases (Table 2) of three-day samples with ZnO and microsilica (Figure 4(a)) are close to the three-day pastes only with nanoparticles for both mixing methods. The hydration of calcium silicates has not taken place at this age. Pastes at seven days (Figure 4(b)) show that calcium silicate hydration has taken place for manual mixing samples but not so for low shear rate dispersion pastes. The total water combined with cement pastes prepared by low shear rate dispersion at seven days is similar to that at three days. Water loss that takes part in the CSH gel (T1) and water combined with portlandite (T2) are low compared with manual mixing pastes according to Table 2. The absence of the peaks corresponding to the water loss that takes part in the CSH gel (T1) and water combined with portlandite (T2) in DTG analysis confirms that the hydration of calcium silicates either has not started or is at an only slightly advanced at seven days for pastes prepared by the dispersion method. There are meaningful differences related to the quantities of water that are part of the CSH gel, as well as water combined with portlandite, that indicate a higher hydration of calcium silicate of manual mixing cement pastes. Pastes only with ZnO hydration at seven days have started in both mixing methods, although higher advanced points for

pastes are obtained by manual mixing. In Figures 4(a)–4(c), two peaks appear before 200°C and 300°C, respectively, that could correspond to the decomposition of hydrated calcium hydroxyzincate $(CaZn_2(OH)_6 \cdot 2H_2O)$ and hydrozincite according to Mellado et al. [54]. In Figure 4(a), these peaks can appear overlapping with the hydration of tricalcium aluminates and therefore have a greater intensity.

At 28 days, the hydration process for cement pastes with NZ and MS obtained for both mixing methods is identical according to DTG analysis and associated water losses (Figure 4(c) and Table 2). At 28 days, the water loss combined with portlandite indicates that about 16 wt.%. of total loss is due to portlandite and the portlandite content for samples obtained by both mixing methods being about 9% of total weight. This minor content, compared with pastes, is only with nanozinc and is a consequence of the pozzolanic reaction between portlandite and the microsilica.

Similar to Figures 1(a)–1(c), calcium carbonate $CaCO_3$ dissociation ensuing form the degradation of the CSH gel takes place between 650°C and 800°C and $CaCO_3$ calcite ensuing from portlandite $Ca(OH)_2$ carbonation dissociated between 850°C and 1000°C according to Figures 4(a)–4(c).

The CSH gel-portlandite ratio is higher for samples with nanozinc oxide and microsilica than for the reference cement paste at 28 days for both experimental methods of obtaining the cement pastes, as occurred in the case of the cement pastes only with nanozinc oxide. Unlike cement pastes only with zinc oxide, this result is a consequence not only of the higher content of water (higher combined water) in the gel (T1) due to the water bound to zinc in calcium hydroxyzincate but also of the formation of the CSH gel because of the pozzolanic reaction due to microsilica addition. As a result of the pozzolanic reaction, the portlandite content (equivalent calcium) for the cement pastes with nano-ZnO regardless of the mixing method is similar and lower than that for reference pastes without ZnO. As for cement pastes with only ZnO, the

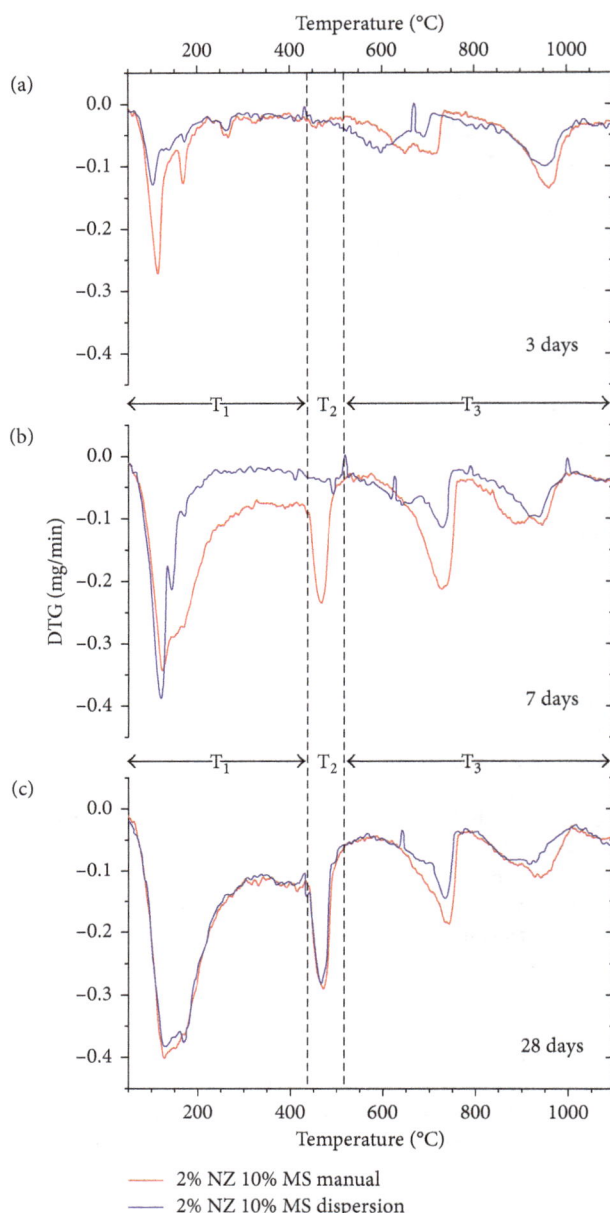

FIGURE 4: DTG-T^a analysis for cement paste with 2 wt.% of nanozinc (NZ) + 10 wt.% microsilica (MS) for (a) 3, (b) 7, and (c) 28 days.

hydration process in the first ages is different for each cement paste with nanozinc oxide and microsilica depending on its mixing method, although in both cases hydration is later, with this being higher for cement pastes obtained at low shear rate dispersion. ZnO does not occur in the hydration in cement pastes prepared by this method in the first ages, and the water bonded is lower than that in pastes prepared by manual mixing. At 28 days, hydration is similar for cement pastes obtained by both mixing methods. Cement pastes obtained by the novel dispersion method show a higher CSH gel-portlandite ratio than the manual mixing ones, indicating a higher Ca/Si ratio and therefore denser CSH gels.

The CSH gel-portlandite ratio is higher for samples with nanozinc oxide and microsilica than for the reference cement paste at 28 days for both experimental methods of obtaining the cement pastes, due in part to the water content in calcium hydroxyzincate and because it is not related to the CSH gel (there is also a formation of secondary CSH gel consequence of the pozzolanic reaction). This result would provide better mechanical properties for samples with nanozinc oxide and microsilica substitutions for both mixing methods compared with the reference pastes. The CSH gel-portlandite ratio for samples with zinc oxide and microsilica is higher than that for samples only with zinc oxide due to the pozzolanic reaction forming a secondary CSH gel.

The Ca/Si ratio in the CSH gel of cement pastes with NZ and MS obtained by SEM-EDX analysis at 28 days shows such a ratio for both mixing methods as similar, although slightly higher for manual mixing, with the values of 2.20 and 2.25, respectively. The Ca/Si ratio decreases with the addition of SiO_2.

As the plot of DTG versus T^a of cement pastes that contain NZ and nanosilica NS is close to that of cement pastes that contain NZ and MS, it is not shown in this article. Table 3 shows the mass loss from the TG data for the interest temperature ranges defined previously (T1, T2, and T3) and associated with the water loss of different nature for cement pastes with NZ and NS for three, seven, and twenty-eight days, respectively.

At three days of hydration, cement pastes with NZ and NS obtained by both mixing methods, as for cement pastes with NZ and MS, the hydration of calcium silicates has either not taken place or at a little advanced point according to the gel water losses. Pastes at seven days show that calcium silicate hydration has taken place for manual mixing samples but not so for low shear rate dispersion pastes. The total water combined with the cement paste prepared by low shear rate dispersion at seven days is similar to that at three days, as shown in Table 3. According to this DTG analysis and the values shown in Table 3, the result would indicate that at seven days for manual mixing pastes, hydration is already advanced and for dispersing samples, it would be starting. According to findings in another study, nanosilica accelerated C_3S hydration [21]; however, in this study, the hydration retarder effect of ZnO prevailed. Water losses that took part in the CSH gel (T1) and combined with portlandite (T2) in the first ages (three and seven days) are similar for samples that do not contain silica or contain micro- or nanosilica. For 28 days, the hydration process for cement pastes obtained for both mixing methods is similar. Water loss that takes part in the CSH gel (T1) at 28 days for cement pastes with NS is higher than that for the MS ones, with this indicating that nanosilica produces a higher quantity of hydrated phase and showing agreement with the literature (Tobón et al. [59]).

It is important to note, in agreement with the literature, that nanosilica has a pozzolanic reactivity higher than microsilica and, that is, produces additional CSH gel (Land and Stephan [21]). According to Table 3, cement pastes with nanosilica for both mixing methods at 28 days show a higher water loss that takes part in the CSH gel (T1) than the water loss for samples with microsilica (Table 3) which correspond

TABLE 2: Water losses associated with different phases in cement pastes with 2 wt.% of nanozinc (NZ) + 10 wt.% microsilica (MS).

	2% NZ + 10% MS manual mixing (wt.%)			2% NZ + 10% MS low shear rate dispersion (wt.%)			52.5R		
	3 d	7 d	28 d	3 d	7 d	28 d	3 d	7 d	28 d
T1	1.8	6.1	8.8	1.3	2.0	9.0	6.6	7.0	7.8
T2	0.4	1.8	2.5	0.4	0.5	2.5	3.4	3.5	3.5
T3	2.1	3.0	3.0	2.0	2.2	2.6	2.6	2.8	2.6
Total combined water (T1 + T2 + T3) wt.%	4.3	10.9	14.3	3.7	4.7	14.1	12.6	13.3	13.9
Equivalent calcium (T2 + T3) wt.%	2.5	4.8	5.5	2.4	2.7	5.1	6.0	6.3	6.1
CSH gel (T1) wt.%/total portlandite (T2 + T3) wt.%	0.72	1.27	1.60	0.54	0.74	1.76	1.10	1.11	1.28

TABLE 3: Water losses associated with different phases in cement pastes with 2 wt.% of nanozinc (NZ) + 10 wt.% nanosilica (NS).

	2% NZ + 10% NS manual mixing (wt.%)			2% NZ + 10% NS low shear rate dispersion (wt.%)			52.5R		
	3 d	7 d	28 d	7 d	28 d	28 d	3 d	7 d	28 d
T1	3.0	8.0	11.2	2.7	3.8	10.5	6.6	7.0	7.8
T2	0.8	1.7	2.0	0.7	1.0	2.1	3.4	3.5	3.5
T3	1.3	3.0	2.8	0.9	1.9	3.0	2.6	2.8	2.6
Total combined water	5.1	12.7	16.0	4.3	6.7	15.6	12.6	13.3	13.9
Equivalent calcium (T2 + T3) wt.%	2.1	4.7	4.8	1.6	2.9	5.1	6.0	6.3	6.1
CSH gel (T1) wt.%/total portlandite (T2 + T3) wt.%	1.43	1.70	2.33	1.68	1.31	2.06	1.10	1.11	1.28

with a higher CSH gel formation. A higher CSH gel formation is important since this phase is responsible for the mechanical properties, adherence, and contribution to the durability of cement-based materials. For 28 days, water loss combined with portlandite indicates that about 9 wt.% of total loss is due to the portlandite (the portlandite content for samples obtained by both mixing methods is about a 6% total weight). This minor portlandite content, compared with pastes only with nanozinc, is a consequence of the pozzolanic reaction. According to the literature, nanosilica promotes the pozzolanic reaction since the pozzolanic reaction is proportional to the specific surface available for the reaction. Micro- and nanosilica, in this work, have a surface area of 28 and $150 \, m^2/g$, respectively (Jo et al. [60]). For samples with nanozinc oxide and with or without microsilica, the CSH gel-portlandite ratio is higher for samples with nanozinc oxide and nanosilica than for the reference cement paste at 28 days regardless of the mixing method. The explanation is the same as that for samples with microsilica. Cement pastes with nanosilica for both mixing methods show the highest CSH gel-portlandite ratios as a consequence of a higher extent of the pozzolanic reaction since such a reaction is proportional to the surface area of silica and is one order of magnitude higher for nanosilica than for microsilica. In all cement paste mixtures, the hydration process at early ages is different for each mixing method; although in both types, the hydration process is retarded (such an effect is greater for cement pastes obtained by low shear rate dispersion). ZnO is not involved in the hydration in cement pastes prepared by this method in the first ages, and the water bonded is lower than that in pastes prepared by manual mixing. At 28 days, hydration is similar for cement pastes obtained by both mixing methods.

The Ca/Si ratio in a CSH gel of cement pastes obtained by SEM-EDX analysis at the age of 28 days shows such a ratio for both mixing methods as similar with values of 1.95 and 1.78 for manual and dispersion mixing methods, respectively. The addition of silica decreases the Ca/Si ratio in the CSH gel, with this decrease being more meaningful when nanosilica (NS) is added.

FTIR analysis (Figure 5) for cement pastes with NZ and MS obtained by both mixing methods shows at three days that the cement paste by mixing oxide could be forecast as the peak corresponding to OH^- groups joined to Zn ions (zinc hydroxide), although in this case it is even weaker. The hydration behaviour is similar to that identified in cement pastes only with NZ: calcium silicate hydration has not started and the hydration of the calcium aluminates has only happened, according to the identification of alite (A) and ettringite (E). Pastes at seven days show that calcium silicate hydration has taken place for manual mixing samples but not so in the case of low shear rate dispersion pastes, in agreement with DTA-TG analysis. As can be observed in Figure 5, in manual mixing samples, the peak corresponding to the unhydrated alite phase has moved from 920 to $995 \, cm^{-1}$, corresponding to the C_3S hydration product (formation of the CSH gel) in addition to the portlandite phase. The hydration advance is higher for cement pastes prepared by a manual mixing method than for low-shear one. This is illustrated by the displacement of the peak corresponding to the CSH gel formation and also by a weak peak identified at $3645 \, cm^{-1}$ corresponding to the O-H stretching of portlandite, formed together with the CSH gel in the hydration of calcium silicates. For cured cement pastes at 28 days, FTIR analysis shows an identical hydration of cement pastes obtained by both mixing methods. When the hydration age is 28 days, there is a meaningful decrease in the peak corresponding to the ettringite phase regardless of the mixing method used in agreement with the literature

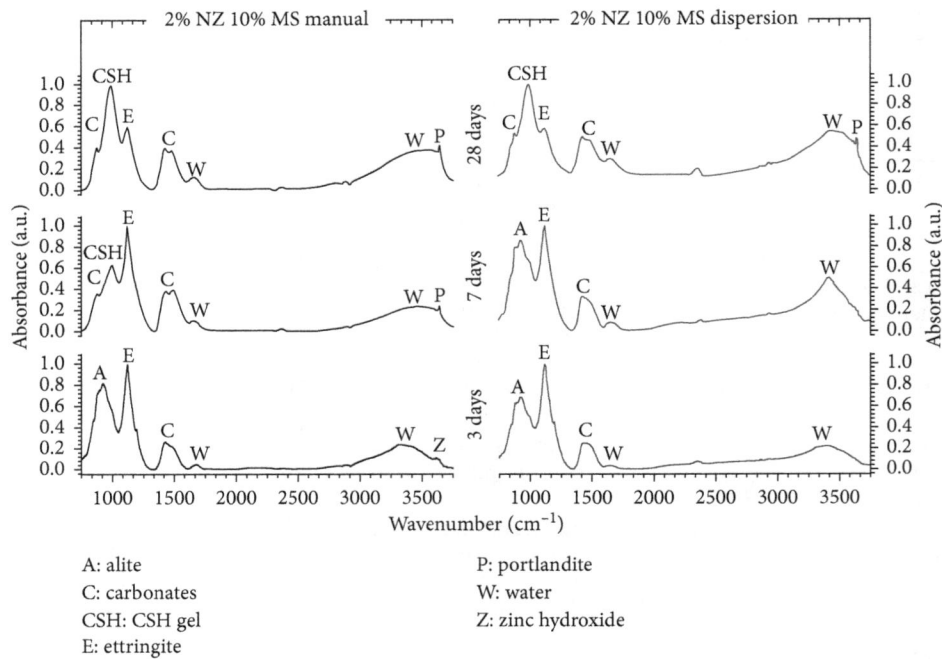

FIGURE 5: FTIR analysis for cement paste with 2 wt.% of nanozinc (NZ) + 10 wt.% microsilica (MS).

[57]. The intensity of the band corresponding to ettringite decreases as its transformation in monosulfoaluminates occurs. The only exceptions are cement pastes with only NZ as an addition, in which this change takes place when the age increases from three to seven days. It is important to note that the peak corresponding to portlandite ($3645\,cm^{-1}$) in samples with microsilica (seven and 28 days) shows a significantly lower intensity than for samples without silica due to portlandite reacting with silica in the presence of water to form a secondary CSH gel according to the following equation given by Neville and Brooks [61]:

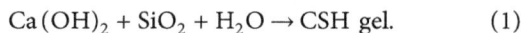

$$Ca(OH)_2 + SiO_2 + H_2O \rightarrow CSH\ gel. \qquad (1)$$

Although the pozzolanic reaction does not lead to a higher water loss T1, if the values provided in Tables 1 and 2 are compared, where in samples only with ZnO, the pozzolanic reaction cannot take place. Secondary CSH gels show lower Ca/Si ratios, in accordance with the literature (Al-Dulaijan et al. [62]).

FTIR analysis for cement pastes with NZ and NS obtained by both mixing methods is similar to cement pastes that contain NZ and MS. For this reason, this is not shown in the article.

At three days, in all cement pastes obtained by both mixing methods, it is possible to observe calcium silicates (alite) and ettringite but not CSH gels. Therefore, the hydration process has been delayed by ZnO. Pastes at seven days show that calcium silicate hydration takes place for manual mixing samples but not so for low shear rate dispersion pastes in agreement with DTG-T^a analysis since for manual mixing samples the displacement of the peak corresponding to the unhydrated alite phase from 920 to $995\,cm^{-1}$ corresponding to the C_3S hydration product (formation of CSH gel) has

taken place. For cured cement pastes at 28 days, FTIR analysis shows an identical hydration of cement pastes obtained by both mixing methods. When the hydration age is at 28 days, there is a meaningful decrease in the peak corresponding to ettringite phase regardless of the mixing method used to obtain the cement pastes. It is important to note that the peak corresponding to portlandite ($3645\,cm^{-1}$) is not identified for any of the samples, as a consequence of a further extension of the pozzolanic reaction according to its higher surface area.

X-ray diffraction characterization (Figure 6) for cement pastes with NZ and MS shows agreement with DTA-TG analysis and FTIR in that hydration at three days is delayed for both mixing methods since only unhydrated phases are identified (calcium silicates and aluminates) in addition to calcium hydrozincite. At seven days, hydration is at a more advanced point in cement pastes obtained by manual mixing than for low shear rate dispersion. At this cured age, manual mixing sample characterization shows peaks corresponding to the portlandite phase (P), unlike dispersing samples in which such a phase is not identified. For pastes obtained by dispersion, as the XRD pattern for ages three and seven days is identical, hydration is delayed. At 28 days, according to XRD analysis, there are no hydration differences between cement pastes obtained by both mixing methods. It should be noted that it is possible to identify the crystalline calcium hydroxyzincate dihydrate $CaZn_2(OH)_6 \cdot 2H_2O$ (CZ) phase to any cured age for both pastes as in the pastes only with ZnO. As in the case of the pastes only with ZnO, given that an alite phase in cement pastes at ages of 28 days is identified, there are unhydrated cement particles.

X-ray diffraction characterization for cement pastes with NZ and NS is identical to cement pastes with NZ and MS (Figure 6). At three days, hydration is delayed for both mixing methods, with unhydrated phases (calcium silicates

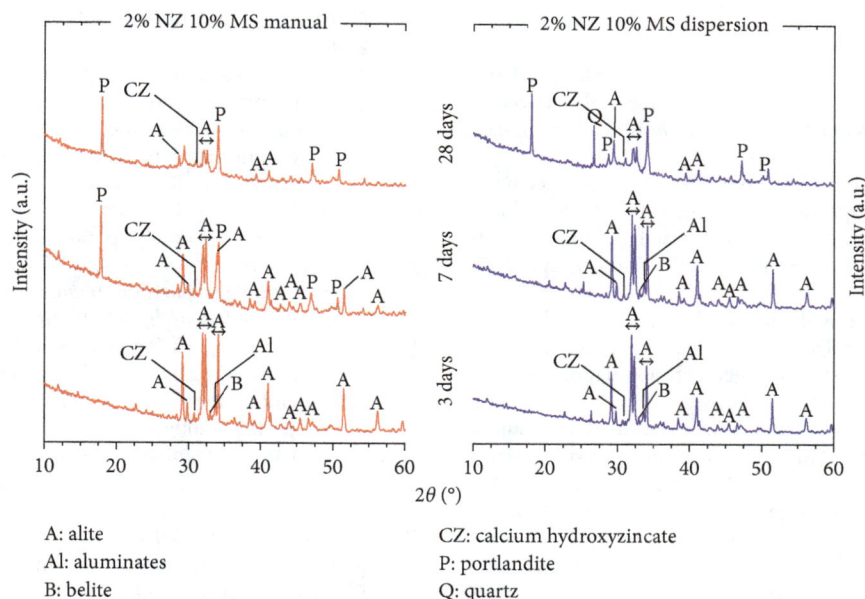

A: alite
Al: aluminates
B: belite

CZ: calcium hydroxyzincate
P: portlandite
Q: quartz

FIGURE 6: XRD characterization for cement paste with 2 wt.% of nanozinc (NZ) + 10 wt.% microsilica (MS).

and aluminates) being identified. At seven days, hydration is at a more advanced point in cement pastes obtained by manual mixing than for low shear rate dispersion. The portlandite phase only can be observed in pastes obtained by manual mixing. At 28 days, hydration for cement pastes obtained by both mixing methods is identical. It should be noted that it is possible to identify the crystalline calcium hydroxyzincate dihydrate $CaZn_2(OH)_6 \cdot 2H_2O$ (CZ) phase to any cured age for both pastes, as in the pastes only with ZnO or with ZnO and microsilica. As in the case of other pastes, a small quantity of alite in cement pastes for both mixing methods at ages of 28 days is identified, indicating that there are unhydrated cement particles as a consequence of the hydration relay by the ZnO addition. By means of XRD diffraction, it is not possible to confirm the greatest extent of the pozzolanic reaction for pastes with NS than for MS (T2 values in Tables 2 and 3 and Figure 5) according to the intensity of the peak associated to the portlandite phase.

4. Conclusions

In this work, cement pastes with simultaneous functional additions (micro- and nanosilica and nanozinc oxide) have been prepared by means of two different mixing methods: a common manual mixing method and a novel low shear rate dispersion method. It highlights the viability of using a new dispersion method based on a low shear mixing to prepare cement pastes with nanometric simultaneous functional additions and limit exposure to the nanoparticles in order to avoid health risks.

At earliest age, hydration for both mixing methods has not started and it is identified by the amorphous zinc hydroxide phase, responsible for the hydration inhibition. When hydration is restarted, it is identified by the crystalline calcium hydroxyzincate phase in cement pastes prepared by both methods. The delay in hydration is higher for cement

pastes obtained by low shear rate dispersion than with manual mixing. Nevertheless, at 28 days, hydration is similar for cement pastes obtained by both mixing methods.

In all the cement pastes with additions, the CSH gel-portlandite ratio is higher than that for the reference cement paste without additions at 28 days regardless of the mixing method, with cement pastes with nanozinc oxide and nanosilica reaching the highest ratio value. This result is a consequence of the higher content of water (higher combined water) in the gel (T1) due to the water bound to zinc in calcium hydroxyzincate for cement pastes with nano-ZnO. In addition, the formation of a secondary CSH gel due to the pozzolanic reaction is greater for cement pastes with nanosilica because of its higher surface area.

The most important result obtained in agreement with the similar hydration behaviour of the different cement pastes prepared by both mixing methods is associated with the feasibility of using the novel low energy dry dispersion method to prepare cement pastes with simultaneous nanometric functional additions avoiding the exposure to the nanoparticles and therefore minimizing health risks.

Conflicts of Interest

The authors declare that there are no conflicts of interest regarding the publication of this paper.

Acknowledgments

This work was supported by the Spanish Ministry of Economy and Competitiveness through the Project MAT-2017-86450-C4-1-R and Spanish National Research Council under Project NANOMIND 201560E068 F. Rubio-Marcos is also indebted to MINECO for a "Ramón y Cajal" contract (Ref. no. RyC-2015-18628), which is cofinanced by the European Social Fund.

References

[1] L. Yang, A. Hakki, F. Wang, and D. E. Macphee, "Photocatalyst efficiencies in concrete technology: the effect of photocatalyst placement," *Applied Catalysis B: Environmental*, vol. 222, pp. 200–208, 2018.

[2] J. Chen, S.-c. Kou, and C.-s. Poon, "Photocatalytic cement-based materials: comparison of nitrogen oxides and toluene removal potentials and evaluation of self cleaning performance," *Building and Environment*, vol. 46, no. 9, pp. 1827–1833, 2011.

[3] P. Sikora, A. Augustyniak, K. Cendrowski et al., "Characterization of mechanical and bactericidal properties of cement mortars containing waste glass aggregate and nanomaterials," *Materials*, vol. 9, no. 8, p. 701, 2016.

[4] P. Sikora, K. Cendrowski, A. Markowska-Szczupak, E. Horszczaruk, and E. Mijowska, "The effects of silica/titania nanocomposite on the mechanical and bactericidal properties of cement mortars," *Construction and Building Materials*, vol. 150, no. 309, pp. 738–746, 2017.

[5] F. Tittarelli, "Oxygen diffusion through hydrophobic cement-based materials," *Cement and Concrete Research*, vol. 39, no. 10, pp. 924–928, 2009.

[6] C. Nunes and Z. Slizkova, "Hydrophobic lime based mortars with linseed oil: characterization and durability assessment," *Cement and Concrete Research*, vol. 61-62, pp. 28–39, 2014.

[7] W. De Muynck, N. De Belie, and W. Verstraete, "Antimicrobial mortar surfaces for the improvement of hygienic conditions," *Journal of Applied Microbiology*, vol. 108, no. 1, pp. 62–72, 2009.

[8] N. León, J. Massana, F. Alonso, A. Moragues, and E. Sanchez-Espinosa, "Effect of nano-SiO_2 and nano-Al_2O_3 on cement mortars for use in agriculture and livestock production," *Biosystems Engineering*, vol. 123, no. 1, pp. 1–11, 2014.

[9] E. Mohseni, B. Mehdizaded, J. Yang, and M. Ali Yazdi, "Singel and combined effects of nano-SiO_2, nano-Al_2O_3 and nano-TiO_2 on the mechanical, rheological and durability properties of self-compacting mortars containing fly ash," *Construction and Building Materials*, vol. 84, pp. 331–340, 2015.

[10] V. Vishwakarma, S. Sekhar Samal, and N. Manoharan, "Safety and risk associated with nanoparticles. A review," *Journal of Minerals and Materials Characterization and Engineering*, vol. 9, no. 5, pp. 455–459, 2010.

[11] M. A. Albrecht, C. W Evans, and C. L Raston, "Green chemistry and the health implications of nanoparticles," *Green Chemistry*, vol. 8, no. 5, pp. 417–432, 2006.

[12] N. C. Mueller and B. Nowack, "Exposure modeling of engineered nanoparticles in the environment," *Environmental Science and Technology*, vol. 42, no. 12, pp. 4447–4453, 2008.

[13] S. Haruehansapong, T. Pulngern, and S. Chucheepsakul, "Effect of the particle size of nanosilica on the compressive strength and the optimum replacement content of cement mortar containing nano-SiO_2," *Construction and Building Materials*, vol. 50, pp. 471–477, 2014.

[14] L. Senff, J. A. Labrincha, V. M. Ferreira, D. Hotza, and W. Repette, "Effect of nano-silica on rheology and fresh properties of cement pastes and mortars," *Construction and Building Materials*, vol. 23, no. 7, pp. 2487–2491, 2009.

[15] J.-X. Lu and C.-S. Poon, "Improvement of early-age properties for glass-cement mortar by adding nanosilica," *Cement and Concrete Composites*, vol. 89, pp. 18–30, 2018.

[16] J. Massana, E. Reyes, J. Bernal, N. León, and E. Sanchez-Espinosa, "Influence of nano and micro-silica additions on the durability of a high performance self-compacting concrete,"

[17] A. A. Ramezanianpour and M. A. Moeini, "Mechanical and durability properties of alkali activated slag coating mortars containing nanosilica and silica fume," *Construction and Building Materials*, vol. 163, no. 28, pp. 611–621, 2018.

[18] H. Cheng-yi and R. F. Feldman, "Hydration reactions in Portland cement-silica fume blends," *Cement and Concrete Research*, vol. 15, no. 4, pp. 585–592, 1985.

[19] J. A. Larbi, A. L. A. Fraay, and J. M. Bijen, "The chemistry of the pore fluid of silica fume-blended cement systems," *Cement and Concrete Research*, vol. 20, no. 4, pp. 506–516, 1990.

[20] J. J. Thomas, H. M. Jennings, and J. J. Chen, "Influence of nucleation seeding on the hydration mechanism of tricalcium silicate and cement," *Journal of Physical Chemistry C*, vol. 113, no. 11, p. 4237, 2009.

[21] G. Land and D. Stephan, "The influence of nano-silica on the hydration of ordinary Portland cement," *Journal of Materials Science*, vol. 47, no. 2, pp. 1011–1017, 2012.

[22] A. Moezzi, A. M. McDonagh, and M. B. Cortie, "Zinc oxide particles: synthesis, properties and applications," *Chemical Engineering Journal*, vol. 185-186, pp. 1–22, 2012.

[23] R. K. Sendi, "Effects of different compositions from magnetic and nonmagnetic dopands on structural and electrical properties of ZnO nanoparticles-based varistor ceramics," *Solid State Sciences*, vol. 77, pp. 54–61, 2018.

[24] D. Xu, K. Song, Y. Li, L. Jiao, and J. Song, "Sc_2O_3 doped Bi_2O_3-ZnO thin film varistors prepared by sol-gel method," *Journal of Alloys and Compounds*, vol. 746, pp. 314–319, 2018.

[25] M. Kahouli, N. Toursi, H. Guermazi, and S. Guermazi, "Enhanced structural and optical properties of ZnO nanopowder with tailored visible luminescence as a function of sodium hydroxide to zinc sulfate mass ratio," *Advanced Powder Technology*, vol. 29, no. 2, pp. 325–332, 2018.

[26] A. M. Hussein, A. V. Lefanova, R. T. Koodali, B. A. Logue, and R. V. Shende, "Interconnected ZrO_2 doped ZnO/TiO_2 network photoanode for dye-sensitized solar cells," *Energy Reports*, vol. 4, pp. 56–64, 2018.

[27] T. V. Torchynska, B. El Filali, G. Polupan, A. L Diaz Cano, and I. Ballardo, "Luminescence, structure and aging c-axis-oriented silver doped ZnO nanocrystalline films," *Materials Science in Semiconductor Processing*, vol. 79, no. 1, pp. 99–106, 2018.

[28] C. Klingshirn, "ZnO: from basics towards applications," *Physica Status Solidi B*, vol. 244, no. 9, pp. 3027–3073, 2007.

[29] V. S. Ramachandran, *Handbook of Thermal Analysis of Construction Materials*, Noyes Publications, New York, NY, USA, 2002.

[30] D. Nivethitha and S. Dharmar, "Effect if zinc oxide nanoparticle on strength of cement mortar," *International Journal of Scientific Engineering and Technology*, vol. 3, no. 5, pp. 123–127, 2016.

[31] G. Arliguie and J. Grandet, "Influence de la composition d'un ciment Portland sur son hydratation en presence de zinc," *Cement and Concrete Research*, vol. 20, no. 4, pp. 517–524, 1990.

[32] M. Yousuf, A. Mollah, T. R. Hess, Y. N. Tsai, and D. L. Cocke, "An FTIR and XPS investigations of the effects of carbonation on the solidification/stabilization of cement based systems-Portland Type V with zinc," *Cement and Concrete Research*, vol. 29, pp. 55–61, 1999.

[33] D. L. Cocke, M. Y. A. Mollah, T. R. Hess, and T. C. Lin, "In aqueous chemistry and geochemistry of oxides, oxyhydroxides and related materials," *Materials Research Society*, vol. 432, pp. 63–68, 1997.

[34] C. A. Johnson and M. Kersten, "Solubility of Zn(II) in association with calcium silicate hydrates in alkaline solutions," *Environmental Science and Technology*, vol. 33, no. 13, pp. 2296–2298, 1999.

[35] W. Lieber and J. Gebauer, "Einbau von zink in calciumsilikathydrate," *Zem.-Kalk-Gips*, vol. 22, pp. 295–303, 1969.

[36] F. F. Ataie, M. C. G. Juenger, S. C. Taylor-Lange, and K. A. Riding, "Comparison of the retarding mechanisms of zinc oxide and sucrose on cement hydration and interactions with supplementary cementitious materials," *Cement and Concrete Research*, vol. 72, pp. 128–136, 2015.

[37] A. Stumm, K. Garbev, G. Beuchle et al., "Incorporation of zinc into calcium silicates hydrates, part I," *Cement and Concrete Research*, vol. 35, no. 9, pp. 1665–1675, 2005.

[38] S. K. Ouki and C. D. Hills, "Microstructure of Portland cement pastes containing metal nitrate salts," *Waste Management*, vol. 22, no. 2, pp. 147–151, 2002.

[39] F. Ziegler, A. M. Scheidegger, C. A. Johnson, R. Dahn, and E. Wieland, "Sorption mechanism of zinc to calcium silicate hydrate: x-ray absorption fine structure (XAFS) investigation," *Environmental Science and Technology*, vol. 35, no. 7, pp. 1550–1555, 2001.

[40] F. Ziegler and C. A. Johnson, "The solubility of calcium zincate ($CaZn_2(OH)_6 \cdot 2H_2O$)," *Cement and Concrete Research*, vol. 31, no. 9, pp. 1327–1332, 2001.

[41] F. Tittarelli and T. Bellezze, "Investigation of the major reduction reaction occurring during the passivation of galvanized steels rebars," *Corrosion Science*, vol. 52, no. 3, pp. 978–983, 2010.

[42] Y. Xie, Y. He, P. L. Irwin, T. Jin, and X. Shi, "Antibacterial activity and mechanism of action of zinc oxide nanoparticles against *Campylobacter jejuni*," *Applied and Environmental Microbiology*, vol. 77, no. 7, pp. 2325–2331, 2011.

[43] A. Sirelkhatim, S. Mahmud, A. Seeni et al., "Review on zinc oxide nanoparticles. Antibacterial activity and toxicity mechanism," *Nano-Micro Letters*, vol. 7, no. 3, pp. 219–242, 2015.

[44] N. Madhusudhana, K. Yogendra, K. Mahadevan, and S. Naik, "Photocatalytic degradation of coralene dark red 2B azo dye using calcium zincate nanoparticles in presence of natural sunlight: an aid to environmental remediation," *International Journal of Chemical Engineering and Applications*, vol. 2, no. 4, pp. 294–298, 2011.

[45] N. M. Gomez-Ortiz, W. S. Gonzalez Gomez, S. C. de la Rosa-Garcia et al., "Antifungal activity of $Ca[Zn(OH)_3]_2 \cdot 2H_2O$ coatings for the presentation of limestone monuments: an in vitro study," *International Biodeterioration and Biodegradation*, vol. 91, pp. 1–8, 2014.

[46] J. F. Hernandez-Sierra, F. Ruiz, D. C. Cruz Pena et al., "The antimicrobial sensitivity of Streptococcus mutans to nanoparticles of silver, zinc oxide and gold," *Nanomedicine, Nanotechnology, Biology, and Medicine*, vol. 4, no. 3, pp. 237–240, 2008.

[47] J. F. Fernández Lozano, I. Lorite Villalba, F. Rubio Marcos et al., "Method for dry dispersion of nanoparticles and production of hierarchical structures and coatings," US Patent WO2010/010220 A1, 2014.

[48] *Aerosil®, Aeroxide® and Sipernat® Products for the Construction Materials Industry, Technical Information 1398*, Evonik Industries, Essen, Germany.

[49] I. Lorite, L. Pérez, J. J. Romero, and J. F. Fernández, "Effect of the dry nanodispersion procedure in the magnetic order of the Co_3O_4 surface," *Ceramics International*, vol. 39, no. 4, pp. 4377–4381, 2013.

[50] D. Alonso-Domínguez, I. Álvarez-Serrano, E. Reyes, and A. Moragues, "New mortars fabricated by electrostatic dry deposition of nano and microsilica additions: enhanced properties," *Construction and Building Materials*, vol. 135, pp. 186–193, 2017.

[51] E. Ghafari, S. A. Ghafari, Y. Feng, F. Severgnini, and N. Lu, "Effect of zinc oxide and Al-zinc oxide nanoparticles on the rheological properties of cement paste," *Composites Part B: Engineering*, vol. 105, pp. 160–166, 2016.

[52] G. Villain, M. Thiery, and G. Platret, "Measurement method of carbonation profiles in concrete: thermogravimetry, chemical analysis and gammadensimetry," *Construction and Building Materials*, vol. 37, no. 8, pp. 1182–1192, 2007.

[53] V. S. Ramachandran, R. M. Paroli, J. J. Beaudoin, and A. H. Delgado, *Handbook of Thermal Analysis of Construction Materials*, Noyes Publications/William Andrew Publishing, New York, NY, USA, 2003.

[54] A. Mellado, M. V. Borrachero, L. Soriano, J. Payá, and J. Monzó, "Immobilization of Zn(II) in Portland cement pastes," *Journal of Thermal Analysis and Calorimetry*, vol. 112, no. 3, pp. 1377–1389, 2013.

[55] M. Y. A. Mollah, J. P. Pargat, and D. L. Cocke, "An infrared spectroscopic examination of cement-based solidification/ stabilization systems-Portland Types V and IP with zinc," *Journal of Environmental Science and Health*, vol. 27, no. 6, pp. 1503–1519, 1992.

[56] M. A. Trezza, "Hydration study of ordinary Portland cement in the presence of zinc ions," *Materials Research*, vol. 10, no. 4, pp. 331–334, 2007.

[57] T. Vázquez Moreno, "Contribución al estudio de las reacciones de hidratación del cemento portland por espectroscopia infrarroja," *Materiales de Construcción-CSIC*, vol. 16, no. 163, pp. 51–63, 1976.

[58] F. Andrade Brehm, C. A. Mendes Moraes, R. C. Espinosa Modolo, A. C. Faria Vilela, and D. C. Coitinho Dal Molin, "Oxide zinc addition in cement paste aiming electric arc furnace," *Construction and Building Materials*, vol. 139, pp. 172–182, 2016.

[59] J. I. Tobón, J. J. Payá, M. V. Borrachero, and O. J. Restrepo, "Mineralogical evolution of Portland cement blended with silica nanoparticles and its effect on mechanical strength," *Construction and Building Materials*, vol. 21, pp. 1351–1355, 2007.

[60] B.-W. Jo, C.-H. Kim, G. Tae, and J.-B. Park, "Characteristics of cement mortar with nano-SiO_2 particles," *Construction and Building Materials*, vol. 21, no. 6, pp. 1351–1355, 2007.

[61] A. M. Neville and J. J. Brooks, *Concrete Technology*, Longman Group, London, UK, 2nd edition, 2010.

[62] S. U. Al-Dulaijan, A.-H. Al-Tayyib, M. M. Al-Zahrani, G. Parry Jones, and A. I. Al-Mana, "Si MAS-NMR study of hydrated cement paste and mortar made with and without silica fume," *Journal of the American Ceramic Society*, vol. 78, no. 2, pp. 342–346, 1995.

19

The Effect of Water Repellent Surface Impregnation on Durability of Cement-Based Materials

Peng Zhang,[1,2] **Huaishuai Shang,**[1] **Dongshuai Hou,**[1] **Siyao Guo,**[1] **and Tiejun Zhao**[1]

[1]*Center for Durability & Sustainability Studies, Qingdao University of Technology, Qingdao 266033, China*
[2]*Institute of Concrete Structures and Building Materials, Karlsruhe Institute of Technology, 76131 Karlsruhe, Germany*

Correspondence should be addressed to Peng Zhang; zhp0221@163.com and Huaishuai Shang; shanghuaishuai@aliyun.com

Academic Editor: Jun Liu

In many cases, service life of reinforced concrete structures is severely limited by chloride penetration until the steel reinforcement or by carbonation of the covercrete. Water repellent treatment on the surfaces of cement-based materials has often been considered to protect concrete from these deteriorations. In this paper, three types of water repellent agents have been applied on the surface of concrete specimens. Penetration profiles of silicon resin in treated concrete have been determined by FT-IR spectroscopy. Water capillary suction, chloride penetration, carbonation, and reinforcement corrosion in both surface impregnated and untreated specimens have been measured. Results indicate that surface impregnation reduced the coefficient of capillary suction of concrete substantially. An efficient chloride barrier can be established by deep impregnation. Water repellent surface impregnation by silanes also can make the process of carbonation action slow. In addition, it also has been concluded that surface impregnation can provide effective corrosion protection to reinforcing steel in concrete with migrating chloride. The improvement of durability and extension of service life for reinforced concrete structures, therefore, can be expected through the applications of appropriate water repellent surface impregnation.

1. Introduction

The development of cement and concrete dates to the mid-1800s, and it proved to be a revolutionary innovation in building materials. Today, reinforced concrete is the single most widely used building material in the world for both entire buildings and key structural elements that need to be able to withstand various substantial loads. Reinforced concrete is used in such large amounts because it shares the characteristics of relatively good durability, low maintenance cost, and convenience. However, it is nowadays generally accepted that service life of many reinforced concrete structures is frequently not sufficient. The cost for early repair measures is often significantly higher than the cost for new construction. The major origin of these problems of maintenance and repair costs and poor serviceability is a lack of durability for reinforced concrete structures [1–3].

Moisture transport in cement-based materials is a crucial physical process for their durability since many effects that influence the durability of the building structure are induced by water itself as well as harmful substances transported by it. If cement-based materials, such as mortar and concrete, are exposed to water, a series of deteriorating processes can take place. One dominant process or a combination of different processes may eventually limit the expected service life of reinforced concrete structures. The corrosive attack of water with respect to concrete can be subdivided at least into three different types. First, pure water in permanent contact with cement-based materials acts as a solvent. The binding matrix consisting of $Ca(OH)_2$ and C-S-H gel is gradually dissolved by hydrolysis. Second, gases of the environment may be dissolved in the aqueous pore solution of concrete. In this way, acids are formed, for instance, by dissolution of CO_2 and SO_2, which could react rapidly with the hydration products of cement. In the third type of corrosive attack water acts essentially as a vehicle and transports dissolved compounds, such as chloride ions, into the porous system of cementitious matrix. Besides the corrosive attacks, water also plays an

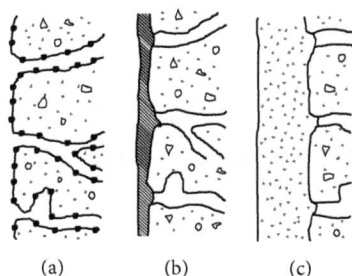

FIGURE 1: Types of surface treatments: (a) impregnation; (b) coating; and (c) cementitious coating surface.

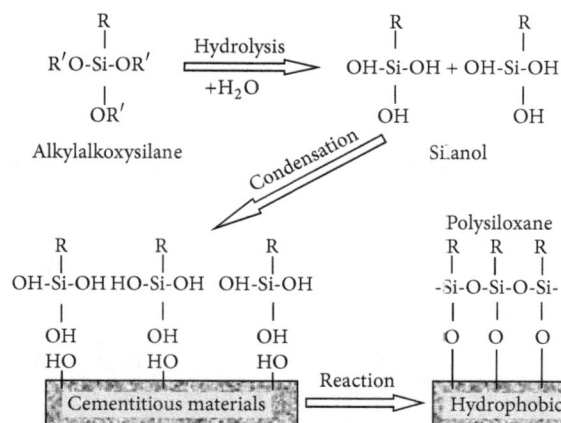

FIGURE 2: Mechanisms for hydrolysis and polymerization of silane in cement-based materials.

important role in some other physical and chemical damage for concrete, such as freeze-thaw action, alkali-aggregate reaction, steel corrosion, and drying shrinkage.

It is obvious that all these three types of aggressive attacks just mentioned act from the surface of concrete. Throughout history, a range of protective materials have been applied to the exposed surfaces of structural concrete elements to prevent the ingress of water, including oils, waxes, or paints. Nowadays big progress has been achieved in the production of water repellent agents and development of water repellent treatment. It has been proved that surface impregnation by water repellent agents should be an effective preventive method for concrete structures [4–9]. For broader information about the studies on water repellent treatment, the proceedings of HYDROPHOBE conference series (Hydrophobes I–VIII) can be known from [10–17].

In this contribution, the basic mechanism of water repellent treatment on cement-based materials has been briefly described. Three types of water repellent agents which are in forms of liquid, cream, and gel have been applied on the surface of two types of cement-based materials. The consequent effects of surface impregnation on reducing water capillary suction, chloride penetration, carbonation, and reinforcement corrosion in concrete will be measured and discussed.

2. Basic Mechanism of Water Repellent Treatment

In general, water repellent surface treatments are mainly classified in three groups according to the mechanism by which the protection is achieved. In Figure 1, types of surface treatments are illustrated according to this classification [18]. Surface treatment by silanes belongs to "impregnation," whose basic mechanisms are given in the next two paragraphs.

The most important silicon-based water repellent agents are those made of silanes and siloxanes which are polymers containing three alkoxy groups, denoted OR′, linked to a silicon atom, with each silicon atom carrying an organic alkyl group, denoted R. The silicon functional alkoxy group reacts with water and yields a reactive silanol group (hydrolysis stage). Further condensation by crosslinking to the hydroxyl groups forms polysiloxane (silicon resin) as the active water repellent product which is linked to the inorganic substrate by way of covalent siloxane bonds, as shown in Figure 2.

The organofunctional alkyl groups will reduce the critical surface tension of the material surface and thus provide hydrophobicity, while the silicon functional groups provide reactivity with the substrate and control the penetration depth.

The effect of water repellents is essentially based on their low surface tension. The behaviour of water when contacting the surface of a material is governed by the surface tension which can be measured by the contact angle, as shown phenomenologically in Figure 3. The intensity of the water repellent property is associated with the contact angle between water and the treated surface. Contact angles of a water droplet of more than 90° represent hydrophobic property with less than 90° hydrophilic property. The higher the contact angle is, the more water repellent the surface becomes. The hydrophobicity of water repellents is in fact realized in two steps. Firstly, the beading effect causes the water droplet to quickly run off and leave the surface. Secondly, when water tends to spread and form a water film over the surface, water absorption is reduced by excluding via treated capillaries.

3. Materials and Methods

3.1. Materials and Preparation of Specimens. Two types of mortar and concrete specimens were prepared for the test series. Ordinary Portland cement type 42.5, crushed aggregates with a maximum diameter of 20 mm and density of 2620 kg/m^3, and river sand with a maximum grain size of 5 mm and density of 2610 kg/m^3 were used. The exact compositions of the concrete used in this project are given in Table 1. The mix with W/C = 0.5 was named concrete C. Mortar with a higher water-cement ratio (W/C = 0.6) was also prepared and called mortar M. Some specimens prepared with both concrete C and mortar M were later surface impregnated with different amount of water repellent agents. The concrete specimens were used for water absorption test, chloride penetration test, carbonation test and steel corrosion test. The mortar specimens were prepared only for neutron radiography test in order to avoid the influence of course aggregate during the image analysis.

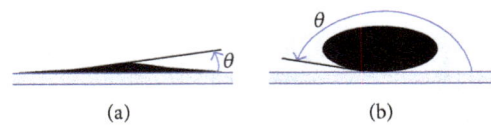

FIGURE 3: The principle of water repellency: (a) untreated concrete, contact angle θ less than 90°; (b) water repellent treated, contact angle θ greater than 90°.

TABLE 1: Composition of the two types of mixtures used in this project, kg/m^3.

Type	W/C	Cement	Sand	Aggregate	Water
Concrete C	0.5	320	653	1267	160
Mortar M	0.6	300	1650	—	180

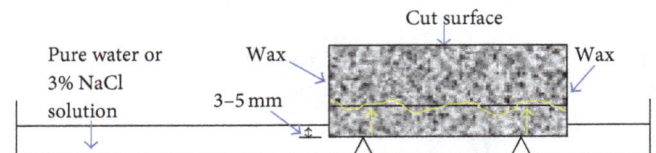

FIGURE 4: Schematic illustration for water absorption and chloride penetration test of concrete.

From all mixes given in Table 1 cubes with side length of 100 mm were produced. Another type of prismatic specimens with dimensions of 280 × 150 × 115 mm with two steel bars was also prepared for steel corrosion test. All specimens were compacted in steel forms and cured for one day before demolding. After that the specimens were moved into a curing room ($T = 20 \pm 2°C$, RH > 95%). At an age of 28 days they were taken out of the curing room for water repellent surface treatment.

3.2. Water Repellent Surface Impregnation. After 28 days of moist curing, the specimens were further stored at RH of 60% for 7 days for drying. Then one of the molded surfaces of cubic specimens and the top surface (280 × 115 mm) of the rectangular parallelepiped specimens have been impregnated with three different types of water repellent agents. The agents' type, usage amount, and the corresponding samples codes are listed in Table 2. After that the specimens were stored again at RH of 60% for another 7 days in order to allow sufficient polymerization of silane. Then the surface impregnated specimens were ready for further tests.

One series have been impregnated with liquid silane. In this case, concrete surface was put in contact with liquid silane for one hour. During this period, liquid silane could be absorbed into specimen due to capillary suction. In second series one of the molded surfaces was covered by silane cream. The amount of usage on the surface was 400 g/m^2. For the third to fifth series, 100, 400, and 600 g/m^2 of silane gel were applied. Both silane cream and gel were covered on the concrete surfaces with a small brush.

From the specimens treated by water repellent agents, layers from the treated surface with a thickness of 1 mm each have been milled consecutively by means of a specially built milling cutter. The powder obtained from this process was collected. The silicon content of these powders was then determined by means of FT-IR spectroscopy. This method has been developed and further refined for this specific application by Gerdes and Wittmann [19].

3.3. Water Absorption and Chloride Penetration. Water absorption of surface treated and untreated specimens has been measured by a standard method [20]. Before the test, the

cubic specimens were cut into two halves and dried in a ventilated oven at a temperature of 50°C for 7 days until mass equilibrium was reached. When the specimens cooled down to the room temperature the treated and untreated specimens were put in contact with water for selected periods of time, as shown in Figure 4. Then the amounts of absorbed water by capillary suction were measured by weighing the specimens after 1, 2, 4, 8, 24, 48, and 72 hours.

In a similar way as described in the last paragraph, chloride penetration test (3% NaCl solution) for water repellent treated and untreated specimens were carried out for 28 days. After the test, powder was milled consecutively starting at the specimens' surface which had been exposed to salt solution. The chloride content of the powder was then determined with the ion selective electrode method. In this way, the chloride profiles in water repellent surface impregnated and untreated specimens have been determined.

3.4. Neutron Radiography. Water repellent mortar specimens and untreated companion specimens were also tested by neutron radiography at Paul-Scherrer-Institute (PSI) in Switzerland. Neutron radiography has been identified as an ideal and unique nondestructive technique to study water movement and moisture distributions in cement-based materials because of their strong attenuation by hydrogen and their insensitivity to the dominant ingredients, such as silica and calcium in cement-based materials. More details about this technique can be found in [21–26].

First, neutron images were taken on samples, which were in hygral equilibrium with the room atmosphere (RH ≈ 60%; $T \approx 20°C$). Then neutron images were taken again on water repellent treated and untreated mortar specimens after contact with water for 0.5 and 2 hours. In this way, the water movement in samples was visualized. In addition, some surface impregnated and untreated samples were placed in water for three days. This period was sufficient to saturate the samples completely. Then neutron images were taken on these water saturated specimens. Both untreated and surface impregnated mortar specimens in the water saturated state

TABLE 2: Three types of water repellent agents used in this project and their uses.

	Type	Amount of use	Note
Ref.	—	—	No treatment, reference sample
L1	Liquid silane	$470\,g/m^2$	Surface absorption
C400	Silane cream	$400\,g/m^2$	Surface brushing
G100	Silane gel	$100\,g/m^2$	Surface brushing
G400	Silane gel	$400\,g/m^2$	Surface brushing
G600	Silane gel	$600\,g/m^2$	Surface brushing

were investigated. From the neutron images the moisture distribution can be analyzed quantitatively.

3.5. Accelerated Carbonation. After drying in lab for 7 days, both surface treated and untreated specimens were submitted to accelerated carbonation for 7 and 28 days. According to the Chinese standard [27], the concentration of CO_2 gas was maintained constant at $20 \pm 2\%$; relative humidity in carbonation box was about 70%; the temperature was $20 \pm 3°C$. Four surfaces except for the treated surface and its opposite surface had been sealed with wax before being placed in carbonation situation. In this way carbonation normal to two opposite surfaces into concrete was imposed. After 7 and 28 days, carbonation depth in the surface impregnated and untreated samples were measured by spraying phenolphthalein solution with 1% in ethanol.

3.6. Reinforcement Corrosion. This test followed ASTM G 109-07 [28]; specimens were $280 \times 150 \times 115\,mm$ with a reservoir of NaCl solution on the test surface. The reservoir with size of $150 \times 75 \times 75\,mm$ was located at the center of top surface. Upper reinforced steel was positioned 20 mm from ponded surface and bottom steels were 25 mm from bottom surface. The ends of steel were protected with electroplater's tape and a 200-mm portion in the middle is bare. During the test, the half-cell potential and the corrosion current density of the steel rebar in surface impregnated and untreated concrete specimens were measured continuously every week.

4. Results and Discussion

4.1. Effect of Water Repellent Surface Impregnation on Water Absorption. Water absorption of both untreated and surface treated concrete specimens has been measured for 72-hour contact with water. Results obtained at different time are shown in Figure 5. Points indicated in Figure 5 are average values of three independent measurements. The variation of the individual measurements is also shown. It can be learned from the results that all the surface impregnated concrete absorbed much less water compared to the untreated concrete. In this case, it is not liquid water but water vapor is trapped by capillary condensation once it has crossed the silane impregnated layer. In addition, capillary condensation can take place in nanopores of concrete, as silane molecules cannot enter these narrow spaces because of geometrical reasons. Therefore, a small amount of capillary condensed water still can migrate into the pores by diffusion. But, compared to the untreated concrete, the amount of absorbed

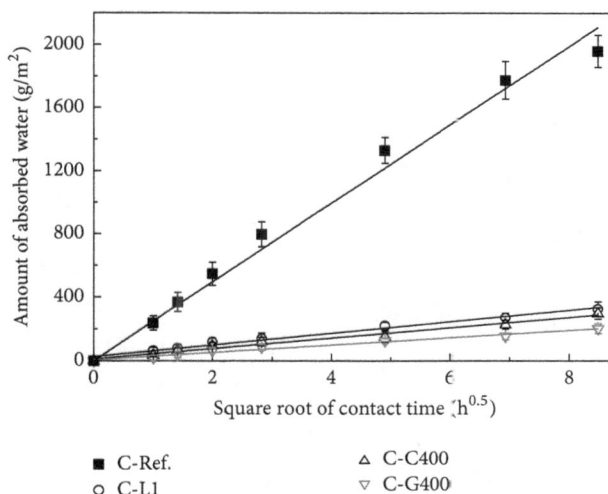

FIGURE 5: The amount of absorbed water in surface impregnated and untreated concrete at different time of water absorption and the linear fitting lines.

water is reduced significantly by surface impregnation with each type of silane.

For a homogeneous porous material, a simple expression can be deduced from theory of capillarity to describe capillary suction as function of time; see (1) [29, 30]. This equation is only a first approximation because the skin effect of concrete will always be the origin of a deviation of measured results from the theoretical prediction.

$$\Delta W = A\sqrt{t}, \qquad (1)$$

where ΔW stands for the amount of absorbed water by capillary suction per unit area and t for the duration of contact. A is the coefficient of capillary suction. The coefficient of capillary suction deduced from Figure 5 for treated and untreated concrete can be calculated. The results indicate that the coefficient of capillary suction for untreated sample is $248.7\,g/(m^2 h^{0.5})$, while for sample L1 (impregnated by liquid silane) it is $40.9\,g/(m^2 h^{0.5})$, approximately one-sixth of untreated sample; for samples C400 (silane cream) and G400 (silane gel) the coefficients are 34.5 and $24.5\,g/(m^2 h^{0.5})$, respectively. They are less than one-seventh and one-tenth of that of untreated sample. This obviously indicates that surface impregnation with water repellent silanes can significantly reduce water penetration into concrete.

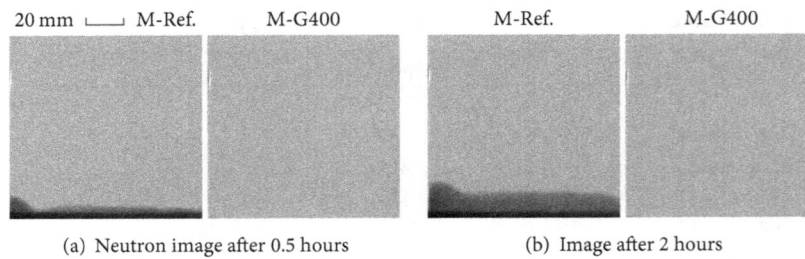

(a) Neutron image after 0.5 hours   (b) Image after 2 hours

FIGURE 6: Observations of water penetration into mortar specimens after 0.5 and 2 hours by means of neutron radiography.

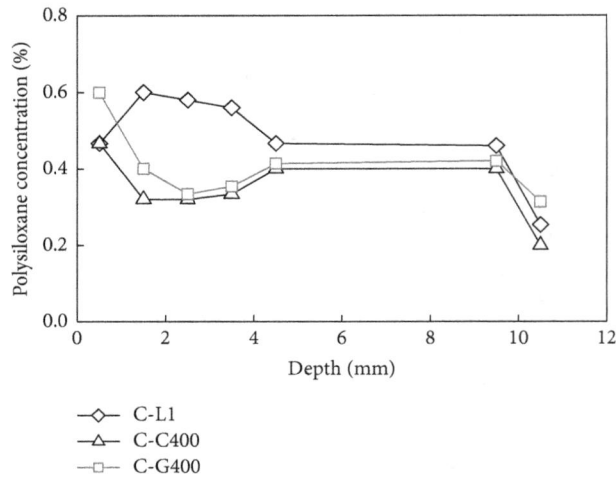

FIGURE 7: Polysiloxane profiles in water repellent treated concrete with silane liquid, silane cream, and silane gel.

Figure 6 shows the visual observation of water penetration into untreated and water repellent surface treated mortar specimens after 0.5 and 2 hours by means of neutron radiography. It can be clearly seen that after half an hour of contact with water a penetration front becomes visible in untreated concrete. This irregular front gradually moves into the porous material with increasing of time. But for the surface impregnated sample water uptake could not be observed with the naked eyes even after two hours because of the polysiloxane film formed from silane, which made the near-surface region hydrophobic.

After being applied on the surfaces of concrete silane penetrated and formed polysiloxane (silicon resin) in the near-surface zone. The polysiloxane concentration in surface impregnated samples has been measured by FT-IR spectroscopy. The results are shown in Figure 7. It can be seen that in each case a penetration depth of about nine millimeters has been reached. This treatment can be called deep impregnation in contrast to simple surface impregnation. In some cases, a simple surface impregnation is sufficient. However, to build up a reliable and durable chloride barrier, a minimum penetration depth of 7 mm is often required [5]. This has to be confirmed in the context of quality assurance after surface treatment in practice. If the penetration depth is too small,

the ingress of aggressive ions with water is slowed down but not prevented for long time.

In addition, neutron images of three types of impregnated and water saturated mortar specimens are shown in Figure 8. The upper impregnated surface is of interest exclusively in this context. It can be clearly seen with the naked eye that the neutron transmission is significantly higher in the outer impregnated layer. The thickness of the impregnated layer can be estimated from the results shown in Figure 8. The average values determined by visual inspection are 2.0, 4.1, and 6.3 mm for samples G100, G400, and G600, respectively.

The moisture distribution was further measured in the near-surface zone as indicated with the rectangular frame shown in Figure 8 (M-G600) from the neutron images obtained from water saturated specimens. Results are shown in Figure 9. As expected, the moisture content in the untreated specimen is essentially homogeneously distributed all over the volume. The observed slight decrease of water content close to the surface may be attributed to a small water loss during handling before taking the first neutron image.

However, on the surface impregnated specimens, the influence of the water repellent near-surface zone can be observed clearly. As expected, the water content in the water repellent zone is significantly reduced. The width of the water repellent zone can also be observed clearly. In samples M-G100 a water repellent layer with a thickness of approximately 2 mm has been established. In samples M-G400 and M-G600 the thickness of the water repellent zone can be estimated to be approximately 4 and 6 mm respectively. What is most important, however, is the fact that in sample M-G100 the water content in the water repellent zone is certainly substantially reduced but still a certain amount of water can be observed in this region. In contrast in sample M-G600 a minimum amount of water can be detected only. From these results, it can again be concluded that deep impregnation is necessary for an efficient chloride barrier.

4.2. Effect of Water Repellent Surface Impregnation on Chloride Penetration. The surfaces of treated and untreated concrete specimens have been brought in contact with an aqueous NaCl solution with concentration of 3% for 28 days. The chloride profiles were determined. The results are shown in Figure 10. It can be seen that a lot of chloride ions penetrated into untreated concrete even up to depth of 30 mm. It already has been shown that capillary suction is a most powerful mechanism for the transport of chlorides into concrete. If

 (a) M-G100 (b) M-G400 (c) M-G600

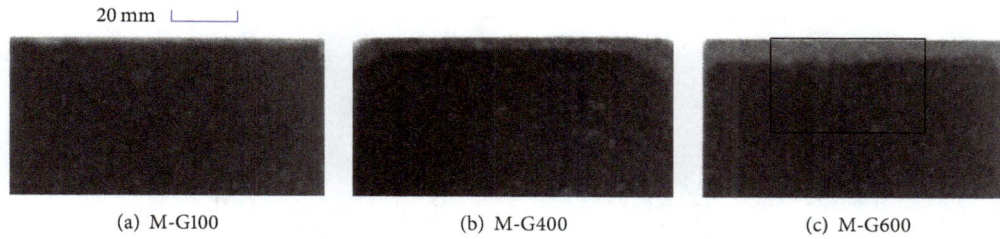

FIGURE 8: Neutron images as obtained on the tree types of water repellent surface impregnated and water saturated mortar specimens. The upper half of the neutron images taken on square slabs is shown only.

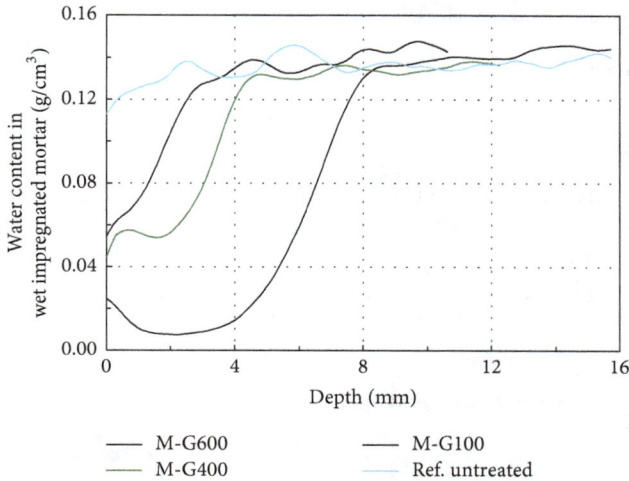

FIGURE 9: Water content in surface impregnated and water saturated mortar specimens G100, G400, and G600. For comparison, the water distribution in untreated mortar is also shown.

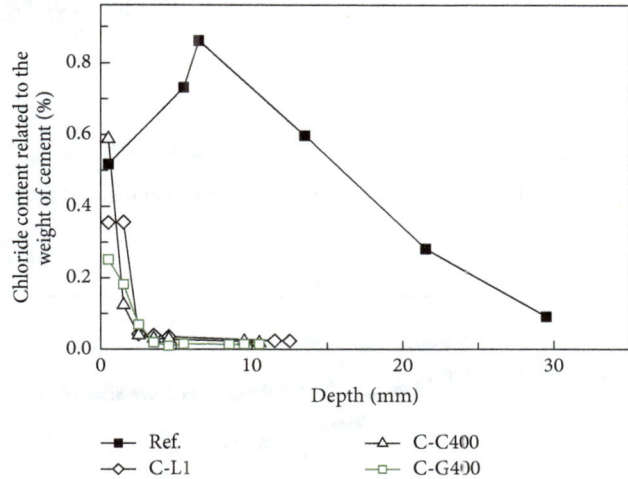

FIGURE 10: Chloride profiles of surface impregnated and untreated concrete after continuous contact with NaCl solution for 28 days.

there is no capillary action, salt solution cannot be taken up by the porous material and if the micropores are not water filled chloride cannot diffuse into the porous structure either. Therefore, by means of surface impregnation with silanes it restrained water from penetrating into concrete and consequently prevented chloride migration. During the exposure period for treated concrete no chloride has penetrated into deep part of the material. The small amount of chloride ions which can be detected in the first 3 mm is due to surface roughness and open big pores in the near-surface zone. Therefore, surface impregnation with silane is an efficient chloride barrier for porous cement-based materials.

4.3. Effect of Water Repellent Surface Impregnation on Carbonation. After 7 and 28 days of carbonation, the carbonation depth of water repellent treated and untreated concrete has been measured. The results are shown in Figure 11. It can be obviously found that surface impregnated specimens have lower carbonation depth than untreated concrete. Among the surface treatments, application of 400 g/m^2 silane cream and silane gel reduces approximately one half of carbonation depth compared to the reference concrete, whose efficiency is much better than 100 g/m^2 covering usage.

By surface impregnation with silanes, the hydrophobic film protects concrete from water penetration, which usually makes the hydrophobic layer almost dry. Very little carbonation action takes place in this area because the neutralization between CO_2 gas and calcium hydrate or C-S-H gel needs water, while this layer also makes the moisture diffusion of concrete very low and consequently makes the area behind the hydrophobic layer moist, under which condition carbonation cannot happen either. However, it must be noticed that the conclusion that surface impregnation reduces carbonation depth by about one half was obtained under RH of 70% in the carbonation box. If the environment is very dry, the untreated concrete would lose water very soon; but, in the treated concrete, the drying rate is slowed down and consequently liquid water in the pores would make carbonation process quicker [31].

4.4. Effect of Water Repellent Surface Impregnation on Reinforcement Corrosion. The half-cell potential (Cu-CuSO$_4$) and corrosion current density of the steel rebar in reinforced concrete have been measured. The results are shown in Figure 12. It indicates clearly that the concrete specimens without surface impregnation exhibits high level of negative corrosion potentials and corrosion current densities, especially after approximately 33 weeks of exposure period. At this stage,

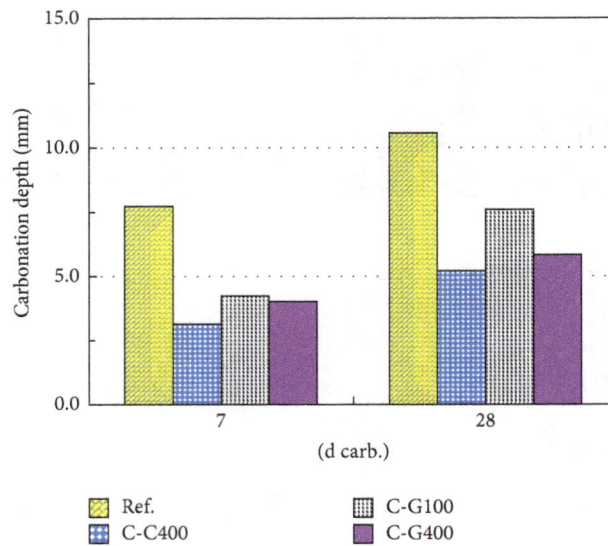

FIGURE 11: Carbonation depth of untreated and surface impregnated concrete after 7 and 28 days of accelerated carbonation.

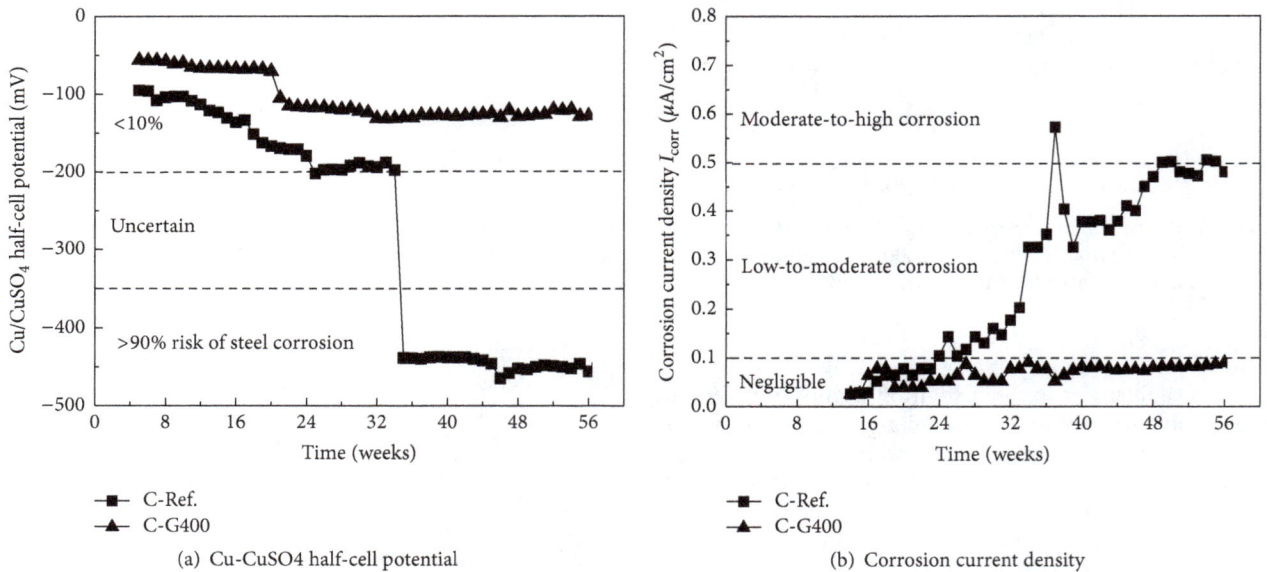

(a) Cu-CuSO4 half-cell potential

(b) Corrosion current density

FIGURE 12: Cu-CuSO$_4$ half-cell potential (a) and corrosion current density (b) of reinforced steel in reference untreated and surface impregnated concrete by silane.

the corrosion potential was about −460 mV. According to the ASTM standard, this means the risk of corrosion was greater than 90% [32]. The corrosion current density was about 0.4~0.5 $\mu A/cm^2$ which means the steel reinforcement started to corrode, while, for water repellent surface treated concrete, both the electric potential and corrosion current density were kept much lower throughout the period measured. The risk of corrosion was maintained lower than 10% from the results of corrosion potential. From the results of corrosion current density, corrosion can be neglected. This shows that corrosion did not happen in water repellent treated

specimens. Therefore, corrosion activities can be reduced considerably by surface impregnation.

5. Conclusions

Based on the results presented herein, the following conclusions can be drawn.

(1) When the surface of water repellent treated concrete is in contact with water, there is no aqueous water penetration; but small amount of water vapor is still absorbed and condenses in the untreated pores of the material. Therefore,

the hydrophobic layer with several millimeters of thickness can reduce significantly water absorption into concrete.

(2) Water vapor, however, does not contribute to ion transport. If the pores of concrete are not water filled, ions diffusion is slowed down effectively. Hence surface impregnation with silane provides an effective chloride barrier. As a consequence, the service life of a concrete structure exposed to sea water or deicing salt can be extended.

(3) Carbonation depth of surface impregnated concrete can be reduced one half under the environmental RH of 70% compared to the untreated concrete.

(4) Surface impregnation with silanes also provides effective corrosion protection to reinforcing steel in concrete contacted with chloride solution. In order to prolong the service life of reinforced concrete structures, water repellent treatment can be taken into consideration to reduce the risk of steel corrosion, provided surface treatment is adequately maintained which can be achieved by appropriate application and deep impregnation (>6 mm) [33]. In addition, the durability of silane impregnation itself and its long-term residual protection has to be studied. In this sense, the efficiency of the protective measure should be controlled in regular intervals. Once the initial requirements are not fulfilled anymore, the treatment should be repeated.

Conflicts of Interest

The authors declare no conflicts of interest.

Acknowledgments

Financial support of ongoing projects by National Natural Science Foundation of China (51420105015, 51278260), Basic Research Program of China (2015CB655100), and 111 Project is gratefully acknowledged.

References

[1] H. S. Müller, M. Haist, and M. Vogel, "Assessment of the sustainability potential of concrete and concrete structures considering their environmental impact, performance and lifetime," *Construction and Building Materials*, vol. 67, pp. 321–337, 2014.

[2] U. M. Angst, R. D. Hooton, J. Marchand et al., "Present and future durability challenges for reinforced concrete structures," *Materials and Corrosion*, vol. 63, no. 12, pp. 1047–1051, 2012.

[3] H. Huang, G. Ye, C. Qian, and E. Schlangen, "Self-healing in cementitious materials: materials, methods and service conditions," *Materials & Design*, vol. 92, pp. 499–511, 2016.

[4] J. Vries, R. B. Polder, and H. Borsje, "Durability of hydrophobic treatment of concrete," in *Proceedings of the 2nd International Conference on Water Repellent Treatment of Building Materials*, pp. 77–90, Aedificatio Publishers, 1998.

[5] P. Zhang, Y. Cong, M. Vogel et al., "Steel reinforcement corrosion in concrete under combined actions: The role of freeze-thaw cycles, chloride ingress, and surface impregnation," *Construction and Building Materials*, vol. 148, pp. 113–121, 2017.

[6] P. Hou, X. Cheng, J. Qian, and S. P. Shah, "Effects and mechanisms of surface treatment of hardened cement-based materials with colloidal nanoSiO2 and its precursor," *Construction and Building Materials*, vol. 53, pp. 66–73, 2014.

[7] Y. Cai, P. Hou, C. Duan et al., "The use of tetraethyl orthosilicate silane (TEOS) for surface-treatment of hardened cement-based materials: A comparison study with normal treatment agents," *Construction and Building Materials*, vol. 117, pp. 144–151, 2016.

[8] C. Schröfl, V. Mechtcherine, A. Kaestner, P. Vontobel, J. Hovind, and E. Lehmann, "Transport of water through Strain-hardening Cement-based Composite (SHCC) applied on top of cracked reinforced concrete slabs with and without hydrophobization of cracks - Investigation by neutron radiography," *Construction and Building Materials*, vol. 76, pp. 70–86, 2015.

[9] J.-G. Dai, Y. Akira, F. H. Wittmann, H. Yokota, and P. Zhang, "Water repellent surface impregnation for extension of service life of reinforced concrete structures in marine environments: The role of cracks," *Cement and Concrete Composites*, vol. 32, no. 2, pp. 101–109, 2010.

[10] F. H. Wittmann, T. A. J. M. Siemes, and L. G. W. Verhoef, *Hydrophobe I*, TU Delft, Delft, The Netherlands, 1995.

[11] F. H. Wittmann and A. Gerdes, *Hydrophobe II*, ETH, Zürich, Switzerland, 1998.

[12] K. Littmann and A. E. Charola, *Hydrophobe III*, University of Hannover, Hanover, Germany, 2001.

[13] J. Silfwerbrand, "Hydrophobe IV," Stockholm, Sweden, 2005.

[14] H. De Clercq and A. E. Charola, "Hydrophobe V, KIK-IRPA," Brussels, Belgium, 2008.

[15] E. Borrelli and V. Fassina, "Hydrophobe VI," Rome, Italy, 2011.

[16] J-M. Mimoso, "Hydrophobe VII," Lisbon, Portugal, 2014.

[17] J. G. Dai, H. Yokota, and T. J. Zhao, "Hydrophobe VIII," Hong Kong, 2017.

[18] H. Kus, *Long term performance of water repellents on rendered autoclaved aerated concrete [Ph.D. Thesis]*, Royal Institute of Technology, Stockholm, Sweden, 2002.

[19] A. Gerdes and F. H. Wittmann, "Quantitation of hydrophobic mass by FT-IR spectroscopy," *Restor. Build. Monum*, vol. 5, pp. 201–210, 1999.

[20] *TC 14-CPV Test Method CPC 11.2 Absorption of Water by Concrete by Capillarity*, RILEM Publications, 1982.

[21] E. H. Lehmann, A. Kaestner, C. Grünzweig, D. Mannes, P. Vontobel, and S. Peetermans, "Materials research and nondestructive testing using neutron tomography methods," *International Journal of Materials Research*, vol. 105, no. 7, pp. 664–670, 2014.

[22] G. Frei, E. H. Lehmann, D. Mannes, and P. Boillat, "The neutron micro-tomography setup at PSI and its use for research purposes and engineering applications," *Nuclear Instruments and Methods in Physics Research, Section A: Accelerators, Spectrometers, Detectors and Associated Equipment*, vol. 605, no. 1-2, pp. 111–114, 2009.

[23] P. Zhang, F. H. Wittmann, T.-J. Zhao, E. H. Lehmann, and P. Vontobel, "Neutron radiography, a powerful method to determine time-dependent moisture distributions in concrete," *Nuclear Engineering and Design*, vol. 241, no. 12, pp. 4758–4766, 2011.

[24] P. Zhang, F. H. Wittmann, M. Vogel, H. S. Müller, and T. Zhao, "Influence of freeze-thaw cycles on capillary absorption and chloride penetration into concrete," *Cement and Concrete Research*, vol. 100, pp. 60–67, 2017.

[25] P. Zhang, Z. Liu, S. Han et al., "Visualization of rapid penetration of water into cracked cement mortar using neutron radiography," *Materials Letters*, vol. 195, pp. 1–4, 2017.

[26] P. Zhang, P. Wang, D. Hou, Z. Liu, M. Haist, and T. Zhao, "Application of neutron radiography in observing and quantifying the time-dependent moisture distributions in multi-cracked

cement-based composites," *Cement and Concrete Composites*, vol. 78, pp. 13–20, 2017.

[27] *GB/T 50082-2009, Standard for Test Methods of Long-term Performance and Durability of Ordinary Concrete*, Ministry of Construction, Beijing, China, 2009.

[28] *ASTM G 109-07 Standard Test Method for Determining the Effects of Chemical Admixtures on the Corrosion of Embedded Steel Reinforcement in Concrete Exposed to Chloride Environments*, American Society for Testing and Materials, Philadelphia, Pa, USA, 2013.

[29] C. Hall, "Barrier performance of concrete: a review of fluid transport theory," *Materials and Structures*, vol. 27, no. 5, pp. 291–306, 1994.

[30] D. A. Quenard, K. Xu, H. M. Künzel, D. P. Bentz, and N. S. Martys, "Microstructure and transport properties of porous building materials," *Materials and Structures*, vol. 31, no. 5, pp. 317–324, 1998.

[31] J. Heinrichs, S. Schmeiser, and A. Gerdes, "Numerical simulation of the influence of water repellent treatment on carbonation of concrete," in *Proceedings of the 4th International Conference on Water Repellent Treatment of Building Materials*, pp. 27–44, Aedificatio Publishers, Stockholm, Sweden, 2005.

[32] *ASTM C 876-15 Standard Test Method for Corrosion Potentials of Uncoated Reinforcing Steel in Concrete*, American Society for Testing and Materials, Philadelphia, Pa, USA, 2015.

[33] S. Meier and F. Wittmann, "Recommendations for water repellent surface impregnation of concrete," *Restoration of Buildings and Monuments*, vol. 17, no. 6, pp. 347–358, 2011.

20

Strength Characteristics and the Reaction Mechanism of Stone Powder Cement Tailings Backfill

Jianhua Hu ⓘ,[1] **Qifan Ren** ⓘ,[2] **Quan Jiang**,[2] **Rugao Gao** ⓘ,[1] **Long Zhang**,[2] **and Zhouquan Luo**[1]

[1]*Professor, School of Resources and Safety Engineering, Central South University, Changsha, Hunan 410081, China*
[2]*Research Assistant, School of Resources and Safety Engineering, Central South University, Changsha, Hunan 410081, China*

Correspondence should be addressed to Qifan Ren; qifanren@csu.edu.cn

Guest Editor: Syed W. Haider

Stone powder cement (SPC) is widely used as a novel cement substitute material in concrete for its good gelling performance and low cost. In order to reduce the backfilling cost and assess the potential of SPC backfilling materials, a series of experiments were conducted to analyze the strength and hydration reaction mechanism of stone powder cement tailings backfill (SPCTB). The analysis was based on SPC and tailings, which were used as the gelling agent and the aggregate, respectively. The results showed that the strength of the backfill was greatly reduced at an early stage and slightly reduced in the final stages. The stone powder content was less than 15%, which met the requirement of mining procedure. The addition of stone powder reduced the content of adsorbed water and capillary water in the early stages, while it increased in the middle stages. The SiO_2 contained in stone powder reacted with the hydration products at later stages, which is the reason why the growth of strength is rapid between the groups with the addition of stone powder. The addition of stone powder improved the microstructure of backfill and produced a denser three-dimensional (3D) network structure; however, the plane porosities of Groups A and B gradually increased with the increase in the content of stone powder. The cement powder mixed appropriately with the stone power could meet the strength requirement and reduce the cost of backfilling materials.

1. Introduction

When manufacturing crushed aggregate, the process creates a stone powder, which could be collected and used to produce SPC. The majority of particles of stone powder range from 1 μm to 100 μm, making them difficult to handle, transport, and recycle. The quantity of stone powder produced from crushed aggregate factories in China is approximately 10 million tons per year, most of which is dumped in soil. The disposal of stone powder is a major environmental problem, and therefore, there is a great interest to find solutions for its safe utilization [1]. Portland cement is used as a gelling agent in traditional backfill in mines, whereas the high cost of cement needs a suitable and sustainable low-cost substitute. In China, most of the mines are in mountainous areas surrounded by quarries where the stone powder is thrown out every year and causes a lot of

environmental pollution [2]. However, it can be collected and utilized in backfill. Figure 1 shows a quarry.

Al-Kheetan et al. [3, 4] introduced crystalline material along with a curing compound in fresh concrete to protect and extend its service life and developed hydrophobic concrete by adding dual-crystalline admixture during the mixing stage. Choi et al. [1] examined the microstructure and strength of alkali-activated systems using stone powder sludge, which had some water content as a replacement material in alkali-activated mixtures that strengthened the concrete's ability to withstand extreme variable temperatures and loads. Compared to ordinary concrete, other properties, including early gain in compressive strength, durability, and high acid and fire resistance, make it an appealing construction material. Compressive strengths of four different natural pozzolans with the replacement level of 10–25% at various ages were studied [5]. The results

FIGURE 1: Quarry around the mine.

showed that, with the increase in natural pozzolans content, natural pozzolans replacements reduced the compressive strengths of concrete due to reduction in cement content in the mixture. However the compressive strength increases with age. Several investigations [6, 7] showed that the compressive strength of mortars with different cements and incorporating 10% silica fume was about 30–50% higher than that of plain cements after 28 days. The strength of silica fume mortars depends on the water to binder ratio of the mixture. The compressive strength of mortars containing 10% silica fume decreases with the decrease in the fineness of parent Portland cement. However, it is always higher than the strength of plain Portland cement after 28 days. In addition, a previous study [7] showed the reduction of chloride penetration of silica fume mortars and concretes in rapid chloride penetration tests (RCPTS). However, there are a few materials used in mine backfilling, and the research on these materials is limited.

The results showed that the stone powder is not completely inert diluent for the cement. A chemical reaction between $CaCO_3$, C_3A, and C_4AF generates $C_3A \cdot X \cdot CaCO_3 \cdot 11H_2O$ and ettringite [8]. As the center of hydration process, stone powder can increase the early rate and degree of hydration [9]. In addition, SPC results in a smaller consumption of water than that of the cement of the same standard; when comparing the bleeding rate of SPC and ordinary cement, Albeck and Sutej [10–12] found that the bleeding rate of SPC is always less than that of the ordinary cement, and it also stops bleeding more quickly. According to El-Didamony et al. [13], the set time of cement decreases with the increase in the stone powder content, which means that the SPC has a high early strength. Consequently, as substituted minerals are added in the cement paste, stone powder can promote the hydration of cement, induce the crystallization of cement hydrate products, accelerate the hydration of cement, and participate in the hydration reactions [14]. Previous studies have found that the powder of silicate minerals could be used as sustainable replacements of cement. Abd Elmoaty et al. [15] used granite dust to replace cement with 5–15%, which could improve concrete's compressive strength and tensile strength. Kannan et al. [16] reported that high-performance concrete can be produced

with significant replacement of between 20% and 40% of Portland cement with ceramic waste powder. Berriel et al. and Akhlaghi et al. [17, 18] showed that a combination of calcined clay, limestone, and gypsum, used as the substitution of Portland cement for up to 50%, can provide economic benefits while maintaining the mechanical properties of the cement. However, if the stone powder content exceeds a certain value, the water demand of the cement paste will increase, and the fluidity will reduce. Therefore, in the current work, the possibility of reasonably using waste stone powder as a partial replacement for cement has been explored. The comprehensive use of stone powder having the same main ingredients as the mother rock conforms to the concept of green building materials and is conducive to sustainable developments in the construction industry [19, 20].

Based on the abovementioned properties of SPC, the experimental study of stone powder cement tailings backfill (SPCTB) can be developed. In this study, the physico-chemical properties and the particle-size distribution of stone powder and tailings were analyzed using X-ray fluorescence (XRF) and laser particle size analyzer (LPSA). In addition, the strength characteristics of backfill with different ratios were analyzed to evaluate the feasibility of SPC used as the cementitious material. The microscopic morphology, composition, water evolution, and pore changes of SPCTB at different ages were examined, and the reaction mechanism of SPCTB was studied using X-ray diffraction spectroscopy (XRD), scanning electron microscopy (SEM), and nuclear magnetic resonance (NMR). The use of SPC as the cementitious material not only reduces the backfilling costs for mountainous mines but also benefits the construction of "green mines" and achieves no-waste mining.

2. Experimental

2.1. Tailings, Stone Powder, and Cement. Tailings were obtained from Gaofeng mine in Guangxi Province, China, and were divided into two types (tailings A and tailings B). The types of tailings were produced by different beneficiation processes. The stone powder was obtained from the

FIGURE 2: Physical appearance of (a) tailings A, (b) tailings B, and (c) stone powder.

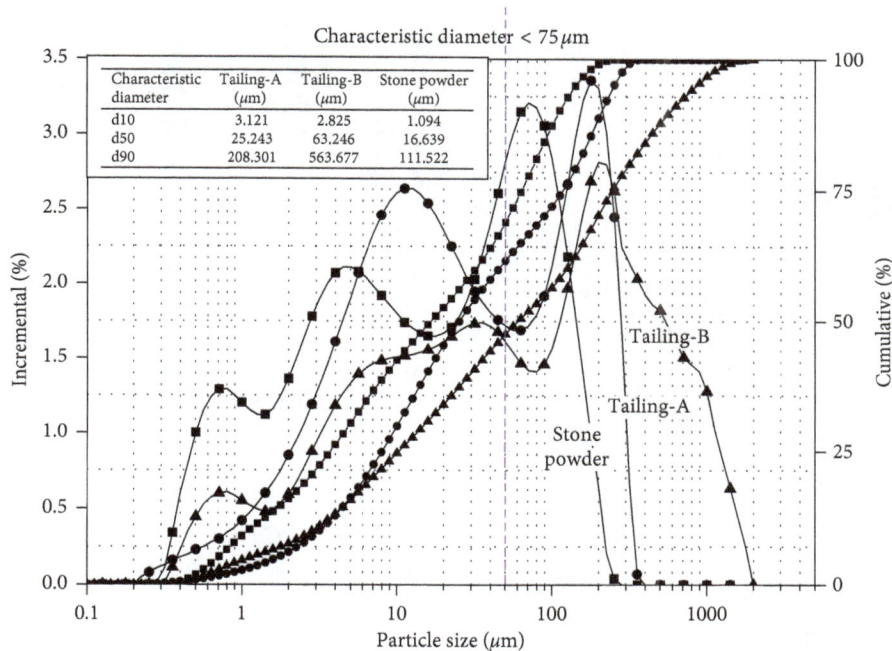

FIGURE 3: Particle-size distributions of tailings and stone powder (including incremental and cumulative values).

quarry around the Gaofeng mine (Figure 2). The particle-size distributions of the tailings and the stone powder were analyzed using a mastersizer 2000 obtained from Malvern Instruments Inc., UK (Figure 3). The characteristic median diameters d_{50} of tailings A and tailings B were 25.243 μm and 31.168 μm, respectively, based on which the tailings could be considered as ultrafine tailings. The median diameter d_{50} of the stone powder was 16.639 μm due to which it belonged to the category of ultrafine particles. The nonuniformity coefficient (Cu) and the curvature coefficient (Cc) of tailings A and tailings B were (14.56, 0.72) and (40.02, 0.71), respectively, indicating that the gradation of tailings A was good, though the tailings B had a wider range.

The apparent density, the bulk density, and the contents of surface water of tailings and stone powder were analyzed according to the standard *GB/T 50080-2016* "Standard Test Method for Performance of Ordinary Concrete Mixtures." The corresponding results are presented in Table 1. The elemental and chemical compositions of tailings and stone powder were obtained using XRF and XRD, and the respective results are provided in Table 2 and Figure 4. According to the chemical composition index [21], the alkaline coefficients of tailings A, tailings B, and stone powders were 3.21, 0.53, and 1.53, respectively. These results showed that tailings A belonged to the alkaline tailings and tailings B belonged to the acid tailings.

The cement from the Xinxing cement factory in Changsha, China, meets the national standard *GB 175-2007* "Common Portland Cement." The mineral composition of the cement is provided in Table 2. Furthermore, tap water was used for the experiments.

TABLE 1: Physical properties of the tailings and the stone powder.

Class	Apparent density	Packing density	Surface moisture content
Tailings A	3.49	1.24	0.128
Tailings B	2.77	1.17	0.135
Stone powder	2.89	0.99	0.162

TABLE 2: Elemental composition of the tailings and the stone powder along with the chemical composition of Portland cement (%).

Elemental composition	O	Fe	S	Ca	Si	Mg	Al
Tailings A	34.699	23.098	15.857	14.025	4.476	2.117	1.113
Tailings B	44.376	12.662	8.877	9.091	16.24	0.403	3.057
Stone powder	58.301	0.942	0.298	23.279	14.901	0.471	1.273
Chemical composition	$3CaO \cdot SiO_2$		$2CaO \cdot SiO_2$		$3CaO \cdot Al_2O_3$	$4CaO \cdot Al_2O_3 \cdot Fe_2O_3$	
Portland cement	52.8		20.7		11.5	8.8	

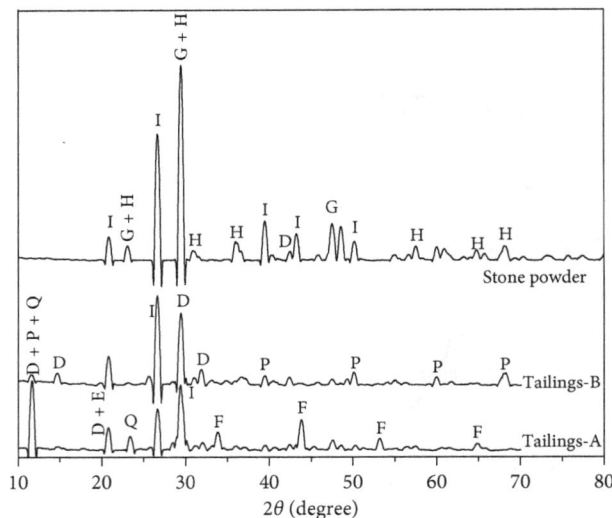

D—gypsum E—brushite
F—pyrrhotite G—magnesium calcite
H—calcite I—SiO_2
P—silicon sulfide Q—hydrotalcite

FIGURE 4: XRD patterns of the tailings and stone powder.

2.2. Specimen Preparation. In this work, SPC and the tailings were used as the cementitious material and the aggregate, respectively. The ratios between the two materials, which were tested in this work, are presented in Table 3. The SPC tailings ratio '(SPCTR), the mass fractions, and the stone powder dosage of Group A using tailings A and those of Group B using tailings B were the same.

The SPCTB components, including stone powder, cement, tailings, and water, were weighted using a high-precision electronic scale having an accuracy of 0.01 g. Mixtures with 10 different formulations (mixing ratios) were mixed in the laboratory blenders for 5 min to ensure homogeneity. Then, the mixtures were casted in plastic cubical molds with the side length of 7.07 cm. A total of 180 sextuplicate specimens (three for UCT and three for NMR) were prepared (Figure 5). The specimens were cured in a humidity chamber at 20°C and 95% relative humidity for different

TABLE 3: Backfilling slurry of stone powder cement tailings ratio (SPCTR), mass fractions, stone powder content, and cement content.

Group	SPCTR	Mass fractions (%)	Stone powder content[a] (%)	Cement content (%)
A/B 1			0	100
A/B 2			10	90
A/B 3	1 : 4	70	15	85
A/B 4			20	80
A/B 5			100	0

[a]The proportion of stone powder in SPC.

durations (3, 7, and 28 days) until the prescribed age reached. After that, the samples were analyzed for their physicochemical properties.

2.3. Uniaxial Compressive Tests. The uniaxial compressive strength (UCS) at a given time is the most important parameter to evaluate the mechanical performance of SPC. Until the predetermined curing time, the SPCTB specimens were analyzed for USC using a computer-controlled universal pressure mechanical device (WDW-2000). The tests were performed following the procedure given in the standard *ASTM D2166/D2166M-16*. The specimens were loaded under a constant vertical displacement rate of 1 mm/min. To reduce the error, the tests were conducted in triplicates, and the average values were used to determine the UCS of SPCTB (accurate to 0.1 MPa) (Figure 6). The individual strengths of three specimens, molded with the same characteristics, should not deviate more than 15% from the mean strength.

2.4. NMR Analysis. During the hardening of backfilling slurry, the water, pore distribution, and strength of the backfilling slurry would change with age, and backfill is formed after hardening. The UCS of the backfill is closely related to the water content in different binding states. The NMR, performed using 1 H relaxation signal, can be used to

FIGURE 5: Production process for the backfill (including metering water and backfilling paste).

FIGURE 6: UCT and destruction process of a 28-day SPCTB specimen containing 10% stone powder.

measure the porous content in the slurry and the backfill. The T2 distribution can be used to analyze the variation in pore distribution, which has the advantages of non-destructive detection and good repeatability [22].

The NMR tests used the MiniMR-60 magnetic resonance imaging (MRI) analysis system, which was manufactured by Shanghai Newmai Co. Ltd., China. The main magnetic field of the device was 0.51 T, and the H proton resonance frequency was 21.7 MHz.

The analysis was performed in triplicates, and the average value for further analysis was calculated. After the SPCTB specimens reached the specified age (3 d, 7 d and 28 d), the SPCTB specimens were taken out from the constant temperature and constant humidity curing box and subjected to NMR relaxation measurements. During the tests, the SPCTB specimens were wrapped in a cling film to reduce the impact of water evaporation.

2.5. SEM and XRD Analyses. At the curing times of 3 d, 7 d, and 28 d, the SPCTB specimen of 1 mm^2 area was taken from the core of the specimen and dehydrated with absolute ethanol to stop the hydration. The specimen was dried at 45°C to constant weight and analyzed using XRD and SEM analyses. The experiments used a Siemens D500 X-ray diffractometer and a TESCAN MIRA3 field-emission scanning electron microscope for these analyses.

3. Strength Characteristics

The (statistical) average strength of each group of specimens and part of the stress-strain curves of these specimens are shown in Figures 7 and 8, respectively.

(1) At the curing time of 3 days (d), the UCS of Group A was 0.4-0.5 MPa, while that of the Group B was

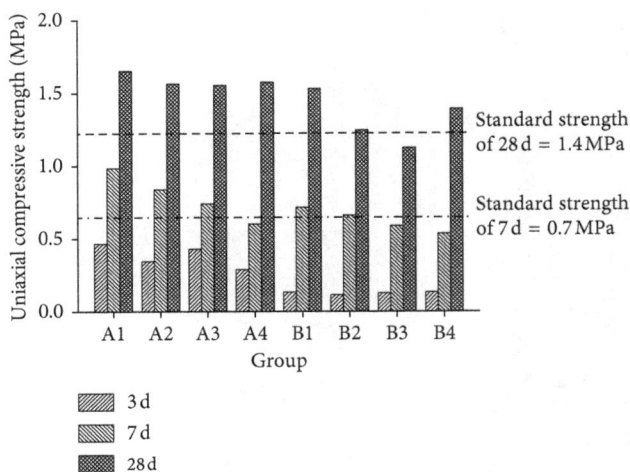

FIGURE 7: Uniaxial compressive strength of each group of specimens at the curing times of 3 d, 7 d, and 28 d (according to the mine information 7 d and 28 d compressive strengths need to reach 0.7 MPa and 1.4 MPa, respectively. A5 and B5 groups are not analyzed for compressive strength).

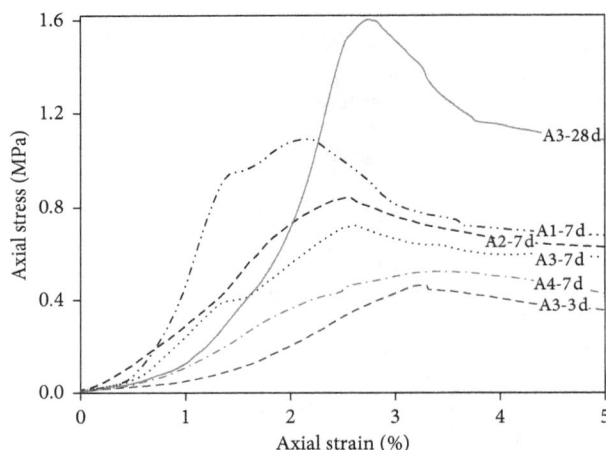

FIGURE 8: Stress-strain curves at the curing time of 7 d of Group A and at the curing times of 3 d and 28 d of Group A3.

0.2 MPa. At the curing time of 7 d, the USC of Group A was 0.6–1.0 MPa, whereas that of Group B was 0.6–0.7 MPa, which gradually decreased with the increase in the content of stone powder. At the curing time of 28 d, the UCS of Group A was 1.6–1.7 MPa, while the USC of SPCTB specimens mixed with the stone powder was basically the same as 1.6 MPa. The USC of Group B was 1.2–1.6 MPa, which tended to decrease with the increase in stone powder's USC. In addition, the USC suddenly increased to 1.5 MPa at the stone powder content of 20%.

(2) The minimum requirements for the backfill strength of mine stopes at the curing times of 3 d, 7 d, and 28 d were 0.25 MPa, 0.7 MPa, and 1.4 MPa, respectively. Group A met the 3 d backfill strength requirements, while the strength's value for Group B was too low. The content of stone powder in Group A was varied to

values of 0%, 10%, and 15%, while Group B1, which was not doped with stone powder, satisfied the requirements of 7 d backfill strength. While for the curing time of 28 d, only Groups B3 and B4 did not meet the backfill strength requirements. Furthermore, Groups A1, A2, and A3 met the requirements of the backfill strength with less than 15% of the stone powder content. The backfill strength of groups that met the requirements had a certain safety factor, which could be changed according to the actual conditions of the stopes and was based on the minimum requirements of the mine. The ratio of the backfill material and the content of the stone powder were adaptable to the environmental parameters of the stopes, which ensured that safe and efficient recovery of backfill material ratio parameters can be selected for the mine.

(3) The intensity ratio after the curing times of 3 d, 7 d, and 28 d for Group A gradually changed from 1 : 2 : 4 to 1 : 2 : 6 with the increase in the content of stone powder. After the stone powder was mixed, the UCS of the SPCTB specimens was reduced, which increased rapidly after 7 days. Additionally, the more the powder added, the faster the increase in the USC value.

(4) With the increase in the content of stone powder, the peak strength and the elastic modulus of the specimen gradually decreased. In addition, the peak strength appeared under the condition of larger strain. The strength and the elastic modulus of the specimens after the curing time of 3 d were low, and the plastic deformation of the specimens was large. The elastic modulus of Group A3 after the curing time of 28 d was lower than that of Group A1 after 7 d. The increase in the content of stone powder reduced the elastic modulus of the backfill. The residual strengths of the six groups of backfill specimens in the postfailure were large and decreased slowly and gradually to stable values, which ensured that the backfill had sufficient strength to keep the backfill and stopes stable.

4. Results and Discussion

4.1. Various States of Bound Water. With the hydration of backfilling slurry, a part of the water in the slurry did not participate in the hydration reaction, which was either secreted or evaporated into the air. The rest remained in the backfill. Water in the hydration backfilling slurry was divided into free water, capillary water, adsorbed water (physical adsorption through hydrogen bonding), interlayer water, and chemically bound water. The loss of water gradually increased as the fluidity of the water gradually deteriorated.

After the NMR sampling, the T2 distribution of the slurry, which was obtained by T2 inversion software, is shown in Figures 9(a) and 10(a). In these figures, the horizontal axis is the relaxation time T2, while the vertical axis is the signal intensity. The peaks from left to right are defined as Peak 1, Peak 2, and Peak 3. Previous relevant studies have shown that different bound states have different T2

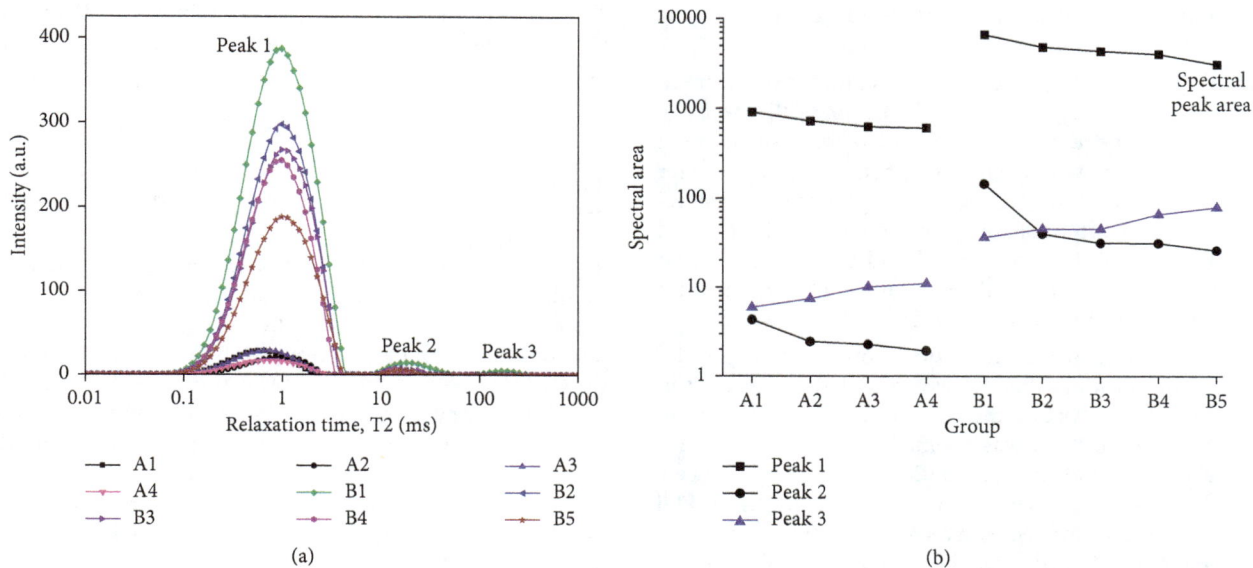

FIGURE 9: (a) NMR T2 spectrum distribution and (b) areas of Peak 1, Peak 2, and Peak 3 after the curing time of 3 d.

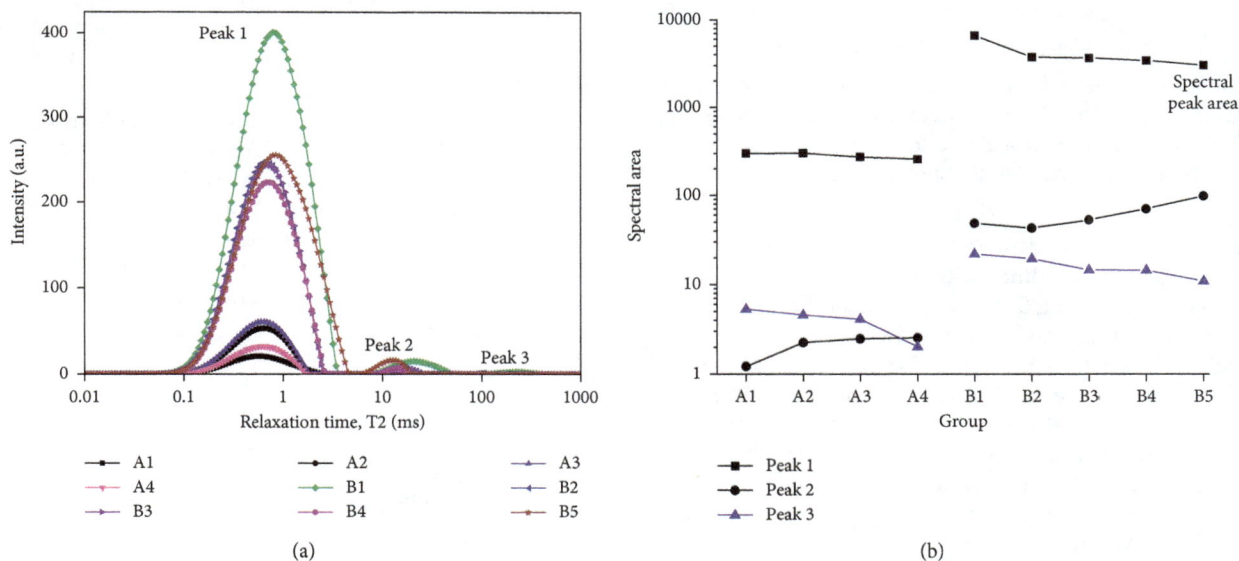

FIGURE 10: (a) NMR T2 spectrum distribution and (b) areas of Peak 1, Peak 2, and Peak 3 after the curing time of 7 d.

distributions of water. Each peak represents a particular bound state of water. The tighter the water bound is, the worse mobility the water has and the shorter the relaxation time T2 is [23–25]. Since the time of relaxation of chemisorbed water in the slurry and interlaminar water is short and the Carr–Purcell–Meiboom–Gill (CPMG) experiment cannot acquire the NMR signal, Peak 1 could be considered as the relaxation signal generated by the adsorbed water. Peak 2 was generated by the capillary water, and Peak 3 was generated by the free water. The relaxation time represented the various states of bound water, whereas the signal intensity represented the content of water and the spectral area represented the content of water of the same type.

(1) The relaxation times of Peak 1 of Groups A and B were nearly the same, followed by those of Peak 2 and Peak 3. In the T2 spectrum, the amplitudes, the signal peak intensities, and the peak areas of Peak 1 and Peak 2 in Group B were larger than those in Group A, which indicated that there were more adsorbed water and capillary water in Group B. With the increase in the curing time, the relaxation time did not change significantly, and the peak amplitude decreased, indicating that the contents of various types of water in the backfill specimen decreased with the increase in curing time. In Groups A and B, the proportion of Peak 1 was more than 97%,

indicating that the majority of adsorbed water was contained in the backfill specimen.

(2) After the curing time of 3 d, the contents of adsorbed water and capillary water in the backfill specimen gradually decreased with the increase in the content of stone powder, while the content of free water increased. Under the same proportioning parameters, the content of bound water in Group B exceeded than that in Group A, which indicated that different tailings had some influence on the change in water content in the backfill specimen.

(3) Since some free water adsorbed on the stone powder particles (SPPs) after 3 d, which caused the decrease in capillary water and the content of adsorbed water in the system, there was no medium for hydration reaction and the rate of hydration reaction was reduced. The adsorbed water and the free water in the backfill specimen decreased with the increase in the content of stone powder, whereas the content of capillary water increased as well. However, with the hydration reaction going on, the free water adsorbed on SPP gradually entered the pores and became the medium of hydration reaction. This was also the reason why the strength of SPCTB increased significantly in the later stage of the addition of stone powder.

The three peak relaxation time ranges and the corresponding water categories are provided in Table 4. Each group NMR T2 spectrum distribution and the peak areas after the curing times of 3 d and 7 d are shown in Figures 9 and 10.

4.2. Analysis of the Pore Structure. The microstructure of the cross section of the specimen with different proportions after 28 d was obtained using SEM with the magnification of 5000 times. Figures 11(a) and 11(b) show the two micromorphologies of Groups A4 and B4. The results show that the gelled structure was fairly compact, though there were some pores around the particles. As can be seen, the structural stability was good when the particles were densely packed. The cementing mesh structure was compact, and the pores were small. The gray areas in SEM images were small, based on which the variation in plane porosity in different proportions can be observed [26, 27].

The images were binarized to ensure the preciseness of the pore results. The process is also known as the threshold segmentation. In view of the differences between the grayscale distribution and the values of the pixels of the particles, the cementing structure, and the pores, the pores in the image could be calculated using the binarized image. Assuming that the size of the SEM image was $M \times N$, $f(x)$ represented the gray area of the pixel located in the line $(x-1)$ and column $(y-1)$ of the image. The principle of binarization of the SEM image is given as follows:

$$f(x, y) = \begin{cases} 0, & f(x, y) < T, \\ 1, & f(x, y) > T, \end{cases} \quad (1)$$

where T represents the threshold grayscale. The SEM image can be binarized and converted to a black-and-white image,

TABLE 4: Relaxation time range of Peak 1, Peak 2, and Peak 3 and the corresponding bound state water.

Peak	Peak 1	Peak 2	Peak 3
Relaxation time range, T2 (ms)	0.3~3	7~60	78~310
Bound-state water	Adsorbed water	Capillary water	Free water

which is represented by a matrix of black-and-white pixels, in which 0 is the white pixel and 1 is the black pixel, and represents the pores and particles, respectively. With the help of Image-Pro Plus software, and for the threshold of 35, the SEM image of every group was processed (Figures 11(c) and 11(d)). The number and the area of pores contained within each group were calculated, and the plane porosity was determined. Each image consisted of four sets of data in four different positions to avoid any accidental error. The plane porosities of Groups A and B were linearly fitted and are shown in Figure 12.

(1) Fitting curves of Groups A and B are represented by the equations: $Y = 0.1594X + 2.033$ and $Y = 0.8894X + 8.178$, respectively. It can be seen that the plane porosities of Groups A and B gradually increased with the increase in the content of stone powder, which led to the decrease in the intensity of the SPCTB specimen.

(2) For the same ratio, the plane porosity of Group B was larger than that of the Group A. Because of this, not only the stone powder but also the nature of tailings will have an effect on the change in pores. The plane porosity of Group B increased faster, indicating that the addition of stone powder exhibited a greater influence on the plane porosity of SPCTB made up of tailings B.

4.3. Analysis of the Hydration Products. The results for the component analysis (done using XRD) are shown in Figure 13. The figure shows the hydration products of backfill specimens after different curing times. Based upon the results, the following conclusions can be drawn:

(1) There was no hydration product in the backfill in the absence of cement, indicating that the stone powder was in an inert state, and therefore, the hydration reaction did not occur. Gypsum existed throughout the hydration process, while in CTB and SPCTB, gypsum almost disappeared after the curing time of 28 d, indicating that the gypsum in tailings and the cement were involved in the hydration reaction.

(2) Some C-S-H gel dispersions and hydroxide diffraction peaks appeared in the SPCTB and CTB after the curing time of 3 d, though they were not very obvious. With the increase in the hydration time, some dispersive peaks with low intensity and discrete diffraction angles appeared after the curing time of 28 days. These were ascribed to C-S-H dispersion. At the same time, there was an increase in the diffraction

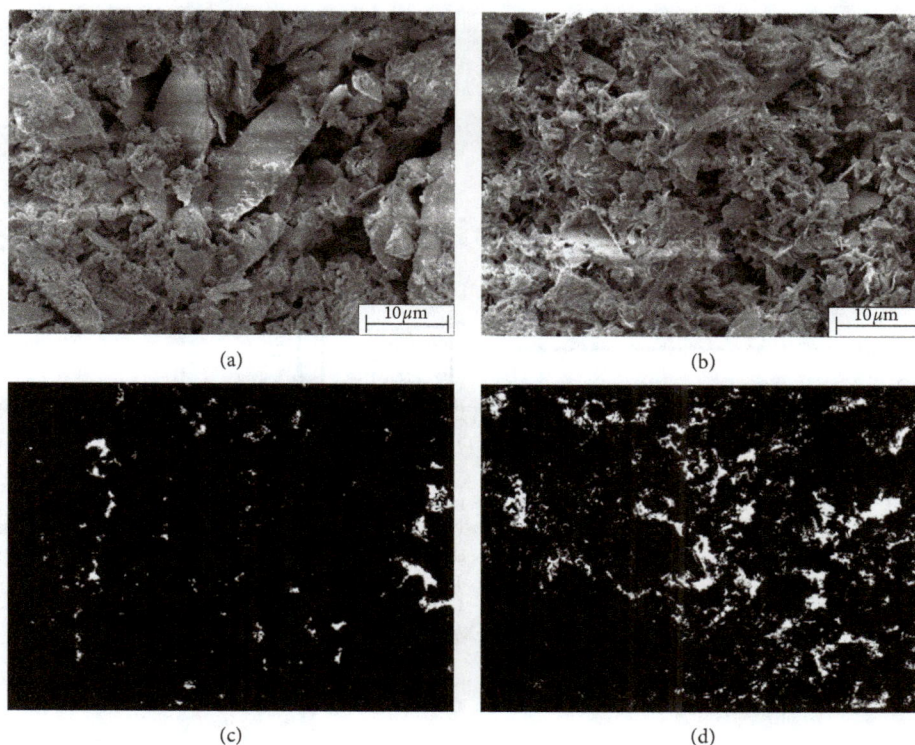

FIGURE 11: Microtopography and its binarized image; (a) magnified SEM image of Group A4; (b) magnified SEM image of Group B4; (c) Group A4 after binarization; (d) Group B4 after binarization.

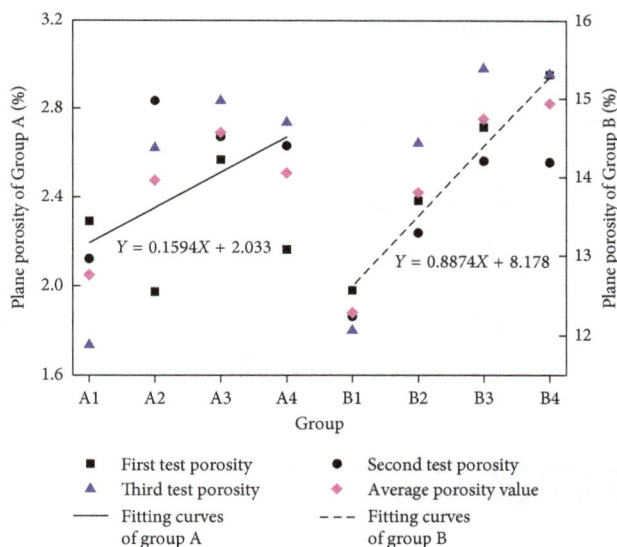

FIGURE 12: Plane porosity of each group and the fitting curves of the plane porosity of Groups A and B after the curing time of 28 d.

peak of the hydroxide, which indicated that the hydration reaction continued with the increase in curing time. AFt and some AFm in hydration products after the curing time of 3 d gradually decreased after 7 d and 28 d, which indicated that the hydration products at the early stage were mainly AFt, while at the later stage, they were the C-S-H gel.

(3) Comparing Figure 13(a) with Figure 13(b), it can be seen that the intensities of AFt and AFm diffraction peaks in SPCTB were weaker at 3 d, while AFt was the main reason for the early strength of backfill. Therefore, the early strength of SPCTB would be lower. At 28 d, $Ca(OH)_2$ diffraction peaks were weaker and C-S-H gel dispersion peaks were basically the same, indicating that the stone powder consumed $Ca(OH)_2$ to generate C-S-H gel after the hydration reaction, which was more stable than AFt. This ensured the poststrength and volume stability of SPCTB.

4.4. Analysis of the Micromorphology. Figure 14 shows the microstructural and micromorphology features of different cementitious materials at different curing times. Based upon the results, the following conclusions are drawn.

(1) For the curing time of 7 d, a large amount of hydration products, including a large amount of acicular AFt crystals, some amorphous C-S-H gel, some stone powder, and tailings particles were formed in the CTB and SPCTB. Acicular AFt crystals grew on the surface and in the pores of the particles, which developed into a three-dimensional (3D) network structure that was covered with SPP and tailings particles, having some early strength. AFt of the SPCTB acicular crystal grew sparser, though the microstructure was more compact than the CTB. This was due to the reason that the

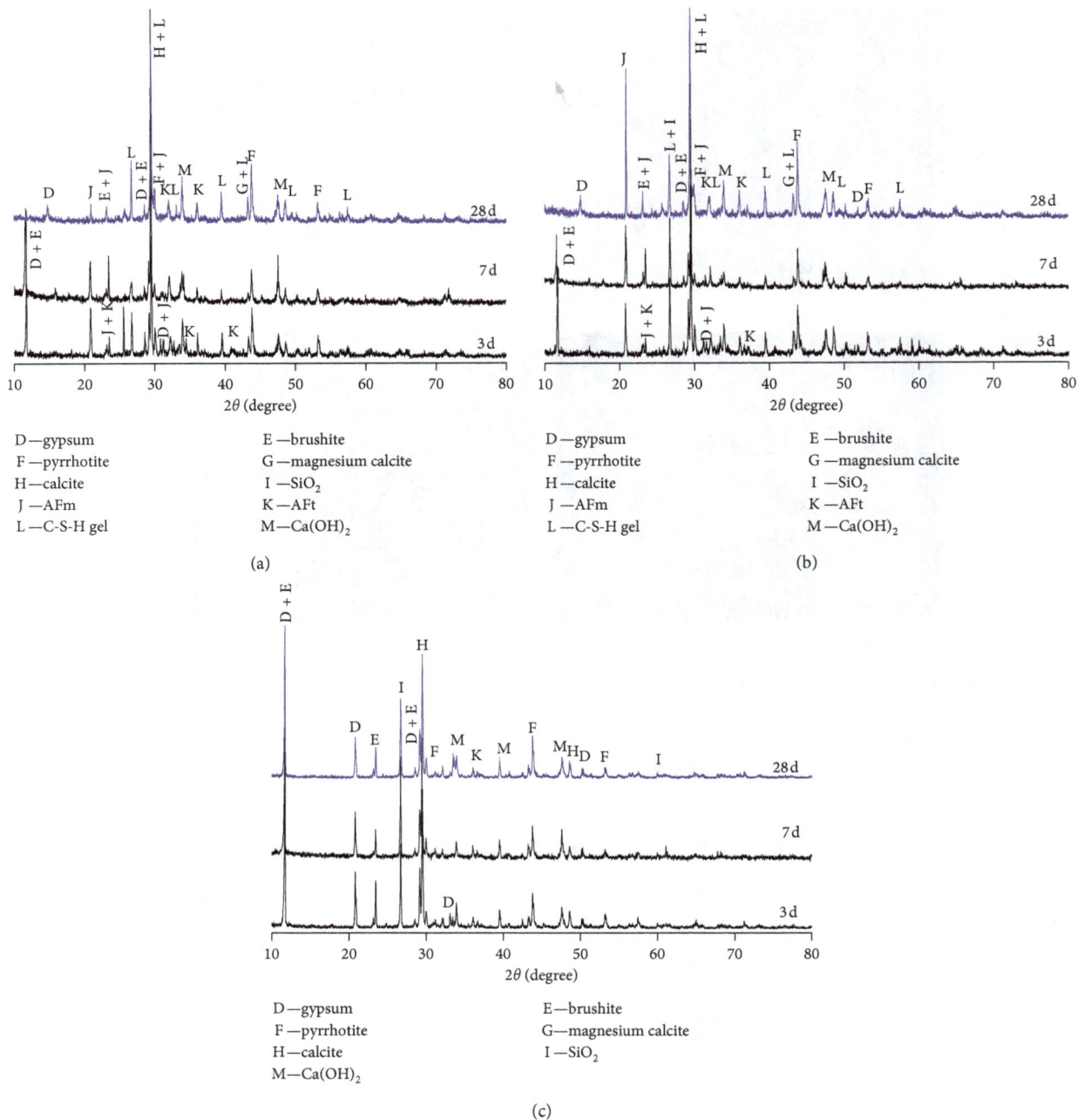

FIGURE 13: XRD patterns of the hydration products; (a) cement tailings backfill (CTB); (b) stone powder cement tailings backfill; (c) stone powder tailings backfill (SPTB). AFt represents ettringite; AFm represents monosulfate hydrate.

shape of SPP can fill the pores formed by the rod-shaped tailings particles.

(2) For the curing time of 28 d, the AFt crystals of acicular structure almost disappeared. The amorphous C-S-H gel grew on the surface of tailing particles and closely combined with it. Compared with the results for the curing time of 7 d, the microstructure was more compact. The strength of backfill and the other aspects were improved for amorphous and rod-like C-S-H gel structure, which grew in the original pores and was tightly combined with the tailings particles. The SPP

became obscure, while the smooth surface became uneven, indicating that the stone powder reacted with the hydration products to generate secondary products. The growth of a rod-like structure in the pores and the overall absence of obvious pores indicated that the microstructure was basically similar to that of the CTB.

4.5. Reaction Mechanism of SPC in Backfill. According to the different times and the characteristics of hydration between the SPCTB and CTB, the hydration process of backfilling material system can be divided into following four stages:

FIGURE 14: SEM image of the hydration product in the backfill; (a) CTB at 7 d; (b) SPCTB at 7 d; (c) CTB at 28 d; (d) SPCTB at 28 d.

dissolution period, condensation period, infiltration period, and hardening period. Figure 15 shows the model of hydration process of SPCTB.

(1) *Dissolution Period.* One mechanism [28] of hydration of Portland cement shows that the through-solution hydration involves the dissolution of anhydrous compounds into their ionic constituents, which results in the formation of hydrates in the solution. This phenomenon happens due to their low solubility and eventual precipitation of the hydrates from the supersaturated solution. From SEM images of 3D hydrating cement pastes (Figure 15), it appears that the through-solution mechanism is dominant in the early stages of cement hydration. The hydration reactions are most violent during the dissolution period. C_3A and C_4AF in the cement particles dissolved first and produced large amounts of $[AlO_4]^-$, $[SO_4]^{2-}$, $[Ca]^{2+}$, and $[OH]^-$ ions. However, both SPP and tailings dissolved in water containing these ions. Water acted as a reaction medium in the violent hydration reaction of these ions.

(2) *Condensation Period.* Depending on the concentration of aluminate and sulfate ions in the solution, the precipitating crystalline product is either calcium aluminate trisulfate hydrate or calcium aluminate monosulfate hydrate. In solutions saturated with calcium and hydroxyl ions, the former crystallizes as short prismatic needles and is also referred to as high sulfate or by its mineralogical name, ettringite (AFt crystals). The monosulfate is also called low sulfate and crystallizes (AFm crystals) as thin hexagonal plates. The relevant chemical reactions may be expressed using the reaction equations [28].

Ettringite:

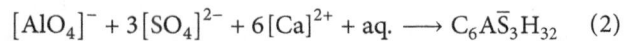

$$[AlO_4]^- + 3[SO_4]^{2-} + 6[Ca]^{2+} + aq. \longrightarrow C_6A\bar{S}_3H_{32} \quad (2)$$

Monosulfate:

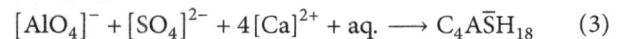

$$[AlO_4]^- + [SO_4]^{2-} + 4[Ca]^{2+} + aq. \longrightarrow C_4A\bar{S}H_{18} \quad (3)$$

The hydration reaction produced a large number of acicular AFt crystals and hexagonal plate-shaped AFm crystals. AFt is the main cause of the early strength of the specimens and represents the generation of AF. It is also the beginning of coagulation period and reaches the final coagulation state until the complete formation of AFt crystals. At the same time, the hydration reaction will also generate a part of amorphous C-S-H gel, which is attached to the surface of tailings particles and SPP.

(3) *Infiltration Period.* At later ages of hydration reaction, when the ionic mobility in the solution becomes restricted, the hydration of residual

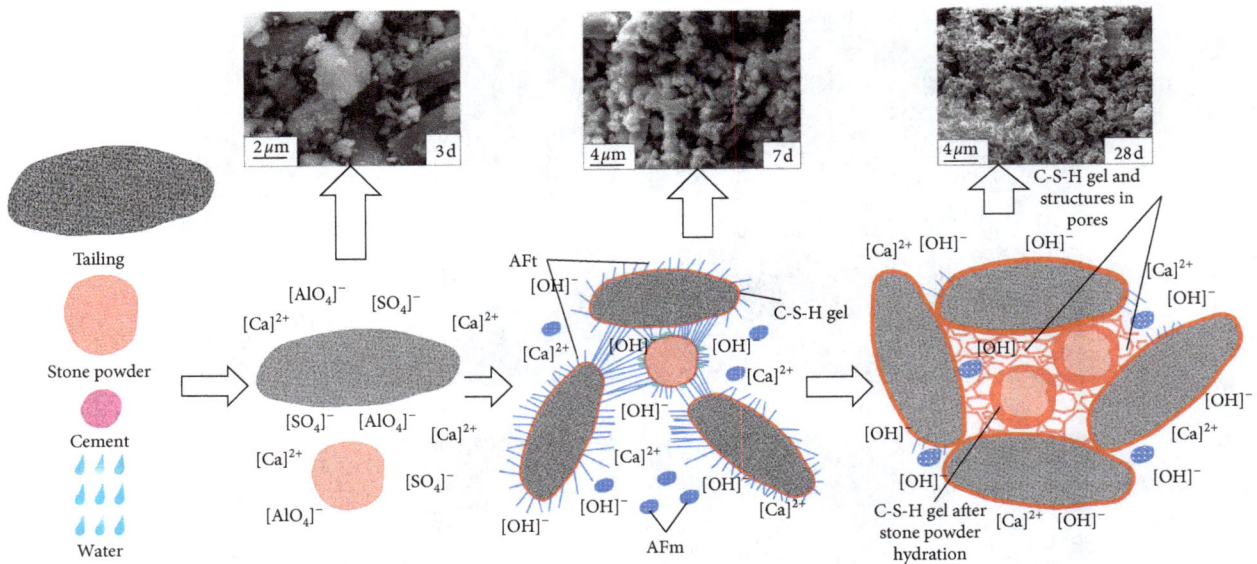

FIGURE 15: Model for the SPCTB hydration process.

cement particle may occur through solid-state reactions. Meanwhile, stone powder acts as nucleation sites for hydration products [29]. It is not surprising that the inclusion of stone powder increases the rate of hydration. At the same time, according to Section 4.3 of hydration products, the proportion of SiO_2 decreased in 7 d and 28 d, which means that SiO_2 reacted with hydration products. In this period, C_3S and C_2S in cement dissolve in water in large quantities and start the hydration reaction. A large amount of C-S-H gel, $[Ca]^{2+}$, and $[OH]^-$ are formed to produce the strong alkaline environment. AFt crystals and AFm crystals formed in the coagulation period under the strong alkaline environment rapidly dissolve and are converted into more stable C-S-H gel. The unstable structure of the surface of stone powder will be destroyed by the strong alkaline environment, and the strong alkali and active SiO_2 of stone powder react to generate C-S-H gel covering the surface of the stone powder. When the osmotic pressure and other factors are not sufficient to drive the continued inward layer reaction, the hydration reaction on the surface of the stone powder would stop. Figure 16 shows the relative amount of hydration products during hydration age.

(4) *Hardening Period.* According to the micromorphology analysis of 28 d, the products of C-S-H gel in the pores and on the surface of the stone powder are gradually formed, whereas the C-S-H gel of every part gradually contacts with other parts to form a stable 3D network-like gel structure, which tends to be stable. In this case, the reaction medium is the little amounts of capillary water and free water. Afterwards, the hydration reaction slows down, and the hydration reaction time continues to increase even for several years.

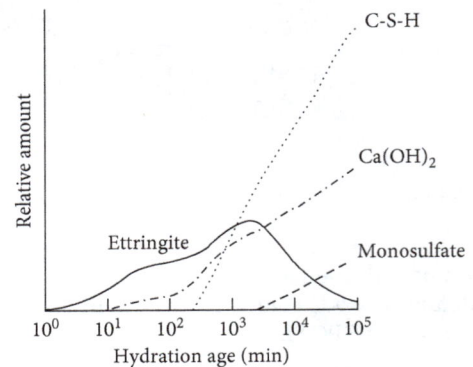

FIGURE 16: Relative amount of different hydration products during hydration age.

During the condensation period, the amount of the cement SPP system is relatively small, and many parts are needed to be cemented. The existing infiltration period slows down the rate of hydration reaction, resulting in low early strength of the SPC system. In the later stage of the infiltration period and the hardening period, the stone powder begins to participate in the hydration reaction, and the strength of the SPC system increases rapidly. At present, the strength has been analyzed only for the curing time of 28 d, while the increasing trend of strength even after 28 d has not been studied. This part of the experiment needs to be studied in a future work to further discuss the reaction mechanism of the SPC in backfill.

5. Conclusions

In this paper, the backfill specimens consisting of stone powder, cement, and tailings were analyzed to study whether the strength of SPCTB meets the requirements of mine backfilling or not. The reaction mechanism of SPC in backfilling slurry hardening process has also been analyzed. The analysis was conducted using techniques, including uniaxial compressive tests (UCTs), nuclear magnetic resonance (NMR), scanning electron microscopy (SEM), and X-ray diffraction (XRD). The main conclusions are as follows:

(1) With the increase in the stone powder content, the strength of each age gradually decreases, whereas different types of tailings also affect the strength of backfill specimens. The strength of Group A with the content of stone powder less than 15% meets the requirements of mine backfilling strength, and therefore, it is more practical for potential applications in mines. The incorporation of stone powder will reduce the early strength of the backfill specimens, while it promotes the strength growth during the hydration reaction in the middle and later stages.

(2) The incorporation of the stone powder will affect the proportion of all kinds of bound states of water in the backfill specimens. With the increase in the stone powder content, the adsorbed water decreases gradually, while the capillary water increases at 3 d, which reduces the rate of hydration reaction. However, it decreases for 7 d period, which results from the free water adsorbed on SPP that gradually enters the pores and becomes the medium of hydration reaction. This was also the reason why the strength of SPCTB increased significantly in the later stage of the addition of stone powder.

(3) The incorporation of stone powder will increase the plane porosity of backfill specimens for 28 d curing period, which will lead to the decrease of strength. Different types of tailings will affect the plane porosity. The backfill specimens consisting of tailings B have larger plane porosity.

(4) When only the stone powder is used as the cementitious material, the hydration reaction would not occur, while the hydration reaction occurred in Groups 2, 3, and 4. The results showed that SiO_2 contained in stone powder reacted with hydration products at later stages of hydration reaction. The hydration products of SPCTB were mainly AFt at the early stages and amorphous C-S-H gel in the final stage. The stone powder particles disappeared in the final stage, indicating that the stone powder reacted with the cement hydration product in the middle and later stages. Additionally, it participates in the hydration reaction, which further promoted the increase in strength of the backfill specimens. Meanwhile, the C-S-H gel and the stone powder hydration products are closely combined, forming a close three-dimensional mesh structure.

(5) The hydration process of the backfilling material system can be divided into following four stages: dissolution period, condensation period, infiltration period, and hardening period. At the third and fourth stages of hydration reaction, the stone powder acts as nucleation sites for the hydration products. At the same time, the unstable structure of the surface of stone powder will be destroyed by the strong alkaline environment, and the strong alkali and active SiO_2 of stone powder react to generate C-S-H gel covering the surface of the stone powder. When the osmotic pressure and other factors are not sufficient to drive the continued inward layer reaction, the hydration reaction on the surface of the stone powder would stop.

Conflicts of Interest

The authors declare that there are no conflicts of interest regarding the publication of this paper.

Acknowledgments

This research was supported by the National Key Research and Development Program of China (2017YFC0602901), National Key Technology Support Program (2015BAB12B01), and National Natural Science Fund (41672298) funded by Ministry of Science and Technology of the People's Republic of China and the Postgraduate Research and Innovation Foundation (2018zzts769) funded by Central South University.

References

[1] S. J. Choi, S. S. Jun, J. E. Oh, and P. J. M. Monteiro, "Properties of alkali-activated systems with stone powder sludge," *Journal of Material Cycles and Waste Management*, vol. 12, no. 4, pp. 275–282, 2010.

[2] G. Marchetti, V. F. Rahhal, and E. F. Irassar, "Influence of packing density and water film thickness on early-age properties of cement pasteswith limestone filler and metakaolin," *Materials and Structures*, vol. 50, no. 2, p. 111, 2017.

[3] M. J. Al-Kheetan, M. M. Rahman, and D. A. Chamberlain, "A novel approach of introducing crystalline protection material and curing agent in fresh concrete for enhancing hydrophobicity," *Construction and Building Materials*, vol. 160, pp. 644–652, 2018.

[4] M. J. Al-Kheetan, M. M. Rahman, and D. A. Chamberlain, "Development of hydrophobic concrete by adding dual-crystalline admixture at mixing stage," *Structural Concrete*, vol. 19, no. 5, pp. 1504–1511, 2018.

[5] A. A. Ramezanianpour, S. S. Mirvalad, E. Aramun, and M. Peidayesh, "Effect of four Iranian natural pozzolans on concrete durability against chloride penetration and sulfate attack," in *Proceedings of Second International Conference on Sustainable Construction Materials and Technologies*, Ancona, Italy, June 2010.

[6] A. Kumar and D. M. Roy, "A study of silica fume modified cements of varied fineness," *Journal of the American Ceramic Society*, vol. 67, no. 1, pp. 61–64, 1984.

[7] A. A. Ramezanianpour, A. Pilvar, M. Mandikhani, and F. Moodi, "Practical evaluation of relationship between

concrete resistivity, water penetration, rapid chloride penetration and compressive strength," *Constrution and Building Materials*, vol. 25, no. 5, pp. 2472–2479, 2011.

[8] S. Tsivilis, E. Chaniotakis, E. Badogiannis, G. Pahoulas, and A. Ilias, "A study on the parameters affecting the properties of Portland limestone cements," *Cement and Concrete Composites*, vol. 21, no. 2, pp. 107–116, 1999.

[9] I. Soroka and N. Stern, "Effect of calcareous fillers on sulfate resistance of Portland cement," *American Ceramic Society Bulletin*, vol. 55, no. 6, pp. 594-595, 1976.

[10] I. Mohammadi and W. South, "The influence of the higher limestone content of General Purpose cement according to high-strength concrete test results and construction field data," *Materials & Structures*, vol. 49, no. 11, pp. 4621–4636, 2016.

[11] H. El-Didamony, T. Salem, N. Gabr, and T. Mohamed, "Limestone as a retarder and filler in limestone blended cement," *Ceramics-Silikáty*, vol. 39, no. 1, pp. 15–19, 1995.

[12] J. Albeck and B. Sutej, "Characteristics of concrete made of portland limestone," *Beton- und Stahlbetonbau*, vol. 41, no. 5, pp. 240–244, 1991.

[13] H. El-Didamony, A. M. Sharara, E. Ebied, and A. El-Aleem, "Hydration characteristics of β-C_2S in the presence of some pozzolanic materials," *Cement and Concrete Research*, vol. 26, no. 8, pp. 1179–1187, 1996.

[14] C. Kemal, S. Hocine, E. H. Y. Yang, and V. C. Li, "Use of high volumes of fly ash to improve ECC mechanical properties and material greenness," *ACI Materials Journal*, vol. 104, no. 6, pp. 620–628, 2007.

[15] A. E. M. Abd Elmoaty, "Mechanical properties and corrosion resistance of concrete modified with granite dust," *Construction and Building Materials*, vol. 47, pp. 743–752, 2013.

[16] D. M. Kannan, S. H. Aboubakr, A. S. El-Dieb, and M. M. Reda-Taha, "High performance concrete incorporating ceramic waste powder as large partial replacement of Portland cement," *Construction and Building Materials*, vol. 14, pp. 435–441, 2017.

[17] S. S. Berriel, A. Favier, E. R. Domínguez et al., "Assessing the environmental and economic potential of limestone calcined clay cement in Cuba," *Journal of Cleaner Production*, vol. 124, pp. 361–369, 2016.

[18] O. Akhlaghi, T. Aytas, B. Tatli et al., "Modified poly(carboxylate ether)-based superplasticizer for enhanced flowability of calcined clay-limestone-gypsum blended Portland cement," *Cement and Concrete Research*, vol. 101, pp. 114–122, 2017.

[19] W. G. Shen, Y. Liu, L. H. Cao et al., "Mixing design and microstructure of ultra high strength concrete with manufactured sand," *Construction and Building Materials*, vol. 143, pp. 312–321, 2017.

[20] K. P. Zhou, R. G. Gao, and F. Gao, "Particle flow characteristics and transportation optimization of superfine unclassified backfilling," *Minerals*, vol. 7, no. 1, p. 6, 2017.

[21] C. H. Hou, Z. Q. Yang, Q. Gao, and D. X. Chen, "Experimental research on a new filling material compound of whole tailings and rod milling sand in Jinchuan mine," *Journal of Shandong University of Science and Technology*, vol. 34, no. 3, pp. 78–84, 2015.

[22] Y. J. Xu, K. P. Zhou, J. L. Li, and Y. M. Zhang, "Study of rock NMR experiment and damage mechanism analysis under freeze-thaw condition," *Rock and Soil Mechanics*, vol. 33, no. 10, pp. 3001–3005, 2012.

[23] J. Y. Jehng, *Microstructure of Wet Cement Pastes: a Nuclear Magnetic Resonance Study*, Ph.D. thesis, Northwestern University, Evanston, IL, USA, 1995.

[24] M. G. Prammer, E. D. Drack, J. C. Bouton et al., "Measurements of clay-bound water and total porosity by magnetic resonance logging," *Log Analyst*, vol. 37, no. 6, 1996.

[25] J. L. Shang, J. H. Hu, K. P. Zhou, X. W. Luo, and M. Aliyu, "Porosity increment and strength degradation of low-porosity sedimentary rocks under different loading conditions," *International Journal of Rock Mechanics and Mining Sciences*, vol. 75, pp. 216–223, 2015.

[26] Q. R. Xu, W. Y. Deng, B. Xu et al., "Quantitative analysis of soft clay three-dimensional porosity based on SEM image information," *Chinese Journal of Rock Mechanics and Engineering*, vol. 37, no. 7, pp. 1497–1502, 2015.

[27] R. G. Gao, K. P. Zhou, and J. L. Li, "Research of the mechanisms of backfill formation and damage," *Materiali in Tehnologije*, vol. 52, no. 2, pp. 163–169, 2018.

[28] P. J. M. Monteiro and P. K. Mehta, *Concrete: Microstructure, Properties and Materials*, Prentice-Hall, London, UK, 2006.

[29] A. A. Ramezanianpour, *Cement Replacement Materials: Properties, Durability, Sustainability*, Springer-Verlag, Berlin, Heidelberg, Germany, 2014.

Permissions

The contributors of this book come from diverse backgrounds, making this book a truly international effort. This book will bring forth new frontiers with its revolutionizing research information and detailed analysis of the nascent developments around the world.

We would like to thank all the contributing authors for lending their expertise to make the book truly unique. They have played a crucial role in the development of this book. Without their invaluable contributions this book wouldn't have been possible. They have made vital efforts to compile up to date information on the varied aspects of this subject to make this book a valuable addition to the collection of many professionals and students.

This book was conceptualized with the vision of imparting up-to-date information and advanced data in this field. To ensure the same, a matchless editorial board was set up. Every individual on the board went through rigorous rounds of assessment to prove their worth. After which they invested a large part of their time researching and compiling the most relevant data for our readers.

The editorial board has been involved in producing this book since its inception. They have spent rigorous hours researching and exploring the diverse topics which have resulted in the successful publishing of this book. They have passed on their knowledge of decades through this book. To expedite this challenging task, the publisher supported the team at every step. A small team of assistant editors was also appointed to further simplify the editing procedure and attain best results for the readers.

Apart from the editorial board, the designing team has also invested a significant amount of their time in understanding the subject and creating the most relevant covers. They scrutinized every image to scout for the most suitable representation of the subject and create an appropriate cover for the book.

The publishing team has been an ardent support to the editorial, designing and production team. Their endless efforts to recruit the best for this project, has resulted in the accomplishment of this book. They are a veteran in the field of academics and their pool of knowledge is as vast as their experience in printing. Their expertise and guidance has proved useful at every step. Their uncompromising quality standards have made this book an exceptional effort. Their encouragement from time to time has been an inspiration for everyone.

The publisher and the editorial board hope that this book will prove to be a valuable piece of knowledge for researchers, students, practitioners and scholars across the globe.

List of Contributors

Mark Bediako
CSIR-Building and Road Research Institute, Materials Engineering Division, Kumasi, Ghana

Eric Opoku Amankwah
Development Office, University of Education, Winneba, Kumasi Campus, Kumasi, Ghana

Jae Hong Kim, Jin Hyun Lee, Tae Yong Shin and Jin Young Yoon
School of Urban and Environmental Engineering, Ulsan National Institute of Science and Technology, Ulsan 44919, Republic of Korea

Liqun Hu, Yangyang Li, Xiaolong Zou, Shaowen Du, Zhuangzhuang Liu and Hao Huang
Key Laboratory of Special Area Highway Engineering of Ministry of Education, Chang'an University, Shaanxi, Xi'an 710064, China

Duan-le Li, Da-peng Zheng, Dong-min Wang, Cheng Du and Cai-fu Ren
School of Chemical and Environmental Engineering, China University of Mining and Technology, Beijing, China

Ji-hui Zhao
Department of Civil Engineering, Tsinghua University, Beijing, China

Zhuangzhuang Liu and Aimin Sha
Key Laboratory for Special Area Highway Engineering, Ministry of Education, Xi'an, China
School of Highway, Chang'an University, Xi'an, China

Wenxiu Jiao, Jie Gao, Zhenqiang Han and Wei Xu
School of Highway, Chang'an University, Xi'an, China

Wuman Zhang, Sheng Gong and Bing Kang
School of Transportation Science and Engineering, Beihang University, Beijing, China

Jin Wook Bang
R&D Center, Tongyang Construction Materials Co., Ltd., No. 2822-1, Gimpo-dearo, Wolgot-myeon, Gimpo-si, Gyeonggi-do 10024, Republic of Korea

Byung Jae Lee
R&D Center, JNTINC Co., Ltd., No. 9, Hyundaikia-ro, 830 Beon-gil, Bibong-myeon, Hwaseong-City, Gyeonggi-do 18284, Republic of Korea

Yun Yong Kim
Department of Civil Engineering, Chungnam National University, Daejeon 34134, Republic of Korea

Xinwei Li
Geological Party 103 Guizhou Bureau of Geology and Minerals, Tongren 554300, China
China University of Geosciences, Wuhan 430074, China

Duoyou Shu, Sui Zhang, Siyang Wang and You He
Geological Party 103 Guizhou Bureau of Geology and Minerals, Tongren 554300, China

E-chuan Yan
China University of Geosciences, Wuhan 430074, China

Yangbing Cao
Fuzhou University, Fuzhou 350108, China

Hui Li
Environmental Geological Prospecting Institute of Hebei Province, Shijiazhuang 050011, China

Chengcheng Fan, Baomin Wang and Tingting Zhang
Institute of Building Materials, School of Civil Engineering, Dalian University of Technology, Dalian 116024, China

Man Yan, Min Deng, Chen Wang and Zhiyang Chen
State Key Laboratory of Materials-Oriented Chemical Engineering, Nanjing Tech University, Nanjing, China

Kwangwoo Wi, Han-Seung Lee and Seungmin Lim
School of Architecture and Architectural Engineering, Hanyang University, Ansan, Gyeonggi-do, Republic of Korea

Mohamed A. Ismail
Department of Civil Engineering, Miami College of Henan Univerisity, Kaifeng, Henan, China

Mohd Warid Hussin
Construction Research Centre (UTM CRC), Institute for Smart Infrastructure and Innovative Construction, Universiti Teknologi Malaysia, 81310 UTM, Johor Bahru, Johor, Malaysia

Jie Yang
College of Water Conservancy and Hydropower Engineering, Hohai University, Nanjing 210098, China

Xin Cai
College of Water Conservancy and Hydropower Engineering, Hohai University, Nanjing 210098, China
College of Mechanics and Materials, Hohai University, Nanjing 210098, China

Qiong Pang and Ying-li Wu
Nanjing Hydraulic Research Institute, Nanjing 210024, China

Xing-wen Guo
College of Mechanics and Materials, Hohai University, Nanjing 210098, China

Jin-lei Zhao
Jiangsu Surveying and Design Institute of Water Resources Co., Ltd., Yangzhou 225127, China

Tomáš Ficker
Faculty of Civil Engineering, Brno University of Technology, CZ-602 00 Brno, Czech Republic

Baoguo Ma, Ting Zhang, Hongbo Tan, Xiaohai Liu, Junpeng Mei, Wenbin Jiang, Huahui Qi and Benqing Gu
State Key Laboratory of Silicate Materials for Architectures, Wuhan University of Technology, Wuhan 430070, China

Qingkun Yu and Liangcai Cai
Air Force Engineering University, Xi'an, Shaanxi Province, 710038, China

Jianwu Wang
Air Force Service Academy, Xuzhou, Jiangsu Province, 221000, China

Ailian Zhang and Linchun Zhang
Department of Civil Engineering, Sichuan College of Architectural Technology, Deyang 618000, China

Li Wang, Hongliang Zhang and Yang Gao
Key Laboratory for Special Area Highway Engineering of Ministry of Education, Chang'an University, Xi'an, Shaanxi, China

M. A. de la Rubia, E. Reyes and A. Moragues
Department of Civil Engineering-Construction, School of Civil Engineering, Polytechnic University of Madrid, C/Professor Aranguren s/n, 28040 Madrid, Spain

E. de Lucas-Gil, F. Rubio-Marcos, M. Torres-Carrasco and J. F. Fernández
Electroceramic Department, Instituto de Cerámica y Vidrio (CSIC), C/Kelsen 5, 28049 Madrid, Spain

Peng Zhang
Center for Durability & Sustainability Studies, Qingdao University of Technology, Qingdao 266033, China
Institute of Concrete Structures and Building Materials, Karlsruhe Institute of Technology, 76131 Karlsruhe, Germany

Huaishuai Shang, Dongshuai Hou, Siyao Guo and Tiejun Zhao
Center for Durability & Sustainability Studies, Qingdao University of Technology, Qingdao 266033, China

Jianhua Hu, Rugao Gao and Zhouquan Luo
School of Resources and Safety Engineering, Central South University, Changsha, Hunan 410081, China

Qifan Ren, Quan Jiang and Long Zhang
School of Resources and Safety Engineering, Central South University, Changsha, Hunan 410081, China

Index

www.ingramcontent.com/pod-product-compliance
Lightning Source LLC
Chambersburg PA
CBHW050443200326
41458CB00014B/5051